Jayantha Katupitiya   Kim Bentley

Interfacing with C++

Programming Real-World Applications

Springer

D1292689

Dr. Jayantha Katupitiya
Senior Lecturer
School of Mechanical and
Manufacturing Engineering
The University of New South Wales
Sydney NSW 2052, Australia
Email: J.Katupitiya@unsw.edu.au

Mr. Kim Bentley

Library of Congress Control Number: 2005937895

ISBN-10   3-540-25378-5 Springer Berlin Heidelberg New York
ISBN-13   978-3-540-25378-5 Springer Berlin Heidelberg New York

Springer is a part of Springer Science+Business Media
springer.com
© Springer-Verlag Berlin Heidelberg 2006
Printed in The Netherlands

Typesetting: by the authors and TechBooks using a Springer LaTeX macro package

Cover design: *design & production* GmbH, Heidelberg

Printed on acid-free paper   SPIN: 11015543   89/TechBooks   5 4 3 2 1 0

Table of Contents

This Book is Written For...

C++ is considered by many to be among the most widely used and powerful object-oriented programming language in industry today. This book is for people who are interested in learning and exploring C++ programming in a fresh and enjoyable environment where programs are developed to interface with real world devices. Other people may leave learning C++ for a later time, instead choosing to interact with various hardware devices by simply running the fully developed programs supplied with this book.

Many readers may already have acquired some knowledge of C++ programming but know little about how to interface a computer to physical devices and want to know more. You might be an engineer, scientist, programmer, technical personnel, hobbyist, student in a technically related field or someone who is simply interested in programming and interfacing a computer to perform real activities.

Inside This Book...

C++ programming is approached in a straightforward, practical and simplified manner using mostly short programs that are clearly explained. You will explore areas of electronics integral to a wide range of modern technologies using an interface board specially developed to support all projects described in this book.

The intertwining of C++ programming and electronics knowledge takes place as we work through interesting and enjoyable real-world projects. These projects encompass the following topics:

- Digital Input and Output.
- Analog-to-Digital Conversion and Digital-to-Analog Conversion.
- DC Motor and Stepper Motor Control.
- Measuring Voltage, Temperature, and Time.

Important concepts are reinforced during the learning and exploration process as we gradually progress from simple straightforward projects to those that are more advanced. Projects on the interface board have been developed as independent modules. This allows readers with C++ programming knowledge to build and play with whichever projects they wish, in any order.

For those readers who want to know how to manage the development of larger programs, a chapter has been specially written to cover the process of program development, demonstrated with the use of a program from an earlier chapter. In this chapter we cover topics such as coding techniques, generating header files and building libraries.

What is C++?

C++ is a language used to program computers to perform specific tasks. There exist many other popular programming languages including C, Pascal, FORTRAN, BASIC, Cobol and Modula II. Computers operate using instructions based on binary format, i.e. on and off states (or ones and zeros). Programming languages allow the programmer to use a language similar to that normally written and then generate computer-based instructions for program execution. Specialised software is used to manage the task of developing programs; in particular converting the program written in its programming language to binary form needed by the computer.

In the recent past the language known as C became very popular and was the most significant commercially used programming language. The C language was developed in response to the need for a good programming language to develop the UNIX operating system. While it is considered a high-level language, it also has many low-level features. This is of great benefit when programs need to work with hardware. On the other hand it was also well suited to performing numerical operations. It can match the capabilities of FORTRAN and Pascal (a language able to handle complex logic). These are some of the reasons for the popularity of the C language.

As the size of programs increased, the benefits of being able to reuse millions of instructions written and assembled by programmers around the world, became apparent. Soon afterwards the concept of object-oriented programming (OOP) was born and the C++ language came into being, evolved from C. C++ can be considered an expanded and better C. In other words, C became a subset of C++. The programmer could now combine associated data and functions to avoid inadvertent misuse. The so-called virtual functions in C++ added extra flexibility allowing decision-making at run time, rather than at compile time. While C++ has gained all this extra power, it has retained other good features of C such as low-level bit and byte operations, easy input and output to ports, etc. In today's world, C++ is the most widely used programming language for sophisticated tasks.

Compiler and Operating System Compatibility

Most programs in this book have been written to carry out some form of interfacing task. An essential feature of such programs is the ability to read from and write to the hardware ports. Some operating systems such as DOS, Windows 3.1, Windows 95/98 allow programs to directly access ports. Other operating systems such as Windows NT/2000/XP and Linux do not allow direct port access. These operating systems will only allow programs to access ports via a piece of software known as a device driver that has the necessary privileges to access ports. The application programs access the ports via the device drivers.

Borland C++ for DOS

Apart from the programs using exception handling (See Chapter 7), all programs in the textbook can be compiled and linked using Borland C++ without any changes to generate executable files. All program listings that are to be compiled using Borland C++ are located in the directory 'BC++' on the companion CD.

GNU C++ for Linux

The programs in the textbook have been modified to request the required privileges to enable them to run under Linux with port access. The modified versions of programs can be found in the directory 'GNUC++' of the companion CD. If a make file is necessary, it is also included in the appropriate chapter subdirectories of the directory GNUC++. Graphics programs, keyboard control programs and PC timer related programs are not available to run under Linux.

Microsoft Visual C++ for Windows

The modified versions of the programs that can be used with Microsoft® Visual C++ can be found in the directory 'VC++' on the companion CD. The programs in the 'Win98' subdirectory can be run under Windows98 without the need of a device driver. The programs in the 'Windows' subdirectory can be run under Windows NT/2000/XP with the use of WinIO, which will act as the driver. These programs have been modified to enable them to access the ports through the use of WinIO. WinIO has not been included in the accompanying CD. Its latest version can be downloaded from http://www.internals.com/. You must first install WinIO in order to be able to run the programs in the 'Windows' subdirectory. The readers of this book who use WinIO are bound by the WinIO licensing agreement published on the web. Graphics programs, keyboard control programs and PC timer related programs are not available to run under Microsoft® Windows.

1

Getting Started

Inside this Chapter

- Developing programs – what is involved?

- Writing and running your first C++ program.

- Program syntax.

- Functions.

- Fundamental data types.

1.1 Introduction

The aim of this chapter is to get you started in writing C++ programs. We will develop a number of simple C++ programs and learn the syntax and typography associated with writing a program. One of the basic building blocks of any C++ program is the so-called function. This chapter will explain the basic concepts behind C++ functions and their use. The C++ language has built-in fundamental data types that can be used to develop complex user-written data types. Some of the fundamental data types will be explained in this chapter.

Towards the end of the chapter we will step through the complete program development process; starting from planning a small program down to using the elements of *program development software* needed to generate a program that can be run on your computer. We will commence with the use of non-object-oriented programming methods because these programs are simpler to understand at this early stage. Object-oriented programming concepts will be explained in Chapter 4 and then used extensively through the remainder of the text.

1.2 Program Development Software

The process of program development includes a number of subtasks. To be able to develop a program you must have an *editor*, a *compiler* and a *linker*. In modern program development platforms, these subtasks are seamlessly integrated and the entire process is very transparent. Such platforms are known as *Integrated Development Environments* (IDEs). Most modern C++ packages (the software that you will use to develop C++ programs) provide some sort of an IDE. Some of the commercially available packages include Turbo C++, Borland C++, C++ Builder and Visual C++. There are also packages referred to as *command line* versions. The command line versions require you to type a command (say at the DOS prompt) to invoke the editor. Then you must use another command line to invoke the compiler and so forth.

Along with the editor, compiler and linker, these packages also provide extensive library support. Sometimes these libraries are referred to as run-time libraries (RTLs). They contain a wide variety of routines or functions we can use within our programs. Regardless of what package we use, it is worthwhile to understand what happens during each subtask. The following sections will describe *editing*, *pre-processing*, *compiling*, and *linking*.

1.2.1 Editing

The first step in preparing your program is to use some kind of editor to type your program. Not every editor is suitable for this purpose. The *edit* program of DOS and the *Notepad* editor of Windows are two suitable editors. Integrated Development Environments (IDE) that are part of C++ packages provide built-in editors known as text editors. At the end of the editing session you must store the contents of the editor into a file. The two editors mentioned above will only store what you type. They will not add extra characters to your file (unlike some editors). What we normally type includes digits, letters, punctuation marks, the space, tab, carriage return and line-feed characters. The line-feed character is used by the editor to position the cursor on a new line. The carriage return character is used by the editor to position the cursor at the start of the next line. A program file must not contain characters apart from those listed above. The file that contains all programming instructions, is known as the *source file*. The source file is said to contain the *source code*, which is nothing more than the programming instructions you typed.

1.2.2 Compiling

The second step is to *compile* the source file. For this purpose, a special program known as a *compiler* is used. As part of the compiler, a program named the *preprocessor* is invoked. This takes place before the actual compilation of your source code. The preprocessor attends to your source code statements that start with the '#' sign. (See the program listings ahead for the lines starting with a '#' sign). These statements are referred to as *compiler directives*. The preprocessor takes action as directed by these statements and will modify your original source file. At the end of preprocessing, all lines starting with the '#' sign will have been processed and eliminated. This process is shown in Figure 1-1. The preprocessor and the compiler are gradually becoming merged - most modern compilers have the preprocessor as a built-in part of the compiler itself.

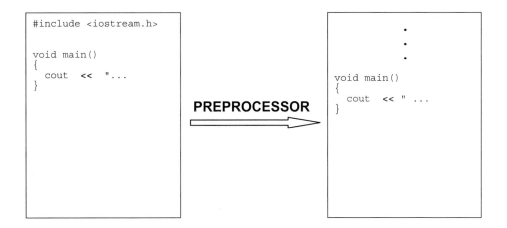

Figure 1-1 Preprocessor attends to all lines starting with '#' symbol.

The compiler in-turn processes the file produced by the preprocessor and produces a file known as an *object file*. The object file contains what is known as object code, which the Central Processing Unit (CPU) of your computer understands, also known as machine code. However, the PC cannot execute the object code since it still has a few parts missing. At this stage your program is in a similar state to an unfinished highway with some stretches complete and others not. As a result, the compiled program cannot yet be executed (i.e. run on your computer).

At this incomplete stage, the object code is said to contain *undefined references*. The undefined references refer to pieces of object code that need to be retrieved from elsewhere to complete the entire program. Just like the highway, the object file does not have a continuous execution path. The compiling process is shown in Figure 1-2.

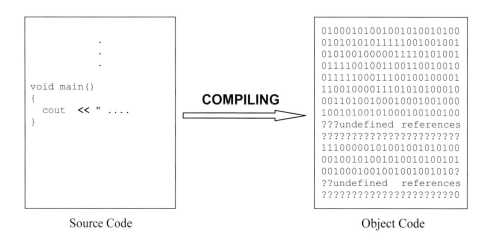

Source Code   Object Code

Figure 1-2 The compiler converts the source code to object code.

The syntax used as part of the program statements is extremely important. As mentioned earlier, syntax refers to the use of punctuation marks within the source file. Most of the time these punctuation marks act as *delimiters*. A delimiter identifies the end of variables, keywords, numbers, statements etc. The space, the comma, the semicolon, the colon, the brace etc., act as delimiters for different contexts of usage. Compilers have limited in-built intelligence. If you miss a semicolon the compiler will detect it and report an error, but it cannot correct the error for you.

As mentioned earlier, the object code is incomplete with many unresolved areas and it cannot be executed. For example, the object code may contain calls to various routines. The object file includes function calls to be made. The actual instructions to be executed during the call are not yet in place. These instructions may be available elsewhere in the object file, or they may need to come from a library file or another object file. Note that finding the missing bits is not part of the compiler's duties – the compiler can be viewed in basic terms as a translator that checks grammatical content!

1.2.3 Linking

The program that bridges all the gaps and completes assembly of the program is known as the *linker*. It will search all the object files and the libraries to find the missing sets of instructions. Sometimes the linker must be told to search certain libraries and object files. These are either third party libraries you may have purchased or the libraries and object files you developed. The linker automatically searches the libraries and object files that come with the C++ software, one being the so-called Run-Time Library (RTL). The linker will insert the missing sets of instructions into appropriate places to form a file that has a 'gaps free' execution path. This process is known as linking. At the end of the linking process, we have a file the PC can execute, known as an *executable file*.

The program must be *loaded* into the computer's memory before execution can begin. This action is carried out by a piece of executable code known as a *loader*. Most linkers append a loader to the start of the executable file. Therefore, when we try to run the program, first the loader will run, loading the program into memory and then actual program execution will begin. Figure 1-3 shows the linking process.

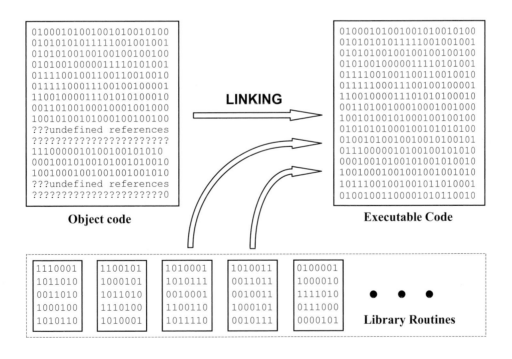

Figure 1-3 Linking forms a gaps-free executable code.

1.3 A C++ Program

A computer sees a program as a set of instructions to be executed. The programmer arranges these instructions in a certain order depending on the tasks the computer is expected to perform. To give you a simple example; if you want to write a program to add two numbers, the numbers must be entered first and then the addition must be carried out. Therefore, the instructions to read the numbers must come before the instructions to add the numbers.

Each programming language has its own unique syntax. Syntax is the typography and the use of punctuation marks. Here, we will learn the syntax that applies to the C++ programming language.

As mentioned earlier, the basic building block of a C++ program can be viewed as the function – a procedure that produces an end result. Therefore, every C++ program that contains a set of executable instructions *must* have a function. One of these functions is special, and is named main. To uniquely identify functions separately from other entities in our text, we use a pair of parentheses () after the function name. Simple programs can be written just with a main() function. When programs become more elaborate and complex, other functions may have to be written in addition to the main() function.

The aim of our first C++ program is to print a text message on the screen of your computer. The lines of this program are given in Listing 1-1.

Listing 1-1 Program to print a text message on the screen.

```
/* This program prints a text message on your screen.
   The program consists of just one function named
   main.*/


#include <iostream.h>

// The main function.
void main()
{
    cout << "Getting Started " << endl;
}
```

If you run this program, you will see the message:

```
Getting Started
```

printed on your screen. The following sections explain the composition of this program.

1.3.1 Comments in Programs

Comments are descriptions included in a program that are used so programmers can document their work. They often describe a program or specific parts of a program and do not form any part of the actual program's instructions that will run on the computer. If you include comments, you must indicate to the compiler that they are not to be considered as actual code when the compiler prepares the final program prior to execution. There are two different ways to include comments:

(i) To include single line or multi-line comments you can use '/*' at the start of the comment and '*/' at the end of the comment.

(ii) If the comment is a single line comment you may use '//' at the start of the comment.

In Listing 1-1, we have a multi-line comment and a single line comment. The multi-line comment is:

```
/* This program prints a text message on your screen
   The program consists of just one function named
   main().*/
```

The single line comment is:

```
// The main function.
```

The text contained within '/*' and '*/' will be ignored by the compiler. Likewise for the text after '//' on that line.

1.3.2 Header Files

The first line after the multi-line comment of Listing 1-1 is an *include statement*:

```
#include <iostream.h>
```

It instructs the preprocessor to replace that statement with the entire contents of the file `iostream.h`. In our program this takes place just before the start of the `main()` function. The files with the file extension '.h' are known as *header files* or as *include files*. A header file can already exist within the C++ development software, or it may be a file created by the programmer. If it is a file provided with the C++ development software, then it resides in a special sub-directory known as the *Include Directory*, as is the case for the `iostream.h` file. Programs can have more than one include statement, resulting in the inclusion of a number of header files.

The header files are text files that contain C++ programming statements, most of which do not form executable program statements. Not all statements in your program are executable. However, the statements in header files play a major role in the preparation of your program. The majority of the statements in a header file assist the compiler to carry out a thorough check of the program statements you write in your program. Once the header files are written and tested, we do not change them. If the compiler issues error or mismatch messages, then we must change our program – not the header file.

Library routines are ready-made pieces of software we can make use of. The programmers who write the library routines must also prepare the header files belonging to the library routines. By programming in strict conformance with the header files, we are conforming with the library routines we have used that are associated with those same header files.

In the program shown in Listing 1-1 we have used `cout`, double left arrows '`<<`', and `endl` within our program. They do not form part of the C++ language in the context we have used them. Unless we instruct the compiler as to the usage of these elements, the compiler cannot interpret their proper use. The header file `iostream.h` contains all the necessary programming statements to inform the compiler how the elements should be used. This information must appear before using `cout`, `<<` and `endl`. Hence, the `include` statement appears before the first use of `cout`, `<<` and `endl` in our program.

The compiler does not need the entire contents of the file `iostream.h` to be able to translate the program shown in Listing 1-1 into code that the computer understands. In our example, it would be sufficient to show the part of `iostream.h` that describes `cout`, `endl`, and the behaviour of the operator `<<`. However, it is very difficult to determine exactly which parts of a header file are necessary for a particular program. Therefore, compilers run through the entire header file. The size of the header files will not have any affect on the size of the executable files, although the time to prepare the program will increase slightly. If necessary, we may need to include more than one header file. In addition, there may be other header files used in each of the ones we include.

In conclusion, the appropriate header file must be included first to provide various definitions of constants and data types, and also to declare various functions before using those constants, data types and functions in a program.

1.3.3 Program Syntax

Syntax refers to the use of punctuation marks in the program. In our program, we have used the # symbol, angle brackets (< >), the pair of parentheses, braces ({ }), semi-colon and <<. These punctuation marks must be correctly inserted at the appropriate places before the compiler can recognise your program as being error-free. The program in Listing 1-1 shows only basic syntax. As programs become more complex, their syntax also becomes more involved.

All lines starting with a hash symbol (#) are instructions to a special part of program development software named the preprocessor, discussed previously.

Our program has just one function – the `main()` function. The start of the *body* of the `main()` function is signified by the open brace ({). The end of the `main()` function is signified by the close brace (}). Between the two braces are the statements to be executed by the program. Program execution always starts at the line that contains `main()`. It ends at the closing brace of the `main()` function's body.

The syntax of the `main()` function can be expressed in a compact form as shown:

```
void main(){statement1; statement2; statement3;}
```

The function has the name `main`. The pair of parentheses that follow the name `main` may or may not be empty - in our simple program they are empty. As can be seen, semi-colons are used to separate the statements of the program. Although it may appear redundant, the semi-colon after the last statement is essential.

1.3.4 Keywords

Keywords are words reserved by the language. They must not be used for purposes other than those specified for them by the C++ language. For example, a keyword cannot be used as an *identifier*. Identifiers are variable names we create to identify various entities such as functions, user created data types and data. So far, the only keyword we have seen is `void`. A list of keywords is given in Appendix B.

1.3.5 The Return Value Type

The word `void` right at the start of our main function describes the *return value type* of the `main()` function. Every function, let it be the `main()` function or some other function, must specify a return value type. The return value can be viewed as the end-product or the output produced by the function. If a function is programmed to return a value, the programmer must specify the data type of the value to be returned (to be issued out). It is also possible to program a function to not return a value. Such functions generally carry out some task but do not produce a value to be issued out. For these functions the return value type is `void`. This is the case in our program. Note that when no values are returned by a function the keyword `void` *must* be used to specify the return value type.

> **NOTE**
>
> If a return value type is not specified for the `main()` function, the return value type will default to that of an integer. This means the function must produce an integer output.

1.3.6 The Body of `main()`

The body of the `main()` function contains just one statement. This line is enclosed within the open brace (`{`) and the close brace (`}`). If there is to be more than one statement forming the body of the `main()` function, they all need to be included within the two braces. The solitary statement in our program reads as:

```
cout << "Getting Started " << endl;
```

The use of `cout` will instruct the computer to *stream* whatever follows to the standard output device, in this case your screen. Streaming has a definition in the C++ language. For now, it's sufficient to understand streaming as directing one entity (such as a group of characters, an integer, etc.) after another to a certain destination. First, `Getting Started` will appear on the screen. Next, `endl` will be streamed to the screen. The effect of this is to position the cursor at the start of a new line on the screen. Execution of the program is now complete.

You can experiment by replacing the previous statement by:

```
cout << "Getting Started";
```

This will only stream `Getting Started` to the screen. It will not stream `endl` to the screen. You will see the cursor blinking at the end of the words 'Getting Started'.

1.4 Use of Functions

As mentioned earlier, functions form an integral, important part of C++ programming. In this section we will learn how to use a function. As explained earlier, a function can be thought of as a procedure that produces some sort of an end result based on the inputs it receives. Some of the inputs the function will receive are known as *parameters* or *formal arguments*. The formal arguments are used at the time of programming a function to indicate the *type* of arguments it can receive. At the time of executing the function in your computer, the formal arguments must be replaced by *actual arguments*. For example, at the time of programming, we may use a formal argument named `a`. At the time of executing the function, the formal argument `a` must be replaced by an actual argument such as the number 3. The same function can be *called* (executed) again replacing the formal argument with a different actual argument, producing a different return value. It is worth mentioning here that, although the formal way of receiving the output of a function is via the return value, there are other ways of receiving the outputs from functions.

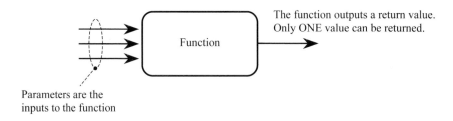

Figure 1-4 General schematic of a function.

What we have described so far is the most general case for a function. This is shown schematically in Figure 1-4. There are a number of special cases. These special cases depend on whether or not the function receives parameters and whether or not it returns a value. The number of parameters received by a function can vary from function to function.

The number of values returned by a function is always one and it must be a *scalar* quantity. A scalar quantity can be loosely defined as a single entity. In other words, functions cannot return *arrays* (groups of entities). For example, a function can produce a result through the return value, which is just one integer. It cannot produce a result that has more than one integer. Figure 1-5 and Figure 1-6 show some typical forms of functions.

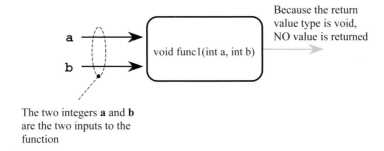

Figure 1-5 A function that takes two arguments and returns NO value.

For the case shown in Figure 1-5 the function will perform a task such as calculate a value and print a message on the screen. If that is all the function needs to do, then there is no need for the function to return a value.

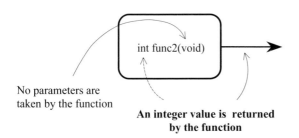

Figure 1-6 A function that takes no parameters but returns a value.

In the case of Figure 1-6, the function may be receiving some data from an external source such as the printer port and returning an integer number. For example, the integer number may indicate the status of paper in the printer; 0 indicating no paper and 1 indicating paper is still present.

1.4.1 A Program with a Function Call

The program shown in Listing 1-2 produces the same output as the program in Listing 1-1. The only difference is that it uses a function to produce the output on the screen. Moreover, the function does not receive any parameters and does not produce any return value. The main emphasis in this section is to explain the concept of *procedure abstraction*. Procedure abstraction means hiding the details of a certain procedure behind a function and then calling the function to have the procedure carried out.

Listing 1-2 A program with a function call.

```
/* This program prints a text message on your screen.
   The program consists of two functions named
   PrintMessage and main.*/

#include <iostream.h>

// The PrintMessage() function
void PrintMessage()
{
    cout << "Getting Started" << endl;
}

// The main function.
void main()
{
    PrintMessage(); // calling the function
}
```

A new function named PrintMessage has been added to this program. The name PrintMessage is our own creation. We have also added the pair of parentheses at the end of the name PrintMessage to signify it as being a function. The pair of parentheses are empty (which is equivalent to placing void there) because the function does not have any parameters. The return value type of the PrintMessage() function is void because the function does not return any value. The definition of the PrintMessage() function is as follows:

```
void PrintMessage()
{
    cout << "Getting Started" << endl;
}
```

A function definition must specify *four* things, being:

1. The return value type

2. The function name
3. Number of parameters and their types
4. The body of the function

The syntax of a function is depicted in Figure 1-7:

```
        Return value type              Number of parameters and their types
            Function name

void PrintMessage()
{
    cout << "Getting Started" << endl;        The body of the function
}
```

Figure 1-7 The syntax of a function definition.

The function definition contains the complete function, informing the compiler what instructions need to be executed. In other words, the body of the `PrintMessage()` function is provided, starting with the open brace and ending with the close brace. Note: a semicolon is not placed after the function name `PrintMessage()`. This allows the following lines containing the function body to be associated with the function name.

The return value type is `void` for the `PrintMessage()` function. The function name is `PrintMessage`. The list of parameters is empty and the body of the function contains the `cout` statement.

A function *declaration* is slightly different (as shown in Figure 1-8). It is sufficient for the compiler to just see the function declaration for it to be able to compile calls to the function. The entire function definition is not needed at this stage. However, in order to execute the function, a compiled version of the function definition is needed. The function declaration has to specify only *three* things:

1. The return value type.
2. The function name.
3. Number of parameters and their types.

```
        Return value type              Number of parameters and their types
            Function name

void PrintMessage();
```

Figure 1-8 Function declaration, also known as the function prototype.

The body of the function is not necessary. However, it must be provided sometime before execution of your program. If the body of the function is obtained from a library, then it will be brought in at linking time. If it is not obtained from a library or another object file, then you must type the code for the function somewhere in your program. In our example, the function declaration would be:

```
void PrintMessage();
```

Note that the line ends with a semi-colon.

In C++, the function *prototype* is exactly the same as the function declaration. However, in a C program the function declaration and function prototype are two different things. See the note in Section 1.6 for an example.

In the `main()` function the body has been changed. The only statement in the body is:

```
PrintMessage();
```

Note that the line ends with a semi-colon and the return value type does not appear. Such a line is termed a *function call*. In a function call, two things must be specified. They are:

1. The function name.
2. The list of actual arguments.

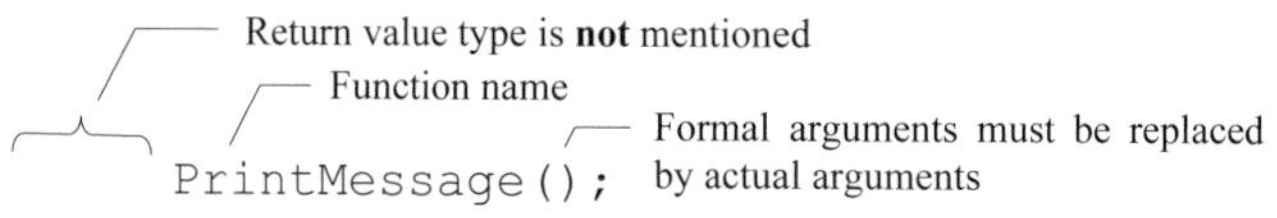

Figure 1-9 An example showing the syntax of a function call.

The actual arguments replace the parameters (or the formal arguments) when it comes to the time of execution. Note the syntax and that the line ends with a semi-colon. An example of a function call that uses a parameter list can be found in Section 1.6.

When the program is executed, as always its execution will begin at the `main()` function. Then the body of the `main()` function will be executed. At this time the computer will encounter the instruction:

```
PrintMessage();
```

This is a function call that results in the execution of the body of the `PrintMessage()` function. Therefore, the message `Getting Started` will be printed on your screen. As mentioned at the start of this section, using the `PrintMessage()` function in the `main()` function enables the details of what it does to be hidden - known as procedure abstraction (explained in Section 1.6).

1.5 Fundamental Data Types

Most of the time programming instructions manipulate data. As such, data plays an important role in programs. Data comes in a variety of *data types* that is sometimes mixed in with other types. It is important to be able to identify data of different types. There are a small number of data types built into the C++ language known as *fundamental data types*. Data types are described by three attributes:

1. The name of the data type.
2. The size of the data type in *bytes* (see Chapter 2).
3. The range of values the data type can handle.

For C++ data types, the size (and therefore the range) differs depending on whether we write 16-bit programs or 32-bit programs. Bits and bytes are explained in the next chapter. For now, it is sufficient to know that 32-bit data types occupy more memory and cover values over a larger range than 16-bit data types.

Data types can be broadly categorised into three types:

1. *Integral* data types.
2. *Floating* point data types.
3. *Pointer* data types.

Integral data types are used to store integral type data (whole numbers) whereas floating-point data types store numbers with a fractional part. Pointer data types are used to store memory addresses. Memory addresses are described in Chapter 2 and pointer data types are discussed in Chapter 7.

The integral data types are further sub-divided into signed types and unsigned types. The signed types can carry both positive and negative numbers whereas unsigned types carry only positive numbers. Floating point numbers can carry both positive and negative numbers. The pointer data types are always positive since there are no negative memory locations.

A programmer can use these fundamental data types to develop custom data types that can be very complex. To start with we will be looking at the following three fundamental data types:

```
char
int
float
```

The first two are integral types and the third is a floating-point type. Table 1-1 shows the three data types mentioned above along with their sizes in bytes and the range of values they can take.

Table 1-1 A few of the fundamental data types.

Data type	Number of bytes	Range of values
char	1	-128 to 127
unsigned char	1	0 to 255
int	2	-32768 to 32767
unsigned int	2	0 to 65535
float	4	$\pm 3.4 \times 10^{-38}$ to $\pm 3.4 \times 10^{38}$
double	8	$\pm 1.7 \times 10^{-308}$ to $\pm 1.7 \times 10^{308}$

Data Type char

This is the data type that is primarily used to store characters. One char type data will occupy one byte in memory. The signed version is simply referred to as char and the unsigned version as unsigned char. This data type can also be used to store small integer numbers that fit into one byte of memory.

Data Type int

This is the data type that is used to store integer numbers. One int type data will occupy 2 bytes of memory in 16-bit programs. The signed version is simply referred to as int and the unsigned version as unsigned int. A synonym for int in 16-bit representation is short int.

Data Type float

This is the data type that is used to store fractional numbers. One float type data occupies 4 bytes in memory. The data type float can handle both positive and negative numbers and so there is no separate data type named unsigned float!

Identifiers

Identifiers are the symbols or variable names we will be using in our programs to identify various entities such as integers, floating point numbers, characters, memory addresses, functions, objects, classes and many more. Identifiers are *case sensitive* and can be of any length.

NOTE

Identifiers *must* start with a letter (upper case or lower case) or the underscore "_" character. They may contain digits (0 to 9), but not as the first character of the identifier.

An identifier must be declared before using it in a program. When declaring an identifier you must specify two things (as shown in Figure 1-10):

1. The data type
2. The identifier name

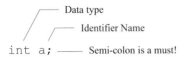

Figure 1-10 An example showing the syntax of a single identifier declaration.

An example of an identifier declaration is:

```
int a;
```

The data type is `int` and the identifier name is `a`. If needed, more than one identifier can be declared in one statement as shown in Figure 1-11:

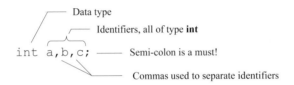

Figure 1-11 An example showing the syntax of a multiple identifier declaration.

Such a declaration *must* be provided before being able to use an identifier in your program.

An identifier can also be declared *and initialised* simultaneously. In such a case, in addition to declaring the variable, we also set the identifier to take up an initial value. An example of such a situation that applies to a single identifier is:

```
int a=0;
```

The data type is `int`, variable name is `a` and it is initialised to have a value of 0 as shown in Figure 1-12.

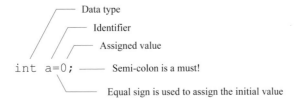

Figure 1-12 An example - syntax of a single identifier declaration and definition.

The identifier declarations we have seen so far can be combined in any manner. Such a declaration is shown in Figure 1-13:

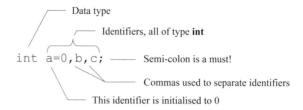

Figure 1-13 An example showing the syntax of a general identifier declaration.

1.6 Functions with Parameters and Return Values

In Section 1.4.1 we learned how to make a function call with a view to understand the concept of procedure abstraction. In this section we will look at a function that can be called repeatedly to carry out the addition of two numbers. We will program the function to receive the two numbers as parameters and to return their sum as the end result produced by the function. This will help us understand the role of function parameters and their return value.

Listing 1-3 Functions with parameters and return values.

```
/* This program calls a function (twice) to add two numbers
together from within the main function and outputs the result
to the screen. */

#include <iostream.h>

float Add (float a, float b)   // The Add() function
{
    float sum;

    sum = a + b;
    return sum;
}

void main()     // The main function.
{
    float p=1, q=2.3, r=3, s=4.5;
```

```
    float Sum1, Sum2;

    Sum1 = Add(p,q); // First call to 'Add' function
    cout << "First Sum " << Sum1 << endl;
    Sum2 = Add(r,s); // Second call to 'Add' function
    cout << "Second Sum " << Sum2 << endl;
}
```

In the program shown in Listing 1-3 we have defined a new function named Add. As mentioned earlier, the definition of a function provides the return value type, the function name, the list of parameters and their types, and the body of the function. Unlike the function we have seen so far in this book, the Add() function's pair of parentheses are not empty, meaning the function receives some parameters. In this case the Add() function receives two parameters of type float. Furthermore, the return value type of the Add() function is float. This means the function *must* produce a return value of type float. The value to be returned must also be specified within the body of the function in a *return statement*. The Add() function is reproduced below to explain its operation:

```
float Add (float a, float b)
{
    float sum;

    sum = a + b;
    return sum;
}
```

NOTE

According to the C language, the declaration of the Add() function is:

```
    float Add();
```

Therefore, a declaration in C does not provide information about the parameters.

The prototype of the Add() function is:

```
    float Add(float, float);
```

This *does* provide information about the parameters. In C++, the declaration and the prototype of a function are exactly the same. Therefore, the prototype (and declaration) of the function Add() is:

```
    float Add(float, float);
```

Within the body of this function we have declared a `float` type identifier named `sum`. Then `sum` is assigned the result of adding a to b. Finally, the `return` statement sends the value of `sum` out of the function. Note that the type of the returned value, i.e. the type of `sum` (which is `float`), is the same as the return value type of the `Add()` function (specified on the first line).

The `main()` function of our program is shown ahead, with its function calls highlighted in bold typeface. In the first call to the `Add()` function, its parameters or formal arguments a and b are replaced by copies of the *actual arguments* p and q, which carry real values. The parameters a and b can be viewed as placeholders. In the second call to the `Add()` function, its parameters are replaced by copies of r and s.

```
void main()
{
    float p=1, q=2.3, r=3, s=4.5;
    float Sum1, Sum2;

    Sum1 = Add(p,q); // First call to 'Add' function
    cout << "First Sum " << Sum1 << endl;
    Sum2 = Add(r,s); // Second call to 'Add' function
    cout << "Second Sum " << Sum2 << endl;
}
```

The value returned by the first call to the `Add()` function is assigned to `Sum1`. Therefore, `Sum1` becomes the summation of p and q. In our case, `Sum1` will have the value of 3.3. Similarly, `Sum2` will have the value of 7.5. Since the `main()` function is making the calls to the `Add()` function, the `main()` function becomes the *caller* and at the same time the *recipient* of any return values. In this case, the `main()` function's body is also known as the calling environment. The other lines in the main function are identifier declarations and/or definitions, and the statement used to print the values of `Sum1` and `Sum2` on the screen.

Figure 1-14 shows an example of the sequences a program goes through. This complete program consists of a `main()` function and a number of other functions and data. The program starts from within the `main()` function where various other functions are called throughout its operation.

The `main()` function and all the other functions are stored in the so-called *code area* of program memory, and are generally not expected to change during the life of the program. The data is stored in the *data area* where its contents are expected to change. Apart from the data in fixed data areas, there may be other data that is created in a temporary area known as the *stack*, and also in a semi-permanent area known as the *heap* (or *free store*).

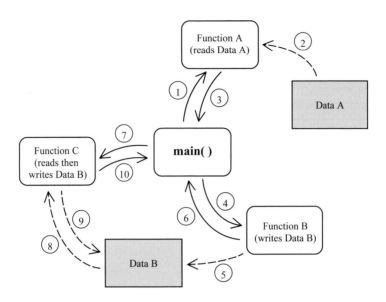

Figure 1-14 Program with main() function, other functions and data.

1.7 Summary

Program development software is typically used to write C++ programs. This software provides an integrated environment for editing, compiling and linking programs. Built-in libraries known as Run-Time Libraries are part of the development environment and contain useful functions. To use these functions, header files are included at the start of the program to provide the respective function declarations.

A C++ program comprises the code written using correct syntax, comments, keywords, identifiers, fundamental data types, user-written data types and header files. Keywords are the reserved words that are part of the C++ language. Fundamental data types are built-in data types +and can be used to develop more complex user-written data types. Identifier names are chosen by the programmer and must not be C++ keywords. Both identifiers and functions must be declared ahead of their use in a program.

C++ programs carry out procedures by using functions that operate on specific data. This simplifies programming since the programmer only calls the function to perform a task and does not need to know how the function implements the call (this is procedure abstraction). A special type of function named main starts and ends program execution. Functions can return a value from within their body after carrying out their assigned operations. The type of this data must be specified at the time of defining the function, therefore, the function has what is known as a return value type. In addition to this, functions often require input data in order to carry out their dedicated operations. This input data is passed into functions with the use of their function parameters.

Early in this chapter we explained a basic C++ program comprising just the `main()` function. An additional function was then added to this program to carry out the same task and demonstrate procedure abstraction. Finally, a program was presented and discussed that added two numbers using a function that had parameters and a return value.

1.8 Bibliography

Kelley, A. and I. Pohl, *A Book on C – programming in C,* Benjamin Cummins, 1995.

House, R*., Beginning with C – An Introduction to Professional Programming,* International Thompson Publishing, 1994.

Deitel H.M. and P.J. Deitel *C: How to Program,* Prentice Hall, 1994.

2

Parallel Port Basics and Interfacing

Inside this Chapter

- Parallel port configuration & functionality.

- Digital logic fundamentals.

- Number systems: decimal, hexadecimal and binary.

- Electronics: port, byte, synchronous, asynchronous, addresses.

2.1 Introduction

A basic understanding of digital logic principles and converting data between number systems is needed before the parallel port can be used effectively. This chapter covers these topics and also describes the configuration of the parallel port itself. Concepts such as binary logic, logic levels, input/output address space and the physical connection to the port will be explained.

Working through this chapter will prime you for programming and connecting to the parallel port. You will use this knowledge in future chapters when developing programs to control and monitor hardware through the port. An understanding of basic electronic logic principles is also beneficial when constructing and testing many circuits on the interface board.

2.2 What is the Parallel Port?

Generally speaking, a port is a portion of electronic hardware that is used as an interface to connect with another electronic device for the purpose of information exchange. This connection allows information in the form of data to flow into, out of, or both into and out of the port.

The parallel port has the facility to transfer data both in and out, between the PC and the outside world. It is normally used for sending information to a printer and also known as the printer port. With older computers, the printer port is made up of circuitry residing on a separate printed circuit board (referred to as a pcb) which plugs into the PC motherboard. Newer computers, however, tend to have the parallel port circuitry integrated along with the rest of the PC motherboard.

Having a basic familiarisation with concepts such as *logic families*, *logic levels* and *noise margins* helps to be able to gain an understanding how electronic devices communicate digitally. This understanding will also prove useful should electrical problems arise when using digital circuitry on the interface board.

2.2.1 Digital Logic

As mentioned earlier, computer programs are executed by hardware which operates using binary logic, also known as *digital logic*. Binary logic has two possible states, ON and OFF. Typically these binary logic states are represented using binary logic notation, where 1's denote the ON state and 0's denote the OFF state.

The ON and OFF states used by the parallel port circuitry and many other digital logic circuits are implemented using voltage levels known as *logic levels* which commonly lie between 0V and +5V. Note that not all types of logic circuits use the same logic voltage levels. These logic circuits are also known as integrated circuits (IC's), containing groups of circuit elements housed on a single piece of semiconductor material known as a "chip". The chip is packaged inside either plastic or ceramic material with metal leads that are bonded internally with wire to the chip to allow external connection.

The two most popular types of logic circuit or *logic families* are TTL (transistor transistor logic) and CMOS (complementary metal oxide semiconductor). Each logic family is fabricated in a unique way, resulting in distinctive electrical operating characteristics. Some basic electrical differences between TTL and CMOS logic families are shown in Figure 2-1. There are several different versions for each family, with characteristic variations in electrical specification.

Note that some CMOS logic families can operate at voltages outside the 0V to +5V range shown. Also, the output voltage level for these circuits does depend on the level of current drawn through each output.

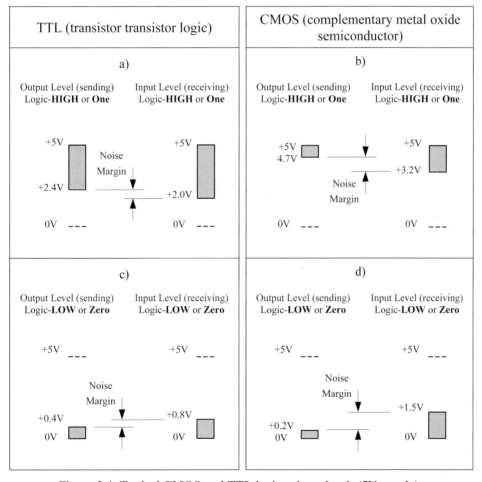

Figure 2-1 Typical CMOS and TTL logic voltage levels (5V supply).

These differences in logic levels from one family to the other are very significant when connecting between them. For example, referring to Figure 2-1 quadrants a) and b); consider the case where a TTL integrated circuit sends a HIGH logic level (+5V to +2.4V) to a CMOS integrated circuit. In this case, the TTL circuit could, at worst, send (output) a logic-HIGH having +2.4V, the lowest output voltage level when operating normally (not damaged or being over-driven). If the CMOS circuit is to correctly recognise a received (input) logic-HIGH level, this received voltage must be at least +3.2V and no more than +5V. The problem with this situation is that the TTL integrated circuit can output a signal down to +2.4V, too low a voltage level for the CMOS integrated circuit to accept as a valid logic-HIGH. The result could be that the CMOS circuit incorrectly mistakes the TTL HIGH level as a LOW level.

Figure 2-1 also shows the voltage *noise margin* when a data signal is sent from one logic circuit to another of the same family. Let us look at the case in which a CMOS circuit outputs a logic-LOW to another CMOS circuit, as shown in quadrant d) of the figure. The sending device will output a signal between 0V and 0.2V during normal operation and the receiving device will accept a signal level between 0V and 1.5V as a valid logic-LOW. If we use an output signal of 0.2V, the worst case for normal operation, then we can have voltage noise of up to 1.3V (1.5V – 0.2V) on this logic signal, and the receiving circuit will still recognise a valid logic-LOW. From this example we can see we have a *noise margin* of 1.3V.

If you examine the same case shown in quadrant c) for a TTL circuit transmitting a logic-LOW to another TTL circuit, you will find that there is a noise margin of only 0.4V. CMOS circuits typically have better noise margin characteristics than TTL circuits. Other differences between TTL and CMOS circuits include their power consumption, their input current requirements, and their output current drive capacity and speed when switching states. For additional information concerning digital logic families, consult the references at the end of this chapter.

2.2.2 Parallel Port Architecture

The parallel port allows print data to be sent from the PC to the printer and data indicating printer status to be received by the PC. This data, sent by the PC, uses eight wires, to transmit a *byte* of information to the printer. A byte is simply a group of eight *bits* used together to make a unit of data. Each wire is used to transmit one bit of data at a time. Each bit of data can have one of two possible logic values, 1 or 0. Another nine wires are used to allow the PC to determine the state of the printer and control the flow of data. These nine lines are broken into a set of five input lines and four input/output lines as shown in Figure 2-2.

The physical connection to this port is through a 25-pin connector known as a 'D25F' connector (where the 'D' refers to the shape of the connector body). The 25 contacts making up this connector are all sockets (female type, hence the 'F' in 'D25F') which mate with the printer cable connector having 25 pins (male type).

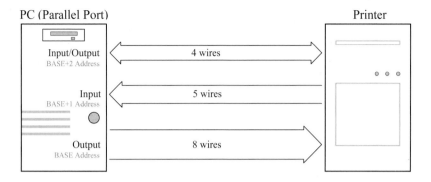

Figure 2-2 Parallel Port Configuration.

The three sets of wires shown in Figure 2-2 show the connection between a PC's parallel port and an external device, in this case a printer. Each group of wires are controlled, or read, by accessing three sequential locations in the PC's *Input/Output address space*, abbreviated to I/O address space. This address space is made up of a number of data storage locations used to allow intercommunication with input/output devices. It is different from the memory generally used by the computer. The PC writes data to particular I/O addresses, where the data is stored and can be accessed by external devices. Other I/O addresses are used to allow external devices to write data into storage for the PC to read, and still other I/O addresses allow bi-directional data transfer.

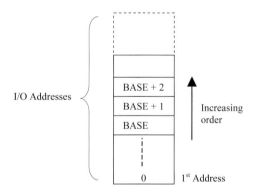

Figure 2-3 I/O Addressing.

The first of the three I/O addresses is referred to as the BASE address as shown in Figure 2-3. It is the lowest address and is used as a reference from which to increment to the other two I/O addresses belonging to the parallel port. Writing to the BASE address will output eight bits of data (a *byte*) from the parallel port (see Figure 2-2), where each bit uses an individual wire.

The next address in this block has a numerical value one more than the BASE address, so we label it the BASE+1 address. The BASE+1 address has access to the five input data bits to the PC. This address can only be used to read the state of these five signals.

The third address of this set is labelled the BASE+2 address, being two addresses past the BASE address. This address location is used to control the four bi-directional data bits of the port. Using this address, we can read and write to these four bits.

> **NOTE**
>
> Beware: the four **BASE+2** lines used for input and output are NOT 'strict' logic outputs. The parallel port interface often has resistors and capacitors connected to these lines to reduce the influence of electrical noise. This causes their states to change much slower than a strict logic output, meaning that erroneous recognition of data can occur when connecting with certain types of logic families.
>
> In addition, due to variation in the individual capacitor values, these signals do not switch at 'exactly' the same time (synchronously). This non-synchronous (asynchronous) switching of **BASE+2** outputs can cause data transfer problems with data interfaces designed to work synchronously.

Table 2-1 provides a summary of the data bits and D25 connector pins that the parallel port connector uses for each of the three port addresses. Each wire in the cable linking the port to the external device (usually a printer), carries the signal of a particular data bit for that port address. The BASE and BASE+2 addresses have their data bits commencing from D0 upwards. The BASE+1 address, however, starts at data bit D3.

Some data bits used by BASE+1 and BASE+2 addresses are inverted by the parallel port circuitry. These inverted bits are marked by a " / " character preceding the letter "D" of that bit. This signal convention is also used on the interface board *schematic diagrams* which show detailed electrical interconnections. When using these data bits, the program must compensate for this inversion in order that signals are output from the port or read in through the port as intended.

If a program needs to send a data bit out as a signal through one of the port's inverted bits, it needs to invert that data bit in software beforehand. This double inversion (once in hardware and again in software) has the effect of correcting the signal back to the intended state. Likewise, when a signal is read through an inverted bit of the port, the now inverted signal must be inverted once more by the program to correct it. The program implements this inversion using one simple line of code, explained in Section 3.6 of the next chapter.

Table 2-1 Parallel Port D25 Connector Pin Assignment.

BASE Address (8-bit *output* data)	BASE +1 Address (5-bit *input* data)	BASE+2 Address (4-bit *input/output* data)
D0 - pin 2		/D0 - pin 1
D1 - pin 3		/D1 - pin 14
D2 - pin 4		D2 - pin 16
D3 - pin 5	D3 - pin 15	/D3 - pin 17
D4 - pin 6	D4 - pin 13	
D5 - pin 7	D5 - pin 12	
D6 - pin 8	D6 - pin 10	
D7 - pin 9	/D7 - pin 11	

Note: "/ " denotes the signal bit is inverted internally by the parallel port circuitry.

D25 pin numbers 18 to 25 are not shown in Table 2-1. They are all connected to the PC electrical 'ground' which is connected to the interface board through the interface cable (Figure 2-4). This cable has a D25 male connector at both ends, connected by individual wires in a "one-to-one" arrangement (D25 pin 1 of one connector to the D25 pin 1 of the other connector; likewise for all remaining pins).

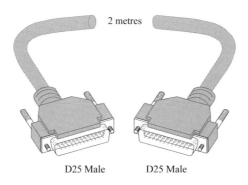

2 metres

D25 Male   D25 Male

Figure 2-4 D25M to D25M Cable.

NOTE

Data bits D0 to D2 of BASE+1 address are not connected to the parallel port circuitry inside the PC. The same holds true for D4 to D7 of the BASE+2 address. Reading these particular bits will produce invalid data.

2.3 Data Representation

As mentioned previously, computers use 'ON' and 'OFF' states (high and low voltages) to store data, termed *binary* data since we have only two states. This leads to the representation of numbers using the binary (on/off) number system. The binary system is based on raising the number two to increasing integer powers to form higher and higher digit values. We can see how such a system works by comparing it with our familiar *decimal* number system. Decimal numbers are based on raising the number ten to higher and higher integer powers.

For example, the decimal number 25 is broken down as follows:

$$\mathbf{25} \quad = \quad \mathbf{2x10^1 + 5x10^0}$$
$$= \quad \mathbf{2x10 + 5x1}$$

Decimal 25 is equivalent to the binary number of 11001 as follows:

$$\mathbf{11001} \quad = \quad \mathbf{1x2^4 + 1x2^3 + 0x2^2 + 0x2^1 + 1x2^0}$$
$$= \quad \mathbf{1x16 + 1x8 + 0x4 + 0x2 + 1x1}$$
$$= \quad \mathbf{25} \text{ (decimal)}$$

NOTE

The binary digit to the far right has the lowest weighting and is known as the *least significant bit* (LSB). Conversely, the left-most binary digit has the highest weighting and is termed the *most significant bit* (MSB).

Binary numbers with many digits are not easy to read. To solve this problem we use a more convenient number representation named *hexadecimal*. This system is based on sixteen number states.

The decimal number system uses ten unique Arabic numerals being 0, 1, 2, ..., to 9. In hexadecimal representation we need *sixteen* unique numerals. The first ten hexadecimal digits use Arabic numerals 0, 1, 2, ..., to 9, however, we must use unique digit representation for the remaining numbers ten to fifteen. This is done by using capital letters A, B, C, D, E and F to represent ten, eleven, twelve, ..., to fifteen.

Table 2-2 illustrates numerical conversion between decimal, binary and hexadecimal numbers.

Table 2-2 Number System Conversions.

Decimal	Binary	Hexadecimal
$0 = 0 \times 10^0$	$0 \quad = 0 \times 2^0$	$0 = 0 \times 16^0$
$1 = 1 \times 10^0$	$1 \quad = 1 \times 2^0$	$1 = 1 \times 16^0$
$2 = 2 \times 10^0$	$10 \quad = 1 \times 2^1 + 0 \times 2^0$	$2 = 2 \times 16^0$
$3 = 3 \times 10^0$	$11 \quad = 1 \times 2^1 + 1 \times 2^0$	$3 = 3 \times 16^0$
$4 = 4 \times 10^0$	$100 \quad = 1 \times 2^2 + 0 \times 2^1 + 0 \times 2^0$	$4 = 4 \times 16^0$
$5 = 5 \times 10^0$	$101 \quad = 1 \times 2^2 + 0 \times 2^1 + 1 \times 2^0$	$5 = 5 \times 16^0$
$6 = 6 \times 10^0$	$110 \quad = 1 \times 2^2 + 1 \times 2^1 + 0 \times 2^0$	$6 = 6 \times 16^0$
$7 = 7 \times 10^0$	$111 \quad = 1 \times 2^2 + 1 \times 2^1 + 1 \times 2^0$	$7 = 7 \times 16^0$
$8 = 8 \times 10^0$	$1000 \quad = 1 \times 2^3 + 0 \times 2^2 + 0 \times 2^1 + 0 \times 2^0$	$8 = 8 \times 16^0$
$9 = 9 \times 10^0$	$1001 \quad = 1 \times 2^3 + 0 \times 2^2 + 0 \times 2^1 + 1 \times 2^0$	$9 = 9 \times 16^0$
$10 = 1 \times 10^1 + 0 \times 10^0$	$1010 \quad = 1 \times 2^3 + 0 \times 2^2 + 1 \times 2^1 + 0 \times 2^0$	$A = 10 \times 16^0$
$11 = 1 \times 10^1 + 1 \times 10^0$	$1011 \quad = 1 \times 2^3 + 0 \times 2^2 + 1 \times 2^1 + 1 \times 2^0$	$B = 11 \times 16^0$
$12 = 1 \times 10^1 + 2 \times 10^0$	$1100 \quad = 1 \times 2^3 + 1 \times 2^2 + 0 \times 2^1 + 0 \times 2^0$	$C = 12 \times 16^0$
$13 = 1 \times 10^1 + 3 \times 10^0$	$1101 \quad = 1 \times 2^3 + 1 \times 2^2 + 0 \times 2^1 + 1 \times 2^0$	$D = 13 \times 16^0$
$14 = 1 \times 10^1 + 4 \times 10^0$	$1110 \quad = 1 \times 2^3 + 1 \times 2^2 + 1 \times 2^1 + 0 \times 2^0$	$E = 14 \times 16^0$
$15 = 1 \times 10^1 + 5 \times 10^0$	$1111 \quad = 1 \times 2^3 + 1 \times 2^2 + 1 \times 2^1 + 1 \times 2^0$	$F = 15 \times 16^0$
$16 = 1 \times 10^1 + 6 \times 10^0$	$10000 = 1 \times 2^4 + 0 \times 2^3 + 0 \times 2^2 + 0 \times 2^1 + 0 \times 2^0$	$10 = 1 \times 16^1 + 0 \times 16^0$
$17 = 1 \times 10^1 + 7 \times 10^0$	$10001 = 1 \times 2^4 + 0 \times 2^3 + 0 \times 2^2 + 0 \times 2^1 + 1 \times 2^0$	$11 = 1 \times 16^1 + 1 \times 16^0$

When developing programs, it is sometimes necessary to output digital signals through a port as one or more bytes of data. The signals to be output form binary bit patterns, which are more conveniently represented within program code as hexadecimal numbers. At other times you will need to represent incoming binary data sent from external devices as hexadecimal numbers. The following examples demonstrate the conversion of a binary number into hexadecimal.

We can obtain the hexadecimal representation for a binary number if we divide the binary number into groups of four digits starting from the least significant digit or bit (LSB, right-most digit of the number). Note that when we break a byte into two groups of four bits, we have what is termed two nibbles of data.

$$10001 \quad = \quad 1 \quad 0001$$
$$\quad\quad = \quad 1 \quad 1 \quad\quad hex \quad \text{(hexadecimal numbers are often written using a 0x prefix , i.e. 0x11)}$$

$$1010001101 \quad = \quad 10 \quad 1000 \quad 1101$$
$$\quad\quad = \quad 2 \quad 8 \quad D \quad hex \quad (0x28D)$$

Alternately, hexadecimal is denoted by using the **$** or **H** symbols, i.e. $11 or 11H.

From the preceding example, if we want our program to store data at the binary address 1010001101, we would manually convert this number into hexadecimal format as 0x28D and use this hexadecimal number as the storage address in our program. Had we converted the binary number into decimal format as 653, we would need to carry out a more involved conversion.

Now we are familiar with hexadecimal representation, we can use this notation for our parallel port address. The BASE address of the parallel port is 0x378 for most PC's, however, in some instances the parallel port uses a BASE address of 0x278, 0x3BC, or 0x300. Working with the more common BASE address of 0x378, we have the BASE+1 address of 0x379 and BASE+2 address as 0x37A.

2.4 Program Demonstrating Hexadecimal to Decimal

The program shown in Listing 2-1 can be used to convert numbers from decimal to hexadecimal and vice-versa. It is often more convenient to use hexadecimal notation in program code when outputting bit patterns through the port than decimal notation. Conversely, it is at times most convenient to display on-screen, the decimal number representation for the bit patterns read in through the port. Use this program to improve your familiarity between decimal and hexadecimal number systems.

Listing 2-1 Program to display numbers in decimal and hexadecimal format.

```
/******************************************************
This program prints a number you would enter in
decimal format in hexadecimal format.
******************************************************/
#include <iostream.h>

void main()
{
    int Number;

    cout << "Enter an integer number -> ";
    cin >> Number;
    cout << "The number is:" << endl;
    cout << dec << Number << " in decimal" << endl;
    cout << hex << Number << " in hexadecimal" << endl;
}
```

In Listing 2-1, the include file `iostream.h` facilitates the use of `cout` and the number conversion argument `hex`. The variable, `Number`, will receive the integer you will pass to the program in response to the prompt `"Enter an integer number -> "`. The same number is then printed on two successive lines, first in decimal format, which is the default number representation, and then in hexadecimal format. The hexadecimal format is activated by the format specifier `hex`.

2.5 Summary

This chapter explained the configuration of the parallel port and the digital logic concepts involved when using the port with external circuitry. These concepts include binary notation, digital logic levels, noise margins and different types of logic families such as CMOS and TTL.

The PC parallel port uses three I/O addresses to transfer data through the port's interface. Each address controls one byte of data, however, for two I/O addresses, several data bits are unused and a few other bits are inverted internally by the port circuitry. The first I/O address is used to output data only, the second address is used to input data through the port and the last address can be used to both input and output data. Furthermore, representation of data using decimal, hexadecimal and binary number systems has been explained. This knowledge will be used when developing programs in the chapters ahead.

2.6 Bibliography

Bergsman, P. , *Controlling The World With Your PC*, HighText Publications, San Diego, 1994.

IBM, *Technical Reference – Personal Computer AT*, IBM Corporation, 1985.

NS CMOS, *CMOS Logic Databook*, National Semiconductor Corporation, 1988.

Wakerly, J. F., *Digital Design Principles and Practices*, Second Edition, Prentice Hall, 1994.

3

Testing the
Parallel Port

Inside this Chapter

- Simple testing of the port.

- Power supply, port interface, logic buffers and driving LEDs.

- C-style programming to use the entire parallel port.

3.1 Introduction

The aim of this chapter is to develop software that will enable us to see the basic input/output operations when using the parallel port of your PC. Program operation is verified using a simple test circuit with LEDs as indicators. Operation of the on-board power supply, parallel port interface and LED Driver circuits is explained before directing the reader to build these circuits. Once these circuits are built (in the order presented), programs can be written and tested.

We will start our program development by writing simple non-object-oriented programs to input and output data through the parallel port. Later on, in Chapter 5, you will be introduced to object-oriented programming (OOP). When you reach the end of this chapter, you will have seen operation of the parallel port for data transfer and you will have gained an understanding of how a simple C++ program is written.

3.2 Interface Board Power Supply

All circuitry on the interface board requires electrical power at a steady voltage in order to function properly. A power supply as part of the interface board generates the individual voltages needed by its different circuits. This section of the board should be assembled and tested first before assembling any other circuitry. You should proceed to assemble the next segment of circuitry only when the power supply is generating correct voltages.

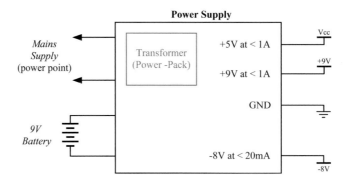

Figure 3-1 Power Supply Block Diagram.

A block diagram of the power supply is shown in Figure 3-1. Most of the power used by the interface board is supplied from the high voltage power point (known as *mains* supply) via the transformer (power-pack). The power supply +5V and +9V outputs can operate at their fixed voltages when supplying currents up to approximately 1A. This maximum current rating applies when the correctly rated heatsink is fitted to each of the voltage regulators. The −8V output does not use mains supply; instead a 9V battery is used to meet its voltage and low current requirements. The −8V output powers analog circuitry which has a wide operating margin for its supply voltage, and hence there is no need for voltage regulation.

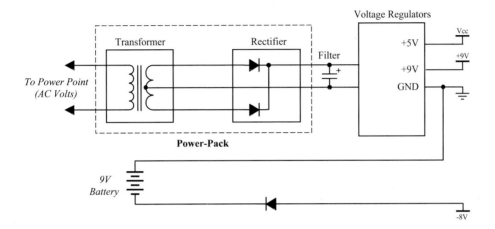

Figure 3-2 Power supply sub-circuits.

Figure 3-2 shows a more detailed functional diagram of the entire power supply. The 9V battery passes electrical energy through a diode that allows current to flow only if the battery is connected with correct polarity. When current is drawn through this circuit, the diode drops approximately 1V, producing a −8V low current power source.

The portion of the power supply using mains voltage is made up of four sub-circuits: a transformer, a rectifier/filter and two voltage regulators (+5V and +9V). The combined effect of these sub-circuits is to take the alternating mains voltage and convert it into stable +5V and +9V DC voltages, free of oscillation.

The alternating mains voltage needs to be reduced in amplitude and then rectified before it can be of use to our 'low voltage' circuits. The transformer inside the Power-Pack carries out this function. It has two diodes fitted inside to allow current to flow in one direction only, rectifying the sinusoidal waveform from the transformer as shown in Figure 3-3.

The voltage regulators will not function properly unless the voltage fed to their input voltage terminals is at least several volts greater than the regulator output voltage. A large value capacitor is added to the output from the power-pack to prevent its output from repeatedly dipping to zero, and instead dipping within acceptable levels as shown in Figure 3-4.

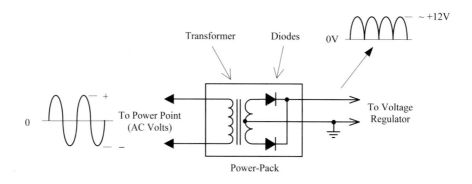

Figure 3-3 Power-Pack Power Supply (without capacitor).

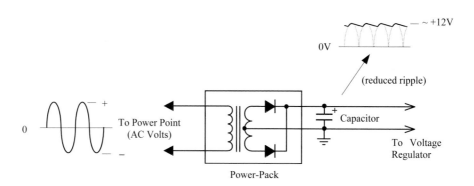

Figure 3-4 Power-Pack Output with Capacitor added.

The voltage regulators accept a rippling input voltage as can be seen in Figure 3-5, and use their internal circuitry to steady their output voltage to within a few percent of their rated voltage. For example, the +5V regulator will produce an output voltage that lies between +4.75V and +5.25V. The voltage regulators on the interface board have several capacitors connected between their input pins and ground, and between their output pins and ground. These capacitors prevent the output of each regulator from oscillating at "high frequencies" and improve the ability of the regulator to respond to fast transient loads.

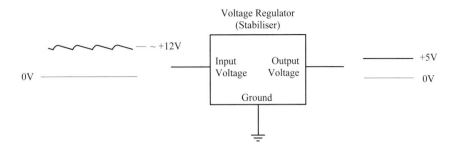

Figure 3-5 Voltage Regulator (without capacitors).

The hardware for the power supply should now be assembled and tested. Appendix A contains guidance for hardware assembly, soldering, schematic diagram conventions, testing and debugging, printed circuit board preparation and checking. This material should be read before commencing with the assembly and testing of the power supply. The power supply schematic diagram, bill of materials and test instructions are also to be found in Appendix A.

3.3 Parallel Port Interface

We will often need to check the correct operation of our programs during their software development cycle. The interface board has circuitry that allows us to test the functionality of the entire parallel port and thus provide feedback on program execution. A block diagram of this circuit is shown in Figure 3-6.

The signals from the parallel port are connected through the interface cable to the interface board D25 connector. From here, eight of the signals connect with a Buffer integrated circuit. These signals are generated by writing data to the BASE address. A second Buffer IC on the interface board is used to send five signals to the PC's BASE+1 Address, via pcb tracks connected to the D25 connector. A further four signals controlled by the BASE+2 address, pass through the D25 connector to individual resistors. This part of the port can both output data or input data.

Permanent damage often occurs when output pins of IC's are accidentally connected together without any means of limiting the resulting currents. Resistors are connected in series with BASE+2 signal lines to limit the currents and minimise damage should any of these lines from the PC be improperly connected to other outputs on the interface board.

Note that the two 8 way Buffer ICs shown in Figure 3-6 also have *pull-up* resistors fitted to their input pins (resistors not shown). Their function is explained in the Parallel Port Interface section of Appendix A.

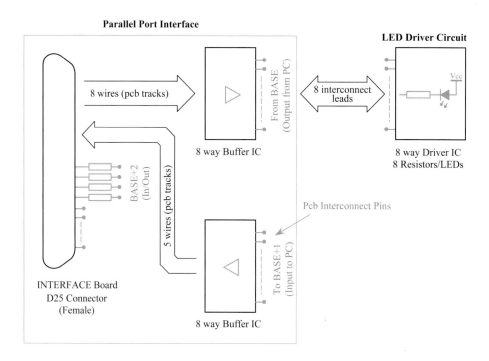

Figure 3-6 Parallel Port Interface & LED Driver Block Diagram.

The circuitry shown in Figure 3-6 has small pins with dots at their ends. These dot symbols represent printed circuit board pins that allow interconnection between other circuits on the board. This is made possible by using an interconnecting lead (shown in Figure 3-7) to connect between pins. You can fabricate these leads as required by following the instructions given in Appendix A.

Each connection made with an interconnecting lead connects an *output* pin of a circuit to an *input* pin of another circuit. DO NOT at any time connect outputs pins to other outputs pins. Doing so will most likely damage the components involved.

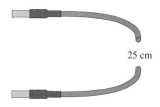

Figure 3-7 Interconnecting Lead.

Also shown on the diagram of Figure 3-6, is a resistor and light emitting diode (LED), representative of a group of eight such devices. This circuit is used to indicate the state of data coming from the PC or the interface board itself. Check the data generated within the interface board by connecting the relevant output pins of the circuit to individual resistor and LED pairs.

To test programs that read data into the PC through the parallel port, the interface board has a number of pcb pins permanently connected to either +5V or GND. This arrangement allows us to test any of the four or five input bits of the PC parallel port that use BASE+1 and BASE+2 addresses respectively.

Proceed to assemble the parallel port hardware and test for correct operation. Appendix A contains the schematic diagram, hardware assembly and test instructions for this circuitry. Several interconnect leads will need to be made for test purposes.

3.3.1 LED Driver Circuit

Figure 3-6 shows eight logic outputs (that originate from the parallel port) connected to a Buffer via pcb tracks. Unfortunately, like most digital logic circuits, a single logic output from the parallel port circuit does not have the capacity to pass sufficient current through a LED; hence the need for the Driver IC. Each output pin from the Buffer is connected to an individual pcb pin. The interconnect lead wires connect the Buffer output pins to the input pins of the LED Driver IC.

The Driver used on this board has the part number ULN2803A. It houses a bank of internal transistors, each one well suited to accept the limited current from a logic output pin and then to drive a LED. Most LEDs require between 5mA and 20mA of current through them to glow with adequate brightness. The ULN2803A Driver switches current flow independently through each of the eight LEDs and resistors, and operates as follows:

1. When a Driver input pin (on the left side of the Driver) is taken to a *high* voltage level, the corresponding output pin is switched internally to ground voltage. This allows current to flow from Vcc (equal to +5V), through the LED, and resistor, and through the Driver output pin to ground, causing the LED to glow.

2. When the Driver input is driven to a *low* voltage level, the corresponding output pin connection path to GND will become highly resistive. This reduces current flow through the LED and resistor to extremely low levels and extinguishes the glow of the LED.

3.3.2 LED Operation

A minimum amount of current needs to flow through a light emitting diode (LED) for it to light. LEDs, like most diodes, conduct significant levels of current in one direction only during normal operation. The current flows from the *anode* (denoted by a triangle) to the *cathode* (denoted by a bar) as shown in Figure 3-8.

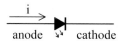

Figure 3-8 Conventional Current through a LED.

For current to flow in this direction, the anode must have a more positive voltage applied than the voltage at the cathode (known as *forward voltage*, V_F). This difference in applied voltage (V_F) typically needs to be approximately 2V for most LEDs and approximately 0.7V for ordinary diodes, if they are to conduct.

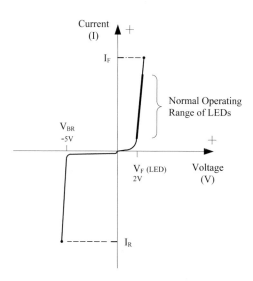

Figure 3-9 Typical LED characteristic curve (without a series resistor).

A characteristic curve for a LED is shown in Figure 3-9. This curve shows the current through the LED and voltage across the LED for its normal range of operation, indicated by the **bold** portion of the curve at the right side of the current axis. When using the LED in its normal operating range, an increase in current through the LED will increase its light output. When the current exceeds the maximum limit for the LED (indicated by the value I_F), the device will be destroyed. A current limiting resistor, shown in Figure 3-10, is used to control the current and prevent failure due to excessive current.

Note that if we reverse the voltage applied across the LED, we will reach the LEDs *reverse breakdown voltage* (V_{BR}, shown as –5V in Figure 3-9). Once the reverse breakdown voltage has been exceeded, reverse current will increase to the point at which the device is destroyed (I_R).

When we generate sufficient forward voltage to make the LED conduct, we say we have *biased* the device to operate. As mentioned previously, the LED will be destroyed if we do not use a resistor to limit the current flowing through it. The amount of current flowing through a resistor depends on the voltage across it. Since the LED and resistor are connected as shown in Figure 3-10, they share the same current. This circuit arrangement is known as a *series* circuit. Current through this circuit is analysed as follows.

If we know the voltage across the resistor we can work out the current flowing through it (and the LED) since Current = Voltage/Resistance.

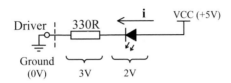

Figure 3-10 Current flow through the Series Circuit.

We know that the voltage across a conducting LED is approximately 2V and we also know that the total voltage across the series circuit from VCC to Ground is 5V. Therefore, the voltage across the resistor is:

$$5V - 2V = 3V$$

The current (I) through a resistor is given by voltage across the resistor, divided by its resistance in Ohms (Ω):

$$\text{i.e. } I = 3V / 330\ \Omega = 0.009\ A \text{ (amperes are denoted by the symbol A.)}$$

The currents flowing in electronic circuits are often small fractions of an ampere and so the units of milliamps (mA) which is $1/1000^{th}$ of an amp are commonly used. Thus we have 0.009 amps which is 9 milliamps flowing through the LED and resistor.

Proceed to assemble and then test the hardware for the LED Driver circuit along with at least eight interconnecting leads, as explained in Appendix A.

3.4 Basic Output Using the Parallel Port

As described in Chapter 2, there are three addresses associated with the parallel port (typically being 0x378, 0x379 and 0x37A). Although we use the term parallel port, this 'port' is really three ports combined together. The simplest of the three ports is the one at address 0x378. In general, this port is only used for output, however, more recent computers have the capability to input data using this port address. Nevertheless, to maintain compatibility, the software we have developed only uses data output to port 0x378. We will write a program that outputs a byte of data via port 0x378 to light up the respective LEDs on the interface board.

To verify the proper operation of our program, we need to connect the interface board to the parallel port of the PC. This is done using the interface board cable described in Chapter 2. The remainder of the connections to be carried out on the interface board must be made according to Table 3-1 below. Note that pin 1 of an IC on the interface board can be recognised by its rectangular shaped pcb pad. Make these connections using the interconnecting leads assembled earlier. When the connections are complete, the signal lines of the port at BASE address (0x378) will be connected to the Driver circuit that lights the eight LEDs via the Buffer IC on the interface board.

Table 3-1 Connections for basic output.

BASE Address (Buffer IC, U13)	ULN2803A Pin No. (Driver IC, U3)
D0	1
D1	2
D2	3
D3	4
D4	5
D5	6
D6	7
D7	8

The program shown in Listing 3-1 has several lines of program statements followed by comments. It uses a library function outportb() to output a byte of data to the port at BASE address. The port address and the data can be specified as actual arguments to outportb() at the time of calling. Since outportb() is a library function, we do not have to provide the body of the function. It comes from the library and will be searched for when linking takes place.

Listing 3-1 Writing to the port at Base address.

```
/*****************************************************
WRITING TO A PORT (output operation)

This program outputs a certain bit pattern to the port at
BASE address to light the respective LEDs on the interface
board.
*****************************************************/
```

```
#include <dos.h>

#define BASE 0x378

void main()
{
    outportb(BASE,255); // in binary, 255 = 1111 1111

// The number 255 can be changed to any value betwen 0
// and 255, causing the eight output signals to
// correspond to the binary value represented by the
// number. For example 65 = 0100 0001.
}
```

As mentioned in Section 1.2.2, the two lines starting with the hash sign (#) are compiler directives. The first compiler directive will include the header file dos.h (which is a source file). This file contains the *prototype* of the outportb() function. The header file dos.h also contains information about many other functions. However, as far as this program is concerned, only the information regarding the outportb() function is needed. The compiler would not be able to process the outportb() function, and the program could not use it if we did not include this header file. Note that the prototype does not specify how the function is to be *executed*. In other words, until linking takes place, the program will not have access to the actual instructions contained in the function.

The second compiler directive is a define statement. It simply instructs the preprocessor to replace any occurrences of BASE in the program by the hexadecimal number 0x378. Using the word BASE instead of 0x378 makes the code more readable since it is easier to relate to the word BASE than a number. In addition, if we ever wanted to change the base address, we only need to modify the define statement and the preprocessor will automatically implement that change in address throughout the program. The define statement can be used when writing larger programs to simplify the task of coding and improve readability.

Next we encounter the main() function (the only function in this particular program) where all C++ programs start their operation. Usually a typical C++ program will have many functions coded (defined). The keyword void indicates that the main() function will not return any value. The body of the main() function starts with the open brace ({) just after the line void main() followed by its instructions. The only executable statement in this program is outportb(BASE,255). This statement is used to output a byte of data from the PC. After a few comments, the body of the main function ends with a close brace (}).

The function outportb() takes in two parameters; a port address and the data to be written to that address. In this program the port address is BASE, which will be replaced with 0x378 by the preprocessor. Therefore, the address of the port where the data is to be sent is `0x378`. The value of the data is 255 (decimal). It must be noted that the size of the data passed to a port is, most of the time, a byte. A byte can take 256 different values. When the value is 0, all eight bits of the byte will have their values as 0. When the value is 255, all eight bits of the byte will have their values set to 1. Other values between 0 and 255 will correspond to different bit patterns. After the interfacing connections given in Table 3-1 are made and the parallel port cable is connected to the interface board, this program can be compiled and executed. It will set all eight bits equal to 1 and as a result you should see all eight LEDs light up. If, for example, you change the `outportb()` line to `outportb(BASE,65)`, then only the LEDs corresponding to bits 0 and 6 will be lit and all other LEDs will be off.

If the program fails to work and light the LEDs, make sure your base address is correct by using the program titled '*base_adr.exe*' included with the accompanying CD-ROM.

Edit the program a number of times replacing the value of 255 with different numbers (less than 255) and observe how the LEDs light up on the board. This exercise will also help you understand how bit patterns relate to decimal data. as explained in Chapter 2. Alternatively, you may replace the number 255 with hexadecimal numbers; for example:

```
outportb(BASE, 0xF0);
```

3.5 Basic Input Using the Parallel Port

In this section we will learn how to write a program to read the port associated with the `BASE+1` address (0x379). This is the only port address of the three port addresses comprising the parallel port that is dedicated to input operations. We will be reading this port and displaying the number on-screen that corresponds to the input data received.

Table 3-2 Connections for basic input.

BASE+1 Address[*] (Buffer ID, U6)	Power Supply +5V and GND pins
D3	+5V
D4	GND
D5	GND
D6	+5V
/D7	+5V

* The signal preceded by a slash (/) is internally inverted by the parallel port hardware.

Once again, program operation can be verified by connecting the PC to the interface board using the interface cable. The remaining connections to be made are on the interface board and shown in Table 3-2. The incoming data can easily be changed by swapping the wiring for the second column of Table 3-2 between +5V and GND.

Use the interconnect leads to make the connections shown in Table 3-2. When the connections are complete, the signals on the interface board are connected to the port at address BASE+1 (0x379). Note that this port has only five signal lines out of eight that can be used. Three of the signal lines, namely the ones corresponding to Data Bits 0 to 2 are unavailable, as they are not dedicated to the port at BASE+1 in the PC parallel port.

The essential steps required in this program are:

1. Read the port.
2. Display the result on the screen.

Both these actions require very simple statements and do not justify coding extra functions. Therefore, just a main function will suffice. To read a port carrying 8 bits of data, we can use the library function inportb(). To display the result on the screen, we can use the library function printf(). It may be convenient to stop the program immediately after displaying the result so we can read the screen. This is especially useful when an Integrated Development Environment (IDE) is used to develop the program. When IDEs are used, the screen will automatically revert to the IDE's editor once program execution terminates. This will prevent the user from observing what happened on the screen at the time the program ran. However, we will be able to see the onscreen results if we make the program wait for a key press just before it ends. We can do this using the library function getch(). Therefore, the program must use the function sequence inportb(), printf() and getch() in that order. In order for us to be able to use these functions we must provide their prototypes, contained in the header files dos.h, stdio.h, and conio.h, respectively. This program is shown in Listing 3-2.

Listing 3-2 Reading the port at address BASE+1.

```
/*****************************************************
   READING THE PORT AT ADDRESS BASE+1

This program will read the port at address BASE+1 and
print the number read on the screen. The number is
formed by combining the five signal lines read in
through the port. Bits 0, 1 and 2 have no valid data as
they are not dedicated to BASE+1 internally in the PC.
Bits 3, 4, 5 and 6 will be read as normal. Bit 7 is
permanently inverted by the parallel port hardware.
*****************************************************/
#include <dos.h>
#include <conio.h>
```

```
#include <stdio.h>

#define BASE 0x378

void main()
{
    unsigned char InputData;      // Declare data type for
                                  // various InputData.

    InputData = inportb(BASE+1);  // Read port at BASE+1.

    printf("%2X\n",InputData);    // Print result to screen
                                  // as a hexadecimal number.

    getch();                      // Wait for key press.
}
```

NOTE

Detailed descriptions of the library functions can be found in the documentation that comes with the C++ development software. More recently, the descriptions are available on-line in help files. The documentation will also provide the names of the header files associated with each library function. This will enable you to determine the name of the header file that must be included in order to be able to use a particular function.

This program has compiler directives; three are include statements and the one is a define statement. The three include statements will include the header files dos.h, stdio.h and conio.h. The define statement will allow us to use the identifier BASE whenever we need to refer to the actual address 0x378.

The main() function begins with the line void main() whose meaning has been explained previously. The main() function like all functions can take in parameters. In this case it does not take any parameters, and so an empty pair of parenthesis are used to follow the word main. This signifies to the compiler that it does not take any parameters and ensures it is recognised as a function.

An example of a function that does take in parameters is the outportb() function used in Listing 3-1. It takes in two parameters - the address the data is to be written to and the data itself. Therefore, its parentheses are not empty.

The next statement (after the brace) in the program is:

```
unsigned char InputData;
```

As mentioned earlier, this is an identifier declaration. It simply informs the compiler the identifier's name (in this case, `InputData`) and the type of data the identifier is allowed to represent (in this case, `unsigned char`).

This variable is declared so it can be used to store the value returned by the `inportb()` function. This is an ideal place to understand the concept of return value of a function. The prototype of the `inportb()` function, as provided by the header file `dos.h` states that its return value is of type `unsigned char`. Therefore, the variable `InputData` must also be declared as `unsigned char` in order to receive the value returned by `inportb()`.

The next statement in the program is:

```
InputData = inportb(BASE+1);
```

This statement makes a call to the function `inportb()` and takes in one parameter, which is an address. The placeholder `BASE` is defined as `0x378`. Therefore, `BASE+1` will evaluate to be `0x379`. This value specifies the location of the port to be read. Therefore, this statement will read the second port associated with the parallel port of your PC. The value read by the `inportb()` function is then placed into the variable `InputData`.

The statement used to display the result on the screen is:

```
printf("%2X\n",InputData);
```

The `printf()` function is a C function – not C++. However, since C is a subset of C++ it can be used in C++ programs. When it comes to printing data on the screen, `printf()` offers good flexibility and ease of use. In our programs we use the good features of C with C++ to develop better applications. The function `printf()` is a relatively unusual and very useful function in that you can pass a *variable* number of parameters to it. Most of the functions you will be writing will have a *fixed* number of parameters. In the current example, the number of parameters passed to the `printf()` function is two. The first argument is the string of characters "`%2X\n`" and the second argument is the value of the variable `InputData`. These two arguments are separated by a comma.

C	Examples of frequently used format specifiers
`%10.3f`	Floating point format with a field width of 10 and 3 decimal places.
`%5d`	Integer format with a field width of 5.
`%c`	Character format.
`%s`	String format (used for a sequence of characters such as a sentence).
`%X or %x`	Hexadecimal format. `X` will print upper case hexadecimal letters and `x` will print lower case hexadecimal letters.

The first argument "%2X\n" is a format specifier that is used when printing the value of InputData on your screen. The characters %2X specify that a hexadecimal format of field width 2 is to be used. A carriage return and a line feed are specified by the two characters \n, the character 'n' known as the new line character.

To represent a byte of incoming data, two hexadecimal digits are needed since each hexadecimal digit represents 4 bits. Therefore, a field width of 2 is appropriate. After printing the number, the cursor on your screen will be positioned at the start of the next line.

The line:

```
getch();
```

is used to make the program wait for a key press and give us time to read what has been printed on the screen. The getch() function waits to receive a character from the keyboard and so the program will not proceed until a key is pressed. When this happens the program will terminate since there are no more statements to execute.

The operation of the program can be verified by interpreting the bit pattern of the hexadecimal value printed on-screen and then checking that this bit pattern corresponds to the actual signals generated on the interface board. You can change the connections on the interface board by changing the connections shown in the right-most column of Table 3-2. That is, you can re-arrange the connection of signals to ground and to +5V. If you run the program again, you should see a different result on the screen.

3.6 Compensating for Internal Inversions

Consider the program shown in Listing 3-2. When we read the port at address BASE+1 (0x379), one of the signals (bit D7) we read from the interface board was inverted by the parallel port hardware. Similarly, some of the signals at port address BASE+2 (0x37A) will be inverted by the hardware when output through this port (bits D0, D1, and D3). In this section we will learn how to modify our software to compensate for such inversions. This compensation can be done in software by simply inverting the affected bits to counteract the inversions that are made by the hardware.

3.6.1 Output Operation

The program shown in Listing 3-3 will write data to the port at address BASE+2. Note that this port only controls bits 0, 1, 2 and 3; bits 4, 5, 6 and 7 are not dedicated for internal use by the port at address BASE+2. Some of these bits that can be controlled are inverted internally by the parallel port electronics when output; bits 0, 1, and 3. Therefore, to nullify this inversion by hardware we must invert bits 0, 1 and 3 in software. Bit 2 is not inverted internally by the parallel port hardware, and so we do not not need to invert it in software.

The that need to be made on the interface board are shown in Table 3-3.

Table 3-3 Connections to the LED Circuit.

BASE+2 Address[†]	ULN2803A (Driver IC, U3)
/D0	D0 (1)
/D1	D1 (2)
D2	D2 (3)
/D3	D3 (4)

[†] The signals preceded with a slash (/) are internally inverted by the parallel port hardware.

Listing 3-3 Writing to the port at BASE+2 with compensation for internal inversions.

```
/********************************************************
WRITING TO PORT @ BASE+2, INTERNAL INVERSIONS COMPENSATED

This program outputs 4 bits of data to the port at address
BASE+2, compensating for the inverted bits 0, 1 and 3.
You can change the value of the actual bit pattern you
want to see output to the interface board.
********************************************************/
#include <dos.h>
#define BASE 0x378

void main()
{
// BASE+2 bits 0,1 and 3 are internally inverted by
// the parallel port hardware before being output. This
// can be compensated in software by carrying out an
// exclusive OR operation with the output data and 0x0B
// (0000 1011).  Bits 4-7 do not matter as they are not
// connected.

    outportb(BASE+2,0x0B ^ 0x0F);

// NOTE: In binary 0x0F = 0000 1111
// The number being output (0x0F) can be changed to any
// value between 0x00 and 0x0F. The four output signals
// will correspond to the binary bit pattern represented by
// the number.
```

```
// Examples:
//  Bit No:  7  6  5  4  3  2  1  0
//  0x0F     0  0  0  0  1  1  1  1
//  0x05     0  0  0  0  0  1  0  1
}
```

In this program, the only line that requires explanation is:

```
outportb(BASE+2,0x0B ^ 0x0F);
```

The `outportb()` function writes data to the port in a manner similar to its use before. In this case, the address of the port is BASE+2. The `define` statement defines BASE to be a placeholder for 0x378. Therefore, the value of the first parameter is 0x378+2, which is 0x37A. The value of the second parameter is the data we want to send out the port. This data is obtained, by evaluating:

```
0x0B ^ 0x0F
```

The operator '^' used in the above expression is known as the Exclusive-OR (XOR) operator. It is one of the many *bit-wise operators* available in C and C++ that is used to operate at bit level. You will have a better understanding of how bit-wise operators work once the operation shown in Table 3-5 has been explained. The operation of the exclusive OR operator will be described with the aid of Table 3-4. This operator requires two operands when used.

Table 3-4 Exclusive OR operation.

Operand A	0	0	1	1
Operand B	0	1	0	1
Result	0	1	1	0

NOTE

In the simple arithmetic operation:

3 + 5

the operator is '+' and the two operands are 3 and 5.

For a bit-wise operator, the operands must be *bits*. In Table 3-4 the two operands are given the names Operand A and Operand B. The results produced by the XOR operation for all four possible combinations of the two operands are listed in the 'Result' row. As can be seen, the result is 1, only when just one of the two operands in a column is 1. When both operands in a column are identical, the result is zero. So when the operands differ, the result is 1. As shown by columns 2 and 4 of Table 3-4, if we hold the Operand B at 1, the result will be the inversion of operand A. Operand B acts as a 'filter' for inverting specific bits of Operand A.

We use this operation to perform software inversions to counteract the internal inversions generated by the parallel port hardware. Table 3-5 explains the result of evaluating:

```
0x0B ^ 0x0F
```

Here Operand A contains the data to be sent out and Operand B is the "filter" used to invert the bits already inverted by the parallel port.

Table 3-5 Evaluation of `0x0B ^ 0x0F`.

Bit No. →	7 6 5 4 3 2 1 0
Operand A (0x0F)	0 0 0 0 1 1 1 1
Operation	XOR
Operand B (0x0B)	0 0 0 0 1 0 1 1
Result	0 0 0 0 0 1 0 0

As explained earlier, bit-wise operators operate on a bit-by-bit basis. In other words, bit 0 of Operand A and bit 0 of Operand B are put through an exclusive OR operation. Likewise, another exclusive OR operation takes place between bit 1 of Operand A and bit 1 of Operand B, and so forth.

The filter comprises data bits that we want inverted set to 1, and bits to be left as is set to 0. Thus, to invert bits 0, 1 and 3 of data to be sent out, the bits 0, 1 and 3 of the filter are set to 1. As can be seen, in the 'Result' row of Table 3-5, bits 0, 1 and 3 are the opposite values of bits 0, 1 and 3 of Operand A. Therefore, when we write the exact bit pattern we want as Operand A, the affected bits will be inverted by software to become the 'Result'. When those affected bits of the 'Result' are then sent to the parallel port hardware and internally inverted, the data arriving at the interface board will correspond to Operand A that we originally want to send out.

To verify operation of the program, you can change the data (i.e. 0x0F) to any value between 0x00 and 0x0F. The LEDs that light up will correspond to the binary bit pattern of the data specified in the program. Note that the filter value 0x0B must not be changed – otherwise not all those specific bits we want to invert (0, 1, and 3) will actually be inverted in software.

3.6.2 Input Operation

In program Listing 3-2, one of the signals being *read in* through the port at address BASE+1 was internally inverted by the parallel port hardware. Note that the port BASE+1 can only input the bits numbered 3, 4, 5, 6 and 7. Of these bits, bit 7 is internally inverted. Similar to the operation performed in the previous section, this internal inversion can also be compensated in software. This is done by performing an Exclusive OR operation using a 'filter' bit to toggle bit 7 as soon as the port is read. The value of the filter to be used is:

```
0x80 = 1 0 0 0  0 0 0 0
```

The required operation is shown in Table 3-6. Note that the unused and therefore invalid data bits D0, D1 and D2 are shown as 'x' in the example. These bits can be in either logic state and therefore the Exclusive OR result will also be indeterminate for these bits.

Table 3-6 Inversion of bit 7.

Bit no.	7 6 5 4 3 2 1 0
Operand A (data received)	1 0 1 1 1 x x x
Operation	XOR
Operand B (the filter, 0x80)	1 0 0 0 0 0 0 0
Result (corrected data)	0 0 1 1 1 x x x

Listing 3-4 Reading the port `BASE+1` with internal inversions compensated.

```
/*******************************************************
READING THE PORT @ BASE+1, INTERNAL INVERSIONS COMPENSATED

This program reads the port at address BASE+1 (0x379). It
compensates for the hardware inversion of bit 7 after
reading the data. The net result is as if the hardware
inversions had not taken place.
*******************************************************/
#include <dos.h>
#include <conio.h>
#include <stdio.h>

#define BASE 0x378

void main()
{
   unsigned char InputPort1;

   InputPort1 = inportb(BASE+1);
   InputPort1 ^= 0x80;
```

```
    printf("%2X\n",InputPort1);
    getch();
}
```

The only line that needs explanation in the program given in Listing 3-4 is:

```
InputPort1 ^= 0x80;
```

This statement is equivalent to the following statement:

```
InputPort1 = InputPort1 ^ 0x80;
```

And is of the form:

Result = Operand A ^ Filter;

Operand A stands for the raw data read from the port. `Result` stands for the compensated value. Consider the statement:

```
InputPort1 = InputPort1 ^ 0x80;
```

`InputPort1` on the right-hand side contains the raw data affected by the internal inversion of the parallel port hardware. The `InputPort1` shown on the left-hand side is the result obtained by carrying out an Exclusive OR operation between the raw data and the filter value 0x80. In other words, the value of `InputPort1` is Exclusive-ORed with the filter value and then this result is stored back into the `InputPort1` variable. The `printf()` statement then prints the compensated value on the screen. As a result, the number appearing on the screen should represent the actual signal levels connected on the interface board.

3.7 Summary

In this chapter we have explained the operation of the interface board power supply, port interface, and LED Driver circuits. These circuits allow the parallel port of the PC to interface with the interface board and test operation of programs.

We learned how to develop C++ programs for sending and receiving bytes of data through the three addresses associated with the parallel port of the PC. These programs printed their results to the screen using either the `cout` object (as we did in Chapter 1), or using the functions of the `printf()` family. We also explained a small subset of the format specifiers that the `printf()` function uses.

The Exclusive OR bitwise operator was used to toggle some of the data bits we transmitted through the parallel port. Bitwise operators are a very useful part of the C and C++ languages and allow us to manipulate specific bits within a byte.

3.8 Bibliography

Bergsman, P. , *Controlling The World With Your PC*, HighText Publications, San Diego, 1994.

IBM, *Technical Reference – Personal Computer AT*, IBM Corporation, 1985.

NS LINEAR *Databook*, National Semiconductor Corporation, 1987.

Savant, C.J., et al., *Electronic Design Circuits and Systems*, Second Edition, Benjamin-Cummings, Redwood City, 1987.

Kelley, A. and I. Pohl, *A Book on C – programming in C*, Benjamin Cummins, 1995.

House, R., *Beginning with C – An Introduction to Professional Programming*, International Thompson Publishing, 1994.

Deitel H.M. and P.J. Deitel *C: How to Program*, Prentice Hall, 1994.

C programming Language – an applied perspective by L Miller and A Quilici, John Wiley Publishing, 1987.

Hanly, J.R., E.B. Koffman and J.C. Horvath, *C Program Design for Engineers*, Addison Wesley, 1995.

Rudd, A., *Mastering C*, John Wiley, 1994.

4

The Object-Oriented Approach

Inside this Chapter

- What exactly is object-oriented programming (OOP)?

- Encapsulation.

- Member data and member functions of an object.

- Inheritance and Polymorphism.

- Constructors and the Destructor.

- Abstract classes.

- Class Hierarchies.

4.1 Introduction

In this chapter we will explain object-oriented programming concepts that apply to C++ programming. Object-oriented programming is the newer way of developing software. The superseded style of developing software is known as *procedural* programming.

In procedural programming, data and functions can be thought of as being 'out in the open'. In this case data may be used and/or altered by any function, and inadvertent misuse is common, often causing detrimental side-effects to a program's operation. Also, any changes or modifications to the existing software can cause problems that are very difficult to debug.

In this chapter we use examples from everyday life to gain a qualitative understanding of the concepts that apply to object-oriented programming. We then take a detailed look at the terminology associated with object-oriented programming and define these concepts. By the end of this chapter, we expect you to have developed a good understanding of the object-oriented concepts and terminology you will need when we commence with object-oriented programming in the next chapter.

4.2 Conceptual and Physically Realisable Objects

Objects in object-oriented programming resemble the objects we encounter in real-life. An object can be viewed as a self-contained entity, which has some sort of a description and some uses associated with it. The description may list all the features of the object. The uses associated with the object may be a set of functions the object carries out for us, or a set of functions we carry out on the object, or a combination of both types of functions. A software representation of such an *object type* in object-oriented programming terms is referred to as an *object class*. The C++ language provides mechanisms to list all the features or properties of an object class and all the functions associated with the object class.

In real-life, we have descriptions of real objects that physically exist and also descriptions of abstract objects that are either imaginary or conceptual. Real objects are tangible whereas conceptual objects are not. While tangible objects can be duplicated, conceptual objects cannot be duplicated for the simple reason that there is no such thing as "two identical concepts". However, it is possible to have "two identical objects". The conceptual objects (abstract objects) can be developed into physically realisable objects by including exact definitions of every detail of the object, at which stage a real object may be produced. Then the object definition has passed through the transition from abstract to real.

To better understand this concept, consider a *vehicle* as an object. The most likely description of a vehicle that comes to mind is an object that is used to transport people or goods, perhaps rolling on wheels, with some form of energy to drive it along (as shown in Figure 4-1 a). A manufacturer cannot proceed to build a vehicle unless they have determined its specific details. Is the vehicle a train, a bus, a car or something else? If it is a car, is it a small car, a medium car or a large car? What is its engine type and capacity? How many doors should it have? This refinement must continue until every detail of the "vehicle to be built" has been defined. At this stage the object is no longer an abstract object. Once completely defined, any number of "cars" can be manufactured as real objects.

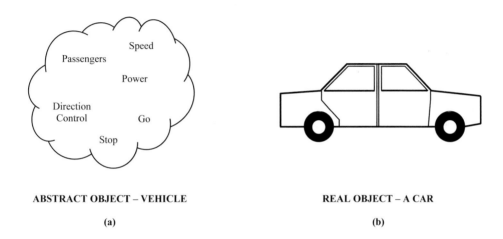

ABSTRACT OBJECT – VEHICLE   **REAL OBJECT – A CAR**

(a)   (b)

Figure 4-1 The concept of abstract objects and real objects.

4.3 Real Objects

Although the cars of this class are identical, when built, they are individual items. Each of these cars will have its own engine, fuel system, braking system, etc. If someone is asked to "Start the engine!", they cannot start an engine without knowing which car is to be started. Accordingly, the word "engine" does not uniquely identify "an engine". To be able to do so, the engine must be *tagged* with a particular car. For example, suppose three cars have been built and they are labelled A, B, and C. Then "Engine of Car A" will uniquely identify an engine. Thus, while "Car" is an object type, "Car A" is an actual object. It is important to understand the difference between the object type and the actual object. The existence of an object type does not necessarily mean that actual objects of that type exist. However, an actual object cannot exist without having its object type in existence.

The terms such as engine, fuel system, braking system, etc. can be used to generally describe those systems. For example, to describe features of the engine of a particular car *type*, we do not have to say the engine of the '*blue*' car has such and such features. We would simply say engines of this particular car *type* have these features. On the other hand, if an engine of a car is faulty, we must specifically refer to that car by saying 'the engine of that *blue* car' is faulty and needs to be repaired. Therefore, when an object type is described, we can use its terms without having to tag them to a physical object.

4.3.1 Public Interface of an Object

A motorcar was a sophisticated object even in early times. There are certain parts of a motorcar that the user is not expected to access; for example, the fuel injection system. The fuel injection system is not directly accessible to the user; nevertheless, it carries out its functions behind the scenes. On the other hand, the driver of the motorcar has direct access to the steering wheel, the brake pedal, the indicator stalk, etc. These can be referred to as the public interface of the object "Car". Similarly, in object-oriented programming, every object must have a public interface for it to be useful. An object without a public interface is like a perfectly built car, which is completely encased and sealed off so that no one can ever get into it to drive it.

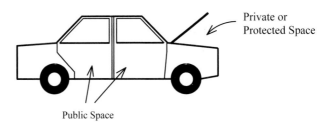

Private or
Protected Space

Public Space

Figure 4-2 The public interface of an object.

The software objects will also have private functions with some designated purpose that are not directly accessed by the users of the object (like the fuel injection system of a motor car). Therefore, in the most general case, an object is an encapsulated entity with a public interface that has restricted access to its hidden details. How objects can be realised in software will be understood as we proceed through the development of object classes in the coming chapters.

4.3.2 Construction and Destruction of Objects

All cars come to life through some kind of a manufacturing process, which we may refer to as the "construction process". If needed there can be slight variations to the cars built according to the 'same' plan – for example, cars that have different colour. Furthermore, manufacturing processes themselves could be slightly different. In one factory, cars may be built starting from sub-assemblies. In another factory, cars may be built from scratch. In either case, the same *type of* car will be produced.

After the car has been brought to physical reality, it can then be driven. At the end of the car's life it will undergo some sort of destruction process which could take place in a car disposal yard. Destruction is an important process, which carries out the disposal of unwanted items to maintain environmental cleanliness and to allow the efficient management of resources. In the case of computer programs, object destruction is necessary for the efficient management of memory.

The techniques available in object-oriented software development provide mechanisms to realise all these aspects. Although we can draw analogies between the objects in our day-to-day life and the software objects, the real power of object-oriented programming in C++ comes from the combination of object-oriented programming techniques and C++ programming techniques.

4.4 Object Classes

An *object class* describes a particular *type* of object. The object class does not refer to a real object but the type of those real objects yet to be created. If we take an analogy from everyday life; a building plan is analogous to an object class – not the building itself. The same building plan can be used to create any number of buildings. Likewise, the same object class can be used to create any number of objects. Each object that will be created according to an object class will reside in the memory of your computer. If more objects of the same type are created, more memory will be used.

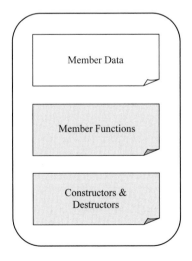

Figure 4-3 Components of an object class.

Each object class will have its own name, also known as the object *type* name. The word `class` is a keyword in C++. It is used to identify a detailed plan with a name. Programmers are free to choose class names when developing new classes.

Each object class must have a *class definition*. Similar to the case for real-life objects, the properties or features of the object and the functions associated with the object are listed in its class definition. The properties or features are given the name *member data* of the class and the functions associated with the object are referred to as *member functions*.

Among the member functions are two special types of functions. The first type can have more than one function per class and are known as *constructors*. Constructors are used by the program to create each individual object (that resides in memory). The second type has only one function per class and is known as a *destructor*. Destructors are used by the program to free the memory that a variable used once the variable is no longer needed. If a constructor is not provided, the C++ compiler will provide a default constructor (invisible to the programmer). Similarly, if a destructor is not provided, the C++ compiler will add a default destructor (also invisible to the programmer).

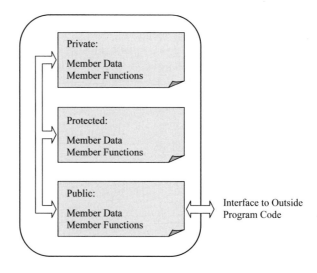

Figure 4-4 Public interface of an object class .

Some of the class members (both data and functions) can be publicly accessed; other members will have restricted access. The publicly accessible members provide the public interface of the object as shown in Figure 4-4. These include the public member data and public member functions. Members with more restricted access are known to have either private or protected access attributes.

4.5 Encapsulation

Grouping the member data and member functions together is known as *encapsulation*. There are some great benefits associated with encapsulation. First of all, encapsulation limits access to the internal details of the object. Access to the data is always through controlled and supervised means. In general, access will be restricted to most data members where data should not be freely accessible or changeable. Some of the functions may also have restricted access.

Access to the object is allowed only through the public interface. This 'restricted visibility' to member data and member functions is also known as information hiding. Sometimes it is not necessary to know the internal details of an object. In this case it is often sufficient to just be able to know how to use it. A good example is a scroll bar. It is sufficient for the programmer to know how to place a scroll bar in an application to make the screen scroll. The scroll bar's internal workings are hidden and normally are of no concernt to the programmer.

While all these restrictions are imposed to functions outside of the class, internally, any function of the class can manipulate or access any other member of the *same* class. This provides efficient operation of object-oriented programs for the following reasons. It eliminates the need to declare member data within the functions and also the need to pass member data as parameters to member functions. While an ordinary non-member function can only return one value, the member functions can return 'more than one value' by being able to change any number of the data members of the class.

4.5.1 Object Instantiation

The process of creating an actual object is known as object *instantiation*. Therefore, to instantiate an object of a particular class, one of the constructors of that class must be called. The constructor resembles "the manufacturing process" in our 'car' analogy. The constructor, like any other function is a function written in C++. It must be called to create a physical object of the object class. While the object class is of type 'Car', the actual objects created, which we refer to as class objects, could be named A, B, C, etc. They are all equivalent, perhaps with minor differences like their colour. In object-oriented programming, a class definition does not occupy memory. However, a class object does occupy memory. Likewise a building plan does not occupy any land, whereas the actual building occupies finite space.

In object-oriented programming, the constructor must be called to create an actual object. At the time of calling the constructor, you can pass parameters to it in a likeness to selecting the colour of a car to be built. In a class definition, there may be more than one constructor. In our car analogy, a car may be constructed from scratch or may be constructed from subassemblies.

4.6 Abstract Classes

Abstract classes can be viewed as initial conceptual class definitions which are insufficiently complete to carry out instantiation. Many good hierarchies of class objects will often start with an abstract class. The real power of abstract classes cannot be properly demonstrated until some advanced concepts of the C++ language are explained in the coming chapters.

An abstract class needs to be developed to become a real class that has working functions if it is to be able to instantiate real objects and use them. There is no such thing as 'an abstract object' for the simple reason that anything abstract does not physically exist. In our 'vehicle' example, there will definitely be a characteristic such as speed. However, we cannot discuss how to increase or decrease speed until we know more details of the actual vehicle. We can also include 'functions' to increase and decrease the value of the speed. However, we cannot define the exact steps these 'functions' should take to increase or decrease the speed without knowing whether the vehicle is a car, or a train, or something else. In an analogous class definition, these types of functions are known as *pure virtual functions*. They are *pure* because their actions (bodies) cannot be defined yet, and *virtual* because their actions (bodies) need to be defined in future derived classes to carry out their required tasks.

4.7 Class Hierarchies

A *class hierarchy* is a set of classes that are developed starting from one or more classes known as *base classes* (located at the root of the hierarchy). The new classes developed from the base classes are known as *derived classes*. The derived classes are more detailed and complete than their base classes. One of the most important benefits of object-oriented programming is *code reuse*. Developing class object hierarchies greatly facilitates the ability to reuse code by not having to rewrite code that new code is based on. Object-oriented programming allows us to re-use already written code as often as we like.

The base class can be an abstract class, although this is not an essential requirement. As mentioned earlier, no real object can be instantiated from an abstract class. However, it is useful to form a general abstract base class by including the essential member data and member functions that will be common to all objects of the derived classes of the hierarchy.

As an example, if you consider a graphics program which works with geometric shapes such as lines, circles, triangles, squares, etc; a base class definition can be formed to move, scale, hide, show, stretch *any* of these objects. Let us call this class the Shape class. The class that we form, without referring to a specific object, can be an abstract class. As mentioned above, base classes do not need to be abstract classes. The abstract class will define the functionality of all objects of the hierarchy without specifying a particular object type.

The derived classes must be more specific than the abstract class. For example, there can be a derived class, which specifically works with a circle. Let us name this the `Circles` class. Then, all functions within the `Circles` class such as `Move`, `Scale`, `Hide`, `Show`, and `Stretch` can be coded specifically to operate on circle objects. As you can imagine, it is very difficult to transfer the functionality of the `Circles` class over to a new class named `Squares` to work with squares! You will need to make so many changes to be able to operate with squares in place of circles. It is easier to directly derive the `Squares` class from the abstract class `Shape`, and then define the functions the `Squares` class needs. Although the abstract classes cannot be used to instantiate new objects, they have a very powerful use associated with *virtual functions* and *late binding*. We will discuss these two concepts in Chapter 8.

4.8 Inheritance

Inheritance is closely associated with class hierarchies. When a new object class is derived from a base class, the new class is said to *inherit* all the member data and member functions of the base class. This is the mechanism that underlies code reuse. The functions and data of the base class automatically exist in the derived class. They do not need to be written again. To provide the new class with functionality beyond that of the base class, we only need to add whatever *additional* data and functions are necessary. It is also possible to modify the inherited functions to suit their more specific tasks. However, there may need to be restrictions for access to some inherited data and functions.

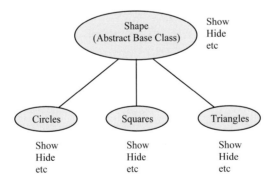

Figure 4-5 Inheriting from an abstract class.

As an example, consider the graphics program we mentioned earlier. We discussed an abstract class named `Shape`. In that class we had graphics functions such as `Move`, `Scale`, `Show`, `Hide`, `Stretch`, etc. If we need a new class to specifically work with circles, then we can use the abstract class `Shape` as the base class and derive the new `Circles` class from it. The derived class `Circles` is said to be a sub-class of the abstract class `Shape` and will inherit all its functions (and data if it had any). The `Shape` class does not have properties specific to a circle object such as centre and radius. These properties will be added as new member data to the derived class.

4.9 Multiple Inheritance

It is also possible for a derived class to have more than one base class. In this case the derived class will inherit all the member data and member functions of all the base classes. For example, we can have a base class named `Colours`, which allows us to set foreground and background colours, choose fill patterns, and carry out filling. We can use the `Shape` class and the `Colours` class to derive our new object class `Circles` (as shown in Figure 4-6). Now the `Circles` class can be made to show colourful circles on the screen!

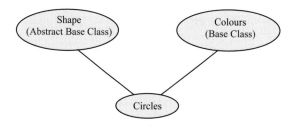

Figure 4-6 Multiple inheritance.

4.10 Polymorphism

In principle, object-oriented concepts are profoundly based on the mechanisms of encapsulation, inheritance, and polymorphism. Polymorphism is a more complex concept than the concepts of encapsulation and inheritance which have already been briefly explained.

As applied to object-oriented programming, polymorphism means the existence of a function with the same name, same return value type, same number of parameters and the same type of parameters in a number of classes of the same hierarchy. The bodies of the functions will differ to suit the requirements for each class. It is not essential for every object in the hierarchy to have this function. Furthermore, there can be any number of polymorph functions in a given hierarchy. Polymorphism allows a common interface for related actions. The most powerful feature of object-oriented programming is associated with polymorphism of virtual functions (discussed in Chapter 8).

Let us consider an example of polymorph functions. Since a polymorph function is the 'same' function throughout the hierarchy, one would imagine that it carries out the same task in each class. Going back to our graphics example, we can consider Show to be a polymorph function. The function Show is in the Shape class, in the Circles class, in the Squares class, and so forth. Another polymorph function is Hide. Suppose the Show function makes the object visible on the screen. If we are using the Show function with a circle object, it will show a circle. If it is used with a square object, it will show a square. Therefore, although the function name is the same, it operates in a context-sensitive manner according to the type of object it is working with.

Polymorphism is the key to harnessing the great benefits of virtual functions. It is possible for a programmer to use a virtual function in a program without knowing which object it will be used with at run-time (when the program is executing). The program will be written generically to suit all classes of the hierarchy. The programmer does not need to write the complex logic for selecting the correct function to suit the object chosen by the user at run-time. This task is passed on to the compiler and linker, simplifying the programming task immensely. This is of most benefit for programs with large numbers of classes and complex hierarchies.

As an example, a programmer can write a generic program in which the virtual function Show is used to show any object in the hierarchy. The user of the program decides at run time, the actual object the function Show will operate on. The programmer has not needed to develop the full range of complex logic needed to handle whatever type of object from the shape class the user will choose at a particular time. Nonetheless, the correct Show function for that object type will automatically be selected during program operation. This concept will be demonstrated in detail in Chapter 8.

4.11 An Example Object Hierarchy

To enable you to relate some of the concepts described previously, we will develop an object hierarchy without using C++ language syntax or its keywords. The class definitions shown below cannot be compiled in an actual C++ program, however, they demonstrate the principles associated with an object hierarchy. We start with the abstract class Vehicle discussed earlier:

```
Abstract Class Vehicle

    Member data:
        Speed
        Power

    Member Functions:
        Stop
        Go
```

This class is the most fundamental of all classes of this example. It has encapsulated the bare minimum that is essential for a vehicle. It must possess `Power` for it to be able to move, and also will have a `Speed` characteristic to describe its motion. The member function `Go` will start the vehicle moving, and the other member function `Stop` will bring it to a stand-still.

New classes with more specific details added to them can then be derived from the `Vehicle` class. We have done this by forming two new class definitions named `Passenger Transport`, and `Goods Transport`. This class hierarchy is shown in Figure 4-7 where there are two branches coming from its root (base class). The class definitions are shown below:

```
Derived Class Passenger Transport: derived from Vehicle class

    Additional member data:
        Number of Passengers


Derived Class Goods Transport: derived from Vehicle class

    Additional member data:
        Load carrying capacity in Kg.
```

Note that these two classes inherit all the member data and member functions of the base class `Vehicle`. For example, if we list everything in the `Passenger Transport` class we will form the class definition show below:

```
Derived Class Passenger Transport

    Member data:
        Speed
        Power
        Number of Passengers

    Member Functions:
        Stop
        Go
```

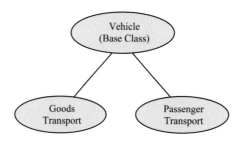

Figure 4-7 Deriving classes from a base class.

Although we did not specifically mention `Speed`, `Power`, `Stop` and `Go`, they are present in the new derived class as a result of inheritance. Furthermore, the `Goods Transport` class is a direct descendant of just the abstract class `Vehicle` and therefore does not inherit anything from the `Passenger Transport` class.

Therefore, there are two branches, right at the root of this object hierarchy as shown in Figure 4-7. Also, the `Passenger Transport` class and the `Goods Transport` class may or may not be abstract classes. Any of these classes can be used to further derive new classes with additional refinements.

In the next class definition, a new class named `Passenger Train` is derived from the `Passenger Transport` base class:

```
Derived Class Passenger Train: derived from Passenger
Transport class

    Additional member data:
        Number of passenger cars

    Additional member functions:
        Doors Open
        Doors Close
        Air Conditioning
```

In this class definition, the members of the `Passenger Transport` class are in effect added to the already existing members of the `Passenger Transport` class that have been inherited (and are invisible). If we managed to see the equivalent complete class definition for the `Passenger Train` class, it would appear as follows - the inherited members are shown in bold italic typeface:

```
Derived Class Passenger Train

    Member data:
        Speed
        Power
        Number of Passengers
        Number of passenger cars

    Member Functions:
        Stop
        Go
        Doors Open
        Doors Close
        Air Conditioning
```

Just as we used `Passenger Transport` as a base class, the `Goods Transport` class can be used to derive further classes. An example of a new `Goods Train` class derived from the `Goods Transport` class is now given:

```
Derived Class Goods Train: derived from Goods Transport class

    Additional member data:
        Number of boxcars
        Number of tank cars
        Number of open cars
```

Motorcars are primarily meant for transporting passengers. Therefore, if we wish to form a new class to represent motorcars, the best place to start is the Passenger Transport class. An example class definition for the new Motorcar class is now given:

```
Derived  Class  Motorcar:  derived  from  Passenger  Transport
class

    Additional member data:
        Engine Capacity
        Body Colour
        Trim Colour
        Number of Cylinders
        Wheel Size
        Number of Doors

    Additional member functions:
        Steer
        Brake
```

If we then need a class definition to represent luxury cars, the best starting point is the Motorcar class. The Motorcar class is chosen instead of the Passenger Transport class because Motorcar objects form a more complete sub-object of a Luxury Car class. If we use the Passenger Transport class as the base class of the Luxury Car class, we will have to re-introduce members such as Engine Capacity, Body Colour, Trim Colour, etc. This involves a lot of unnecessary work, is error prone, and does not take advantage of code reuse.

```
Derived Class Luxury Car: derived from Motor Car class

    Additional member data:
        Inside Air temperature
        Global Position

    Additional member functions:
        Air Conditioning Control
        Power Mirror Control
        Power Window control
        Cruise Control
        Antenna Control
        CD Control
```

In a class hierarchy, future changes that need to be carried out can be done with minimum reprogramming. If the necessary changes are very specific, then the changes are more likely to be made in the most recently derived classes. If the required changes are more general in nature, it is most likely that the changes will be carried out closer to the root of the hierarchy. For example, if all luxury cars are to have automatic navigation in the future, we will add a new member function named `Navigate` to the `Luxury Car` class. On the other hand, if *all* vehicles are to be fitted with automatic navigation facilities in the future, we will add the `Navigate` member function to the abstract class `Vehicle`.

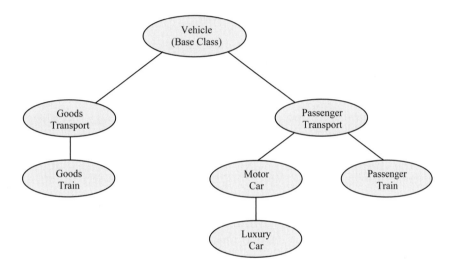

Figure 4-8 The example class hierarchy.

A set of object classes that fit into a class hierarchy always has an *expansive* nature rather than a *multiplicity* nature. In an expansive situation, additional member data and member functions will be added to the new derived class. In a multiplicity situation more members of the same object type are added to the new class.

For example, an object class representing a house and another object class representing a better house fit nicely into a class hierarchy. On the contrary, the class representing a house and another class representing many houses of the same type as the other class do not fit into a class hierarchy. New, completely different member objects need to exist in the new class.

Although you may have understood how classes are formed, their use may still be unclear. In the coming chapters, we will develop classes and begin to use them. For now, it is important to gain an understanding of what a class hierarchy is and how it is formed.

4.12 Advantages of Object-Oriented Programming

A large part of the programming community has already embraced object-oriented programming as a better way to program. One of the main advantages is the robustness of the programs, a direct result of encapsulation. The changes carried out within an object class have no side effects on other parts of the program, and the internal details of an object class are well insulated from the outside world. This significantly simplifies the maintenance of programs. If in the future, the functionality of a class needs to be enhanced, the additional coding needed will be localised to the class itself and will not affect the functionality of unrelated classes. The changes may not even affect the public interface of the object itself. If we take a real-life example, drivers operate motorcars exactly the same way they did in the past. However, the fuel system has changed from carburisation to fuel injection. While the performance of the object is enhanced, the motorcar is driven exactly the same way through the public interface (the accelerator pedal).

Inheritance permits us to reuse the code over and over again. This reduces re-programming time and the associated debugging time. It allows us to reduce our time-to-market and lower the cost of software development. The natural relationship between real-life objects and software objects makes it easier to understand the class structures. This is the strategy we used in this chapter to give you a good insight into object classes.

The most powerful and the most useful feature of object-oriented programming is associated with virtual functions and object hierarchies. Using virtual functions enables the program to select the correct function to operate on the objects that are specified at run time. This relieves the programmer from having to write lengthy code to cater for individual objects that may be specified at run time by the user of the program.

4.13 Disadvantages of Object-Oriented Programming

In general there is reluctance among programmers who are familiar with procedural programming to embrace object-oriented programming. Object-oriented concepts are quite foreign and require some adjustment in thinking, especially so for novices.

Object-oriented programming is often not usually justified when programs are very small. Also, object-oriented programming may not be the best choice for programs requiring time-critical program execution. However, with the increasing speed of computers this is becoming a less significant issue. Operating systems that burden the computer are typically more of a concern than the object-oriented programs themselves.

4.14 Summary

In this chapter we used real-life examples to promote understanding of object-oriented concepts. We started by differentiating the two programming methods; procedural programming and object-oriented programming. Procedural programming exposes data and functions for inadvertent misuse and it can lead to unexpected side-effects and difficult debugging. Object-oriented programming imposes data hiding and protects data from inadvertent misuse by encapsulating data and functions together to form object classes. The public interface of an encapsulated object class has been explained using real-life examples.

A qualitative explanation was given to explain the abstract object classes and actual object classes. This has then been consolidated using object-oriented terminology to explain abstract classes and real classes. Also, inheritance has been exploited in an example object hierarchy to show how a class hierarchy can be developed.

The use of constructors and destructors has been briefly introduced and will be explained in greater detail in the coming chapters. Instantiation has a close association with constructors, and a given class may have any number of constructors.

The important concepts of polymorphism and virtual functions have been discussed in limited detail due to their relative complexity. They will be further explained and used extensively in Chapter 8. Finally, advantages and disadvantages of object-oriented programming have been discussed.

4.15 Bibliography

Meyer, B., *Object Oriented Software Construction*, Prentice Hall, 1988.

Firesmith, D.G., *Object Oriented Requirements, Analysis and Logical Design*, John Wiley, 1993.

Staugaard A. C. (Jr), *Structured and Object Oriented Techniques*, Prentice Hall, 1997.

Gray, N.A.B, *Programming with Classes*, John Wiley, 1994.

5

Object-Oriented Programming

Inside this Chapter

- Creating an object class.

- Developing objects for ports.

- Access attributes.

- Developing the `ParallelPort` object class.

5.1 Introduction

Our aim in this chapter is to teach you how to develop object classes for use in C++ programs. The object-oriented concepts introduced in the previous chapter will be expanded upon and C++ syntax will be added to the class definitions. The object classes developed in this chapter will be used in the development of future programs that interact with various devices on the interface board through the parallel port of your PC.

We will start the chapter by developing a fundamental parallel port object type named `ParallelPort` that will be developed in three separate stages. Each stage of developement will give the `ParallelPort` object additional functionality, with the final class capable of data transfer to and from the computer using most features of the parallel port.

5.2 Naming Convention

In order to improve the readability of our programs, we will establish a naming convention to be used when assigning names to functions and other identifiers. As explained in Chapters 2 and 3, the parallel port of the PC occupies three consecutive addresses. In most cases, the hexadecimal values of these consecutive addresses are 0x378, 0x379 and 0x37A. The term BASE will be used to refer to the first of the three addresses which we call the base address (0x378 in this case). Table 5-1 summarises the naming convention.

Table 5-1 Naming convention for identifiers in programs.

Suffix for Identifiers	Offset with respect to Base Address (BASE)	Physical Address (most widely used)
Port0	0	0x378
Port1	1	0x379
Port2	2	0x37A

NOTE

The port addresses you will be using may not necessarily be 0x378, 0x379 and 0x37A. However, they will be three consecutive addresses. The object classes we develop will have the flexibility to specify the BASE address that applies to your particular case.

According to the naming convention shown in Table 5-1, all member data and member functions ending with `Port0` will use address BASE. For example, a function with the name `WritePort0()` writes data to the port at address BASE. Similarly, `WritePort2()` writes data to the port at address BASE+2.

5.3 Developing an Object Class

By the end of this chapter we will have developed a complete `ParallelPort` class that encompasses most functional aspects of the port. This will take place in three stages; in the first stage we will develop a class using the BASE address. This class will be expanded in two following stages to include the functionality of the BASE+1 address, and lastly the BASE+2 address.

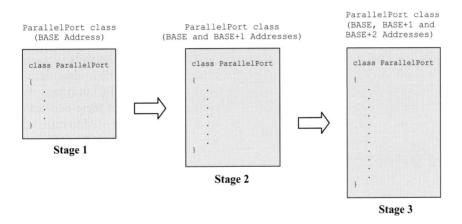

Figure 5-1 Developing the ParallelPort Class.

The port at BASE address is the easiest to use and can send one byte of data at a time. Note that newer computers can both send and receive data through this port. We will only use the BASE address as an output port to maintain compatibility with older computers.

Designing an object class:

1. Select an appropriate name for the object class. It should be concise, yet descriptive enough to suit the purpose and content of the class.
2. Determine class member data (features or properties associated with the object).
3. Determine class member functions (uses of the object).
4. Use the appropriate access attributes to encapsulate the data and functions.
5. Establish constructor(s) and the destructor for the class.

In the previous chapter, member data were identified as features or properties associated with the object. Member functions were identified as the uses of the object. If we use everyday language to describe a port at address BASE as an object class, it will read something like this:

"The port is at a certain address and can output one byte of data"

According to this description, the port is identified by an address. The object is used to output a byte of data to an external device using this port. Now that we have established the features and the uses, we can try to define the object class. We will encapsulate this member data and member function to form our fundamental object class ParallelPort.

5.3.1 Member Data

The ParallelPort object class we are developing must know the base address of the port in order to operate. This data needs to be stored in a data member that we have appropriately named BaseAddress. Most parallel port addresses will fall in the range of 0 to 0x3FF. An 8-bit number (0 to 0xFF) will be too small to store these values, and so a 16-bit number (0 to 0xFFFF) must be used. Therefore, we need to declare a data type of unsigned int to the data member BaseAddress so it can store the required range of possible port addresses:

```
unsigned int BaseAddress;
```

5.3.2 Member Functions

Since the ParallelPort class needs to be able to write a byte of data to the port at BASE address, a function named WritePort0() will be created for this purpose. One byte is represented by the data type unsigned char. The function only outputs data, so it does not need to return a value. This means its return value type should be void. The resulting function is:

```
void WritePort0(unsigned char data)
{
    outportb(BaseAddress, data);
}
```

The function WritePort0() takes in one parameter, namely data of type unsigned char. The body of the function contains just one statement; it outputs the value of the parameter passed as data to the port at the address specified by the value of BaseAddress.

WritePort0() has a return value specified as void. In functions that do not return any value, the return value type must be specified as void. If the functions are to return a value, then there must be a return statement within the body of the function.

5.3.3 Access Attributes

The access attributes control accessibility to the members of the class by functions anywhere else in the program. There are three different types of access attributes; *private*, *protected*, and *public*, described as follows.

private

Functions and data listed under the `private` access attribute can only be accessed by member functions of the same class. Other functions from elsewhere cannot modify or even see any of the private member data, nor can they call any private member functions. Importantly, if an access attribute is not specified for any of the members in a class definition, the access attribute defaults to be `private`.

protected

Functions and data listed under the `protected` access attribute can be accessed by the member functions of the class itself and member functions of all other classes *derived* using this class as the base class. Classes, which have no relationship to this class, cannot access the protected member data or protected member functions. Class derivation is further explained in Chapter 6.

public

Functions and data listed under the `public` access attribute can be accessed by any function regardless of its relationship to the class.

5.3.4 Defining a Class

The more attractive way to develop object classes that offers greater protection is to use *class definitions*. A class definition starts with the keyword `class` followed by the name given to the object type (in this case `ParallelPort`). It is entirely up to you to choose this identifier for the object type. It is also possible in C++ to define an object using aggregates of related data known as `structures`.

As shown in Figure 5-2, the body of the class definition starts at the first open brace (`{`) and it ends when the last close brace (`}`) is encountered. Remember to place a semicolon (`;`) after the close brace to complete the class definition.

There are two segments within the body of the class definition. The first segment relates to member data, and the second segment relates to member functions. Although the member data and member functions can be intermingled, it is good practice to keep them separate.

As mentioned earlier, if an access attribute is not specified, then it defaults to `private`. Therefore, the keyword `private` in the class definition is not essential. If it was dropped, the variable `BaseAddress` would still be private. However, it is good practice to include the keyword `private` to make the code more explicit.

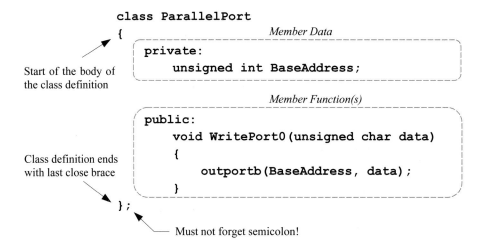

Start of the body of
the class definition

Class definition ends
with last close brace

Member Data

Member Function(s)

Must not forget semicolon!

Figure 5-2 Class definition.

The class has a default constructor and a default destructor provided by the compiler that are not visible. The constructor creates objects for use in a program and the destructor relinquishes objects from memory once the program no longer needs them.

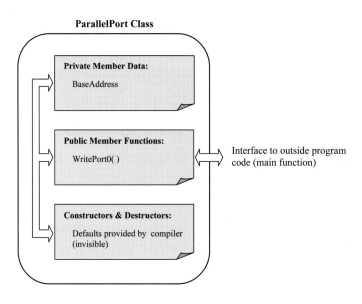

Figure 5-3 Schematic of the `ParallelPort` Class.

In the case of the object class `ParallelPort`, `BaseAddress` is a private data member. Therefore, it can only be accessed by the function `WritePort0()` belonging to this class. This function is declared to be `public` and as such can be called by any other function in your program. This provides an interface between the object and the outside world. If the function `WritePort0()` was declared `private`, it could not be called by any function outside the class (including the `main()` function). As a result the object of type `ParallelPort` could not be used (recall the sealed off car that no one can get in to drive!).

5.3.5 The Constructor

The purpose of a *constructor* is to create objects for use in a program. Simply because there is an object type definition, it does not mean that there is an actual object that resides in memory. For an object to hold data and execute functions, both data and functions must reside in memory. This occurs when the constructor is called. It creates an object and allocates memory to all its data members in a process known as *instantiation*. At the time of instantiating an object all statements in a constructor's body will be executed. The programmer may code the constructor to suit the needs of the objects at the time of their creation.

Constructors always have the same name as the object type – in this case, `ParallelPort`. A constructor, although a function, *never* has a return value. This is a unique feature of constructors. They are one of only two types of functions in C++ which have no return value type, not even `void`. The second type of function with no return value is the destructor.

5.3.6 Default Constructor

The strict definition of the *default constructor* is a constructor that does not take any parameters (i.e. the pair of parentheses are empty). In the absence of a constructor provided by a programmer, the compiler will provide a constructor and that constructor is always a default constructor. The default constructor provided by the compiler has one main job, to *allocate memory* for all data members of the object. It will also call the default constructor of its base class if it has one – this occurs *before* the derived class constructor calls allocates memory for its own data members.

It is also possible for a programmer to provide a default constructor. The default constructor written by the programmer also takes no parameters, but *can* have some statements in its body. If the programmer provides a constructor, be it default or otherwise, the compiler *will not* provide a constructor. The programmer-supplied default constructor will also call the default constructor of its base class *before* executing its own duties.

The compiler also provides a special constructor named the copy constructor, and an overloaded assignment operator (the = sign). We will defer the discussion of copy constructors and overloaded assignment operators until Chapter 12.

5.3.7 Overloading of Constructors

The term *overloading* is used to describe the situation when many functions exist under the same name but carry out different tasks. For the case of overloaded constructors, there is more than one constructor in a class definition, all having the same name (the name of the class), and as for all constructors, not having return value types. Any number of overloaded constructors are allowed in a class. Therefore, the only difference between the constructors should be in the number and types of parameters passed to each of them and the statements included in each of their bodies.

5.3.8 Destructors

The destructor serves the opposite purpose of the constructor. A constructor brings an object into life by creating it in memory. A destructor removes the object from memory, freeing the space. The name of the destructor must always be the name of the object class and preceeded by the tilde symbol (~). For example, the destructor for the `ParallelPort` class would be named `~ParallelPort()`. The program developer can choose to include a destructor in the class definition. If the developer does not include a destructor, the compiler will generate a default destructor.

Unlike constructors, destructors are not explicitly called. They will be called automatically when a locally defined object ceases to exist or when the `delete` operator is used to delete the object from memory. For class hierarchies, the destructor of the derived class is executed *before* the destructor of its base class.

While there are no virtual constructors, destructors can be virtual functions. Virtual destructors will be described in detail in Section 8.5. We will now begin the development of the `ParallelPort` class in the three stages we mentioned at the start of Section 5.3 and as shown in Figure 5-1.

5.4 Parallel Port Class – Stage I

This first stage in the development of the `ParallelPort` class will produce a class having the simplest features of the three stages. This stage will only use the capabilities of the `BASE` address component of the port, being the ability to write a byte of data (8 bits, D0-D7 inclusive) to external devices.

Initially a default constructor will be used as part of the class (generated by the compiler). This design has some deficiencies discussed ahead. The class will be developed further to overcome these shortcomings.

5.4.1 Class Definition

As described earlier, `WritePort0()` is a public member function. Therefore, any function in the program can call this function. The `WritePort0()` function in-turn calls the familiar `outportb()` function to send data to the port at address given by `BaseAddress`.

We can have a default constructor generated automatically by the compiler if we do not write a constructor in the class definition.

C++   Syntax of member functions

There are two ways the compiler can be informed of member functions:

(i)   Member function definition placed outside the class.

In this case the syntax used is the same as that for normal functions, except the object class name followed by two colons (`::`) is placed before the function name as shown in Listing 5-1.

(ii) Member function definition placed within the class.

In this case the syntax used in (i) is not necessary as shown in Listing 5-2.

The member function definition shown in Listing 5-1 for the `WritePort0()` function is defined outside the class.

Listing 5-1 Definition of the member function(s) for the `ParallelPort` class.

```
void ParallelPort::WritePort0(unsigned char data)
{
    outportb(BaseAddress,data);
}
```

When the `WritePort0()` member function is defined within the class definition itself, the class definition will be as shown in Listing 5-2. Any member functions that are defined within the class definition and are small, will be treated as *in-line* functions (discussed ahead). When functions are declared outside the class definition, they will be treated as normal functions.

Listing 5-2 Defining member functions within the class definition.

```
#include <dos.h>

class ParallelPort
{
    private:
        unsigned int BaseAddress;

    public:
        void WritePort0(unsigned char data)
```

```
            {
                    outportb(BaseAddress,data);
            }
};
```

C++   In-line member functions

The compiler allows small member functions *defined* in the class definition to be in-line functions. Whenever an in-line function is called in a source file, the compiler may replace that call with the actual instructions within the body of the in-line function.This will happen if the in-line function is small in size and does not involve complicated (lengthy) parameter passing.

This means every time an in-line function is called in the code, the compiler will add a new instance of the function to the executable code which increases the size of the program. However, there is a benefit that program execution flows directly into the in-line function, avoiding the overheads associated with a function call and therefore promotes greater speed.

Normal functions use only one instance of the function stored in memory. In this case, a function call results in the current program status being temporarily saved before the program jumps to the memory area containing the function. The code for the function is executed followed by a return jump and recovery of prior program status. These extra steps of saving, jumping and recovering add execution time, and slow performance compared with using in-line functions (although the executable programs are smaller in size).

NOTE Listing 5-1 and Listing 5-2 both have a deficiency. Examining the parameters of the WritePort0() function, we see that we can pass an actual argument (such as 0x7F) in place of the parameter data, however, there is no way for us to specify the BaseAddress.

If we do not have a mechanism to specify a numerical value for BaseAddress, the outportb() function will not work. One poor solution is to fix the address at a predetermined value. For example:

```
void WritePort0(unsigned char data)
{
    outportb(0x378,data);
}
```

The above approach will only work for those users having 0x378 as the BASE address of their parallel port. Other users would need to edit and *re-code* the member function to suit whatever BASE address their PC's parallel port uses. Additionally, this solution does not use the data member BaseAddress.

Class Definition – Improvement I

We need to provide the user with the ability to specify the base address according to hardware requirements. This is best done at the time of creating the ParallelPort object – ie at the time of calling the constructor. Our default constructor will need to be replaced with a constructor that allows us to set the member data BaseAddress to a desired value at the time of creating the object.

New constructor: We pass the parameter baseaddress to the constructor, which then assigns this value to the member data BaseAddress. Now the original outportb function will work as intended. You will have noticed in Listing 5-4 that the member function is defined separately from the class definition. The reason for this is explained in the box presented previously titled "In-line member functions".

Listing 5-3 The class definition for ParallelPort - with a constructor.

```
class ParallelPort
{
    private:
        unsigned int BaseAddress;

    public:
        ParallelPort(int baseaddress);  // constructor
        void WritePort0(unsigned char data);
};
```

Listing 5-4 Definitions of member functions for the improved ParallelPort class.

```
ParallelPort::ParallelPort(int baseaddress)  // constructor
{
    BaseAddress = baseaddress;
}

void ParallelPort::WritePort0(unsigned char data)
{
    outportb(BaseAddress,data);
}
```

> **NOTE**
>
> The member data `BaseAddress` is different to `baseaddress`. The parameter `baseaddress` is a placeholder for the actual argument one would pass at the time of calling the constructor and plays a role only at the time of calling the constructor. On the contrary, the member data `BaseAddress` will reside in memory storing data for the entire life of the object; in most cases, for the whole life of the program.

Class Definition – Improvement II

Because most users will operate their parallel port at address 0x378, ideally we would like to have a constructor that by default sets the BASE address to 0x378. We still need the option of being able to set the port to other address values as was implemented in the previous class definition.

We achieve this by adding a second constructor (overloaded constructors) that sets the member data `BaseAddress` to 0x378 by default. This programmer-defined default constructor takes no parameters (ie. a default constructor) but assigns the member data `BaseAddress` to 0x378 inside its body. The class definition and member function definitions are given in Listing 5-5 and Listing 5-6.

Listing 5-5 Class definition for `ParallelPort` with a default constructor.

```
class ParallelPort
{
    private:
        unsigned int BaseAddress;

    public:
        ParallelPort();  // default constructor
        ParallelPort(int baseaddress);  // constructor
        void WritePort0(unsigned char data);
};
```

Listing 5-6 Member function definitions for the class definition in Listing 5-5.

```
ParallelPort::ParallelPort()  // default constructor
{
    BaseAddress = 0x378;
}

ParallelPort::ParallelPort(int baseaddress)  // constructor
{
    BaseAddress = baseaddress;
}

void ParallelPort::WritePort0(unsigned char data)
{
    outportb(BaseAddress,data);
}
```

If in our program we do not pass a value to the constructor, the default constructor will be called and the BASE address will default to 0x378. Alternatively we can set the BASE address to say 0x3BC by passing this value when creating a ParallelPort object as shown:

```
ParallelPort(0x3BC);
```

In this case the second constructor (takes a parameter) will be called to create the ParallelPort object. Although constructors commonly initialise some or all of the member data, a constructor can be programmed to carry out any actions a regular function may perform.

An important point to note is that the member function WritePort0() has not been specifically informed of the member variable BaseAddress – that is, BaseAddress has not been declared within this function. This is not necessary since BaseAddress and WritePort0() are both members of the same class, hence WritePort0() has unrestricted access to BaseAddress as explained in Section 5.3.3.

5.5 Using Class Objects in Programs

We will now use the ParallelPort object type developed in the previous section to output a byte of data to the port at BASE address. The object-oriented program that performs this task seems lengthy, however, the advantages of object-oriented programming will be realised as we progress through later chapters.

Listing 5-7 Writing to port at address BASE using object-oriented approach.

```
/********************************************************
WRITING TO A PORT (object-oriented approach)

The program uses the fundamental ParallelPort object
class to output a byte of data to the interface board.
********************************************************/
#include <dos.h>

class ParallelPort
{
    private:
        unsigned int BaseAddress;

    public:
        ParallelPort();
        ParallelPort(int baseaddress);
        void WritePort0(unsigned char data);
};

ParallelPort::ParallelPort() // default constructor
{
    BaseAddress = 0x378;
}

ParallelPort::ParallelPort(int baseaddress) // constructor
{
    BaseAddress = baseaddress;
}

void ParallelPort::WritePort0(unsigned char data)
{
    outportb(BaseAddress,data);
}

void main()
{
    ParallelPort OurPort; // object instantiation

    OurPort.WritePort0(255); // calling a member function
}
```

The class definition and function definitions are the same as developed earlier and have already been explained. What has not been explained is the body of the `main()` function. This is where a programmer uses the object classes to develop an application. The first of the two lines in `main()` creates an object named `OurPort` of type `ParallelPort`. The second line calls `OurPort`'s member function `WritePort0()` and outputs the value `255` to the port. The actual port address used by `WritePort0()` is assigned when we create the `OurPort` object by calling one of the two constructors. This process will now be explained in more detail.

Object Creation (instantiation)

The first line in `main()` is:

```
ParallelPort OurPort;
```

This is a declaration line and is similar to:

```
int a;   // Data type is int, variable name is a.
```

Similarly; `ParallelPort` is the data type, which is an object type. `OurPort` is the name of the variable. The only difference is that `int` is a fundamental built-in data type whereas `ParallelPort` is a data type we developed.

In C++, the declaration lines not only inform the compiler the name (such as `OurPort`) and type of the object (such as `ParallelPort`), but also call the constructor of that object class to create the object. The creation of the object `OurPort` is what we mentioned earlier as instantiation – it brings into life a real usable object that will reside in memory. It is important to know the difference between 'an object' (also known as a class object) and 'an object type' (also known as the object class). The object is `OurPort` and the object type is `ParallelPort`. Thus, we have created an *object* named `OurPort` of *object type* `ParallelPort`.

Referring to the class definition given in Listing 5-5, we see that there are two constructor functions available. They are:

```
ParallelPort();
ParallelPort(int baseaddress);
```

Which of these two constructors were used by the declaration line to instantiate the `OurPort` object? The compiler will automatically select the correct constructor based on the parameters passed (or the absence of them). Since we did not specify a value for the parameter `baseaddress` at the time of instantiating the `OurPort` object, the compiler will call the first of the above two constructors, being the default constructor.

On the other hand, the declaration line in the `main()` function could have been:

```
ParallelPort OurPort(0x3BC);
```

In this case, the compiler *will not* call the default constructor, since a value for a parameter has been specified. Instead, the other constructor (which accepts a parameter) will be called to instantiate the object `OurPort`:

```
ParallelPort(int baseaddress);
```

In the example above, when creating a `ParallelPort` object, the parameter `baseaddress` is replaced by the actual argument `0x3BC`. The value `0x3BC` will then be assigned to the data member `BaseAddress` (see the body of the constructor in Listing 5-6).

Returning to the program in Listing 5-7, the created object named `OurPort` will have all the features given in the class definition. It has its own private data member named `BaseAddress` and three public member functions. Two of the functions are constructors with the name `ParallelPort` and the other function is named `WritePort0`.

Sending data to the Port

The object we created named `OurPort`, makes use of its member function `WritePort0()` to send the value `255` out the port. The syntax for accessing this member function uses a period placed after the *object name* followed by the member name:

```
OurPort.WritePort0(255);
```

The other members of the object are accessed in exactly the same manner, although their access attributes may restrict access. For example, despite the correct syntax, any function outside the class trying to execute the following instruction will fail:

```
OurPort.BaseAddress = 0; // Will not work!
```

This is because `BaseAddress` is a private data member of the class and functions outside of the class cannot change it (the `main()` function is not part of the class). This illegal attempt to access the member data `BaseAddress` will be detected during compilation and a compilation error will be reported. The next section will present additional examples of access attributes used in our program.

Note that the `WritePort0()` function takes in one parameter. This parameter has been replaced by the actual argument `255`. In binary, the decimal value of `255` will be represented with all eight bits set to 1.

The operation of the program can be checked by connecting the PC to the interface board as shown in Table 3-1. Passing the parameter `255` to the `WritePort0()` function should generate eight lit LEDs. You can change this number to any value between 0 and 255 inclusive. If you re-run the program you should see the LEDs light up accordingly.

5.5.1 Examples using Access Attributes

In the preceding section, an illegal attempt was made to assign the value `0` to data member `BaseAddress`, as now shown again:

```
OurPort.BaseAddress = 0; // Will not work!
```

In that program, BaseAddress was declared as a private data member, which prevented the main() function from gaining access to change its value.

Instead of declaring BaseAddress as a private member, it can be declared as a public member. This being the case, it can be accessed by any function in the program and the compilation error will no longer appear. Listing 5-8 shows how this is done. Note that if you do not have a second parallel port (LPT2:) in your computer, the address 0x3BC does not exist, and the program will *not* function. However, it will compile error-free. Also, ensure the data cable that connects with the parallel port is plugged into the D25 connector of the correct port.

Listing 5-8 Declaring BaseAddress as a public data member.

```
/******************************************************
Note that the member variable BaseAddress is now
declared under the public access attribute.  Therefore
it can be changed from within the main function and
there will be NO compilation errors.

If your computer DOES NOT have a second parallel port,
attempting to use the port address 0x3BC will FAIL!
******************************************************/
#include <dos.h>

class ParallelPort
{
    public:      // Private access attribute has been
                 // changed to public.
        unsigned int BaseAddress;

    public:
        ParallelPort();
        ParallelPort(int baseaddress);
        void WritePort0(unsigned char data);
};

ParallelPort::ParallelPort()
{
    BaseAddress = 0x378;
}

ParallelPort::ParallelPort(int baseaddress)
{
    BaseAddress = baseaddress;
}
```

```
void ParallelPort::WritePort0(unsigned char data)
{
    outportb(BaseAddress,data);
}

void main()
{
    ParallelPort OurPort;

    OurPort.BaseAddress = 0x3BC; // Will NOT cause a
                                 // compilation error
    OurPort.WritePort0(255);
}
```

Declaring member data as public is considered poor programming practice. The main purpose of using object-oriented programming is to form encapsulated objects, that are to some extent protected against misuse. This will not be the case if member data are declared to be public.

If a member variable needs to be changed, then it can be changed through a member function specifically designed for that purpose. This concept is demonstrated by the ChangeAddress() function shown in Listing 5-9.

Listing 5-9 The acceptable way to change a data member of an object.

```
/****************************************************
WRITING TO A PORT (Adding a member function to change
the private member data).

Note: attempting to use the port address 0x3BC will fail
if your computer doesn't have a second parallel port.
****************************************************/

#include <dos.h>

class ParallelPort
{
    private:   // Access attribute has been changed
               // back to private.
        unsigned int BaseAddress;

    public:
        ParallelPort();
        ParallelPort(int baseaddress);
```

```
              void WritePort0(unsigned char data);

     // New public member function added.
          void ChangeAddress(unsigned int newaddress);
};

ParallelPort::ParallelPort()
{
     BaseAddress = 0x378;
}

ParallelPort::ParallelPort(int baseaddress)
{
     BaseAddress = baseaddress;
}

void ParallelPort::WritePort0(unsigned char data)
{
     outportb(BaseAddress,data);
}

// New public member function defined.
void ParallelPort::ChangeAddress(unsigned int newaddress)
{
     BaseAddress = newaddress;
}

void main()
{
     ParallelPort OurPort;

//   The correct way to manipulate a private data member
     OurPort.ChangeAddress(0x3BC);
     OurPort.WritePort0(255);
}
```

The program statement shown below from the previous listing will legally change the value of the private data member BaseAddress to 0x3BC:

OurPort.ChangeAddress(0x3BC);

In this illustrative example we have shown how a private data member can be changed using a public member function. The public member function ChangeAddress() has complete access to the private data member BaseAddress since it is a member function of that same object class. We will not use the ChangeAddress() function in our proper ParallelPort class.

Instead, the user can pass the desired value for the base address to the constructor so the `BaseAddress` can be set to a different value than its default value of 0x378 (set by the default constructor).

5.6 Parallel Port Class – Stage II

In the first stage we created an object class named `ParallelPort` which provided the required functionality to use the port associated with address `BASE`. We now need this object to also use the port at address `BASE+1`. Note that the port at address `BASE` is used as an output port and the port at address `BASE+1` is an input port. The new object's intended functionality is:

- Ability to specify the base address of the parallel port.
- Send data through port at address `BASE`.
- Receive data through port at address `BASE+1`.

Adding further functionality to an existing object is an ideal situation for using class derivation. However, there is no justification to develop a hierarchy of classes for each part of the parallel port, since there is no great use of parts of the parallel port. It is most appropriate to develop the entire parallel port as *one* object. Therefore, in this second stage we will add the extra functionality to the `ParallelPort` class so it can also use the port at address `BASE+1`.

The class definition for the new `ParallelPort` class is given in Listing 5-10, with additions shown in bold text. It contains the declarations for the member data and the member functions. All data members of the class `ParallelPort` are declared as `private`. All the member functions are declared as `public`. As before, `BaseAddress` is one of the data members and the function `WritePort0()` is included so that data can be sent out to port at address `BASE`.

Listing 5-10 New class definition for the object `ParallelPort`.

```
class ParallelPort
{
    private:
        unsigned int BaseAddress;
        unsigned char InDataPort1;
    public:
        ParallelPort(); // default constructor
        ParallelPort(int baseaddress); // constructor
        void WritePort0(unsigned char data);
        unsigned char ReadPort1();
};
```

The port at address BASE+1 is an input port (data into the PC). The function ReadPort1() has been introduced to the ParallelPort class to read data through this port. The private data member InDataPort1 is declared to store the data read from this port. The number assignments used for all the members of the class represent the offsets from the base address. For example, WritePort0() function will be writing to an address with offset 0 with respect to the base address – in this case BASE+0, being the BASE address. Similarly, the ReadPort1() function will read from an address with offset 1. Therefore, it will read the port at address BASE+1.

The definitions of all the functions belonging to this expanded class are contained in Listing 5-11.

Listing 5-11 Function definitions of the ParallelPort object.

```
ParallelPort::ParallelPort()   // default constructor
{
    BaseAddress = 0x378;
    InDataPort1 = 0;
}


ParallelPort::ParallelPort(int baseaddress)   // constructor
{
    BaseAddress = baseaddress;
    InDataPort1 = 0;
}

void ParallelPort::WritePort0(unsigned char data)
{
    outportb(BaseAddress,data);
}

unsigned char ParallelPort::ReadPort1()
{
    InDataPort1 = inportb(BaseAddress+1);
// Invert most significant bit to compensate for
// internal inversion by printer port hardware.
    InDataPort1 ^= 0x80;
// Filter to clear unused data bits D0, D1 and D2 to zero.
    InDataPort1 &= 0xF8;
    return InDataPort1;
}
```

The only change to the constructors is the extra statement that initialises the data member of the BASE+1 address, InDataPort1 to 0. If InDataPort1 is not initialised it will store some unknown value. However, initialising this variable to 0 is not essential. It may be initialised to any other value or left un-initialised provided precautions are taken to prevent its use until `InDataPort1` holds an actual value read from the port.

The function `ReadPort1()`, reads the port at address BASE+1 and returns a value of type `unsigned char`. This requires the body of this function to have a `return` statement, which is `return InDataPort1`. Therefore, while this function stores the results of input operations in the data member `InDataPort1`, at the same time it provides an interface to other functions outside of the class to receive the value of this data member. This will enhance the flexibility of the object. In the coming chapters the `ParallelPort` object will be used when writing many programs. It is advantageous to have full flexibility in the `ParallelPort` object so that it can be used to write good and efficient programs.

The function `inportb()` is called within `ReadPort1()` and carries out the task of reading the data from the port at the specified address, in this case BASE+1. Note that only bits 3 to 7 are free to be read through this port. Also, bit 7 is internally inverted by the parallel port hardware. The `ReadPort1()` function is coded to compensate for the inversion (explained in Section 3.6) and also clear the unused bits D0-D2 to zero by using the logical AND operator (&). The hexadecimal number F8 represents a bit pattern of 1111 1000 and will clear bits D0-D2 of any number it is ANDed with. The value produced from this correcting operation will be stored in the data member `InDataPort1`. The last line of the `ReadPort1()` function contains the `return` statement which returns the value of `InDataPort1`.

The complete program is shown in Listing 5-12. Check operation of the program by connecting your interface board to the PC according to Table 3-1 and Table 3-2.

Listing 5-12 Write data to port at BASE and read data from port at BASE+1.

```
/ * * * * * * * * * * * * * * * * * * * * * * * * * * * * * * * * * * * * * * * * * * * * * * * * * * * * * *
The fundamental object class ParallelPort is expanded to
include the input port at address BASE+1. The combined
object is still named ParallelPort and is used to write
to the port at address BASE and to read data from the port
at address BASE+1.
* * * * * * * * * * * * * * * * * * * * * * * * * * * * * * * * * * * * * * * * * * * * * * * * * * * * * * /
#include <stdio.h>
#include <dos.h>

class ParallelPort
{
    private:
```

```
            unsigned int BaseAddress;
            unsigned char InDataPort1;

        public:
            ParallelPort();
            ParallelPort(int baseaddress);
            void WritePort0(unsigned char data);
            unsigned char ReadPort1();
};

ParallelPort::ParallelPort()
{
    BaseAddress = 0x378;
    InDataPort1 = 0;
}

ParallelPort::ParallelPort(int baseaddress)
{
    BaseAddress = baseaddress;
    InDataPort1 = 0;
}

void ParallelPort::WritePort0(unsigned char data)
{
    outportb(BaseAddress,data);
}

unsigned char ParallelPort::ReadPort1()
{
    InDataPort1 = inportb(BaseAddress+1);
// Invert most significant bit to compensate
// for internal inversion by printer port hardware.
    InDataPort1 ^= 0x80;
// Filter to clear unused data bits D0, D1 and D2 to zero.
    InDataPort1 &= 0xF8;
    return InDataPort1;
}

void main()
{
    unsigned char BASE1Data;
    ParallelPort OurPort;

    OurPort.WritePort0(255);
    BASE1Data = OurPort.ReadPort1();
```

```
    printf("\nData Read from Port at BASE+1
                            %2X\n",BASE1Data);
}
```

Listing 5-10 and Listing 5-11 (explained earlier) are incorporated unchanged in Listing 5-12 which has a `main()` function added.

The first line in the `main()` function is:

```
unsigned char BASE1Data;
```

This line declares a variable named `BASE1Data` to store data of type `unsigned char`. In strict C++ terms, this line has instantiated an object of type `unsigned char` and given it the name `BASE1Data`. As such, `BASE1Data` will now reside in memory. The purpose of `BASE1Data` is to store the value read from the port at address `BASE+1`. How this is done will become clear as we work through the rest of the statements of the `main()` function.

The next line in the `main()` function is:

```
ParallelPort OurPort;
```

This line instantiates the `OurPort` object, which is of type `ParallelPort`. Therefore, `Ourport` will have two data members, namely `BaseAddress` and `InDataPort1`. When the above line is executed, the default constructor will be called (no argument used for the base address). As a result the variable `BaseAddress` will be set to 0x378 and the variable `InDataPort1` will be set to 0.

The two member functions are called in the next two lines:

```
OurPort.WritePort0(255);
BASE1Data = OurPort.ReadPort1();
```

The first line writes a byte of data (255 in this case) to the port at address `BASE`. This will cause all eight LEDs to light. The second line will read the port at address `BASE+1` and compensate for inverted bit D7. `ReadPort1()` stores this result for later retrieval in the data member `InDataPort1` and returns the value of `InDataPort1` to the `main()` function. This value received by the `main()` function is stored into its variable `BASE1Data`.

The last line of the `main()` function displays the value of `BASE1Data` on the screen in hexadecimal format with a field width of 2. A carriage return and line feed is inserted before the value is displayed by using the new line character combination `\n`. In this example `main()` function, its variable `BASE1Data` was assigned the value returned from the `ReadPort1()` function. Our future programs will not always be programmed to do operate this way. For these cases where the `main()` function

Also note we need to have the data member `InDataPort1` so the value read from the port can be stored in our object. Without having such a storage variable, the program must rely on the `main()` function's variable `BASE1Data` to be assigned the value returned from the `ReadPort1()` function. It will not always be desirable for our future programs to be programmed to use a `main()` function variable in this way. In these cases, if the `ParallelPort` object did not have the data member `InDataPort1` to store the value returned from `ReadPort1()`, then this value would be lost once `ReadPort1()` completes its execution.

5.7 Parallel Port Class – Stage III

In this final stage we will further develop the object class `ParallelPort` to encompass all input/output functionality of the parallel port of the PC - with one exception. This being the absence of input through `BASE+2` as it can be unreliable on some computers. This class will output data through the port at address `BASE`, input data through the port at address `BASE+1`, and output data through the port at address `BASE+2`. It will also compensate for internal inversions that occur within the parallel port hardware.

5.7.1 Full function Object Class `ParallelPort`

The functionality required for the final object class `ParallelPort` is:

- Ability to specify the `BASE` address of the parallel port.
- Output data through the port at address `BASE`.
- Input data through the port at address `BASE+1`.
- Output data to the port at address `BASE+2`.

The definition for the final `ParallelPort` class is shown in Listing 5-13.

Listing 5-13 The definition for the `ParallelPort` class.

```
class ParallelPort
{
    private:
        unsigned int BaseAddress;
        unsigned char InDataPort1;

    public:
        ParallelPort();
        ParallelPort(int baseaddress);
        void WritePort0(unsigned char data);
        void WritePort2(unsigned char data);
        unsigned char ReadPort1();
};
```

In the class definition, a function is included for each of the requirements in the list. The definitions of the member functions are given in Listing 5-14. Additions made to the earlier `ParallelPort` object class are shown in bold font in Listing 5-13 and Listing 5-14.

Listing 5-14 Definitions of member functions of the class `ParallelPort`.

```
ParallelPort::ParallelPort()
{
    BaseAddress = 0x378;
    InDataPort1 = 0;
}


ParallelPort::ParallelPort(int baseaddress)
{
    BaseAddress = baseaddress;
    InDataPort1 = 0;
}

void ParallelPort::WritePort0(unsigned char data)
{
    outportb(BaseAddress,data);
}

void ParallelPort::WritePort2(unsigned char data)
{
// Invert bits 0, 1 and 3  to compensate for
// internal inversions by printer port hardware.
    outportb(BaseAddress+2, data ^ 0x0B);
}

unsigned char ParallelPort::ReadPort1()
{
    InDataPort1 = inportb(BaseAddress+1);
// Invert most significant bit to compensate for
// internal inversion by printer port hardware.
    InDataPort1 ^= 0x80;
// Filter to clear unused data bits D0, D1 and D2 to zero.
    InDataPort1 &= 0xF8;
    return InDataPort1;
}
```

The `ParallelPort` object class is used in the program shown in Listing 5-15 to carry out data transfer operations using all three ports of the parallel port of your PC. The operation of the program can be checked with the interface board. The connections to be made on the interface board are those given in Table 3-1 and Table 3-2. Note that before stepping through the program to test the operation of the port at address BASE+2, remove connections from the BASE address outputs to the LED Driver IC and reconnect the LED Driver IC to the BASE+2 address outputs as per Table 3-3.

Listing 5-15 Input and Output operations using `ParallelPort` class.

```
/***************************************************
The object class created to use ports at addresses
BASE and BASE+1 has been expanded to include output
through the port at address BASE+2. The combined object
class is still named ParallelPort.
***************************************************/
#include <dos.h>
#include <conio.h>
#include <stdio.h>

class ParallelPort
{
    private:
        unsigned int BaseAddress;
        unsigned char InDataPort1;

    public:
        ParallelPort();
        ParallelPort(int baseaddress);
        void WritePort0(unsigned char data);
        void WritePort2(unsigned char data);
        unsigned char ReadPort1();
};

ParallelPort::ParallelPort()
{
    BaseAddress = 0x378;
    InDataPort1 = 0;
}

ParallelPort::ParallelPort(int baseaddress)
{
    BaseAddress = baseaddress;
```

```
        InDataPort1 = 0;
    }


    void ParallelPort::WritePort0(unsigned char data)
    {
        outportb(BaseAddress,data);
    }


    void ParallelPort::WritePort2(unsigned char data)
    {
        outportb(BaseAddress+2,data ^ 0x0B);
    }


    unsigned char ParallelPort::ReadPort1()
    {
        InDataPort1 = inportb(BaseAddress+1);
    // Inverting Most significant bit to compensate
    // for internal inversion by printer port hardware.
        InDataPort1 ^= 0x80;
    // Filter to clear unused data bits D0, D1 and D2 to zero.
        InDataPort1 &= 0xF8;
        return InDataPort1;
    }


    void main()
    {
        unsigned char BASE1Data;
        ParallelPort OurPort;

        OurPort.WritePort0(0x55);
        printf("\n\nData sent to Port at BASE\n");
        getch();

        BASE1Data = OurPort.ReadPort1();
        printf("\nData read from Port at BASE+1: %2X\n",
                                            BASE1Data);
        getch();

        OurPort.WritePort2(0x00);
        printf("\nData sent to Port at BASE+2\n");
        getch();
    }
```

The first line of the main() function's body instantiates one object of type unsigned char named BASE1Data used to store data read from the port at address `BASE+1`. The next line calls the default constructor of the object class `ParallelPort` to instantiate the object named `OurPort`. This object has member data and member functions to write or read data to and from all three ports of the parallel port.

The remaining statements of the `main()` function carry out a number of output and input operations. The `getch()` functions are used to make the program wait for a key press to allow the user to read the screen. If the `getch()` statements were omitted, the screen would scroll up or revert back to the IDE before the user could see the results. The `getch()` function is not a member function of the object `OurPort`. Therefore, it is not attached to this object and as such is called as a normal function.

5.8 Summary

At the start of this chapter we developed an object class with the name `ParallelPort`. This class contained only sufficient data members and member functions to give us basic use of the port. We applied particular access attributes to the class members and explained the importance of making proper use of these access attributes.

Several programs were used to explain the relationship between multiple constructors and the default constructor. The `ParallelPort` class was then expanded to include use of the `BASE+1` and `BASE+2` addresses. The operation of objects instantiated from this expanded class was demonstrated using a program which transferred data to and from the interface board. Now that we have a fully functioning `ParallelPort` class, we will be able to use it extensively in future chapters.

5.9 Bibliography

Borland, *Borland C++ Getting Started*, Borland International, 1991.

Winston, P.H., *On to C++*, Addison Wesley, 1994.

Lipman, S.B., *C++ Primer*, Addison Wesley, 1991.

Dench, D. and B. Prior, *Introduction to C++*, Chapman and Hall, 1994.

Etter, D.M., *Introduction to C++ - For Engineers and Scientists*, Prentice Hall, 1997.

6

Digital-to-Analog Conversion

Inside this Chapter

- Digital-to-analog conversion explained.

- Generate voltages using a Digital-to-Analog Converter (DAC).

- Operational amplifier basics.

- Inheritance and derived classes.

- Object class for the DAC.

- Access specifiers.

6.1 Introduction

Digital-to-Analog Converters (DAC) form an integral part of many automated control systems. This chapter describes the principle of operation of a DAC and its typical use to produce an analog voltage or current.

You will learn how new classes are derived from existing classes during the development of software to drive a DAC. The new classes will inherit the existing functionality, have extra functionality added, and the inherited functions will be modified. In the course of this process, access attributes and access specifiers will be further explained.

6.2 Digital-to-Analog Conversion

Digital-to-analog conversion is the generation of an analog voltage or current by using a group or sequence of digital logic levels to set the state of an analog output as shown in Figure 6-1. Some types of digital-to-analog converters (DACs) use digital data presented to the DAC in a serial format, and others use digital data presented in a parallel format.

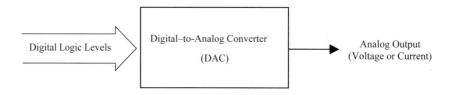

Figure 6-1 Digital-to-Analog Conversion.

There are several types of digital-to-analog converter in use. Some take in a digital pulse-train and use integration to give an analog voltage (for example, frequency to voltage converters). Other DACs use the popular method of digital-to-analog conversion discussed in this text like the DAC in this chapter's project. In order to appreciate how the DAC operates, we need to understand a little about its main building block, the *operational amplifier.*

6.2.1 Operational Amplifier Basics

The operational amplifier is a fundamental building block for many analog electronic systems. The schematic symbol for an *operational amplifier* (op-amp) is shown in Figure 6-2.

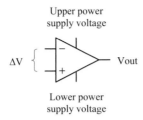

Figure 6-2 Operational amplifier (op-amp) schematic representation.

The operational amplifier component is a very high internal gain amplifier that only needs a minute amount of current to flow through its input pins (marked with a + and − sign), in order to function. The −ve input is referred to as the *inverting input* and the +ve input is referred to as the *non-inverting input*. The device also has power supply rails (normally hidden) that connect to the upper and lower power supply voltages to power the device. When the polarity of the input voltage differential ΔV is positive, with the +ve input voltage higher than the −ve input voltage, the op-amp output voltage (Vout) will be positive. Likewise the reverse holds true when the polarity of ΔV is negative.

Since the op-amp internal gain is very high, say 1 million, then if a 10V output signal was to be generated, only 10/millionth of a volt is required between the two input pins (ΔV = 10μV, μ represents micro, which is 10^{-6}). The current entering the op-amp is extremely low, and might be say 200nA (n represents nano, 10^{-9}).

The op-amp is a building block for various amplifier designs. The two rules concerning its very high gain and practically zero input current allow us to evaluate many op-amp circuits. The op-amp shown previously in Figure 6-2 has no external components connected to it. In this state it is of no practical use, so we must connect external components from the output to the input and take advantage of what is known as a *feedback* configuration. The term feedback is used because we are feeding the output signal back into the input of the op-amp. Now let us look at a current-to-voltage circuit as shown in Figure 6-3, since its function is fundamental to the operation of the DAC on the interface board.

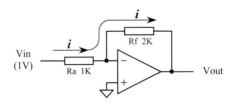

Figure 6-3 Current-to-voltage op-amp circuit.

The input voltage Vin, generates a current *i* which flows through the resistors Ra and Rf as shown, generating voltage Vout. We calculate the output voltage Vout by knowing the value of the current *i* and also the value of the voltage at the –ve input pin (and of course knowing the values of the two resistors Ra and Rf).

Vout cannot fall outside the op-amp's power supply voltage range, since the op-amp doesn't have any special internal circuitry to generate output voltages exceeding that of the power supply. Most op-amps are powered from upper supply voltages of +15V or less and lower power supply voltages of –15V or higher. Bearing this in mind, along with the fact that the op-amp has huge internal gain, then no matter what the output voltage of the op-amp might work out to be, the difference in voltage between the +ve and –ve input pins (ΔV) must be less than say fifteen microvolts.

$$\text{Vout (max)} = \Delta V \times \text{Internal Gain}$$
$$\text{+/- 15V} = 15\mu V \times 10^6$$

In the above example, the output voltage, Vout, cannot exceed either supply voltage at +/-15V, therefore ΔV will always be less than ~15μV, assuming the op-amp internal gain is one million.

The voltage at the –ve input pin is equal to the voltage at the +ve input pin plus ΔV. Since the voltage at the +ve input pin is equal to analog ground (0V, shown by the hollow triangle symbol), and we know that the difference in voltage (ΔV) between the +ve and –ve input pins will always be less than 15μV, the voltage at the –ve input pin will be less than 0V + 15μV = 15μV. This –ve input pin voltage is so close to zero volts that we might as well call it zero volts or *virtual ground*. Now that we know the –ve input pin voltage, we can calculate the current flow *i* and determine the output voltage Vout.

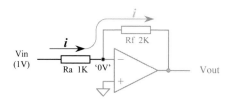

Figure 6-4 Calculating input current i.

Current flows from higher voltages to lower voltages, therefore, the current *i* will flow as shown in Figure 6-4 since Vin is at a greater voltage than the –ve input pin of the op-amp at virtual ground ($\cong$ 0V). With 1V at one end of resistor Ra and 'zero' volts at the other end, the current *i* is as follows:

$$
\begin{aligned}
\text{Current } i \ &= \ \text{Volts / Resistance} \\
&= \ (1V - 0V) / 1K \\
&= \ 1mA
\end{aligned}
$$

Note that the symbol 'K' represents one thousand ohms.

Since effectively zero current flows into the op-amp input pins, all of current i must flow through Rf, into the output pin and to the op-amp load (not shown).

Vout is equal to the voltage at the −ve input pin plus whatever the voltage is across resistor Rf. Using V = I x R, the voltage across resistor Rf is equal to current i multiplied by the resistance of Rf.

$$
\begin{aligned}
V_{Rf} \ &= \ 1mA \text{ x } 2K \\
&= \ 2V
\end{aligned}
$$

As mentioned previously, current flows from higher voltages to lower voltages, therefore the voltage at the left end of Rf is at a higher voltage than the voltage at the right end of Rf. Since the left end of Rf is at 0V, the voltage difference V_{Rf} is 2V, the right end of Rf must be at −2V. Knowing that the right end of Rf is connected to Vout, Vout must be equal to −2V.

The current i flowing through the feedback circuit must enter the junction at the op-amp output and split, part-of this current flowing into the op-amp load (not shown, towards Vout) and the remainder flowing into the op-amp output. How the op-amp draws the right amount of current into its output will be understood once the operation of *negative feedback* is explained.

In principle, the op-amp will draw (*sink*) or supply (*source*) sufficient current to ensure that current i remains at a level to force the −ve input pin to '0V'. This process happens automatically when a negative feedback configuration is used, as shown in Figure 6-3 and Figure 6-4. To configure negative feedback, connect the output voltage Vout back to the −ve input, either directly or through the use of external components, usually resistors. Negative feedback works as follows.

If say, the input voltage Vin increases, the voltage at the −ve input will increase slightly, increasing the small negative voltage difference between the op-amp's +ve and −ve inputs (ΔV). This increased negative voltage (ΔV) will generate an increasingly more negative output voltage Vout, which will in-turn have the *negative* effect of reducing the voltage at the −ve input pin. This generates a decrease in ΔV that will eventually settle to equilibrium. This process takes place automatically and quite rapidly in the op-amp when using negative feedback.

Positive feedback, on the other hand is altogether different and does not produce a self-correcting output voltage. Instead, the output voltage swings to the appropriate voltage supply. Later when the polarity between the +ve and −ve inputs is reversed, the output will swing to the opposite voltage supply. This type of feedback is used inside voltage comparators.

Armed with an understanding of how a current-to-voltage converter circuit works, we can move on to analyse and understand how practical DAC circuitry works.

6.2.2 DAC Circuit Principles

Two DAC methods will be discussed that use a group of parallel digital inputs to generate an analog output voltage. Both methods use current-to-voltage conversion as discussed in the preceding section. The two methods differ in the configuration of their logic input resistor arrays. The first method discussed is the summing amplifier DAC.

6.2.2.1 Summing Amplifier DAC

This circuit (shown in Figure 6-5) operates under the same principle as the current-to-voltage amplifier circuit shown in Figure 6-3 and Figure 6-4 earlier. Previously one input resistor was used to generate the current i_T, whereas in this case the summing amplifier uses four input resistors, each resistor generating its own input current. For the summing amplifier, the individual currents flowing out of each input resistor add together to form i_T.

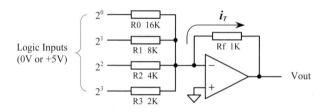

Figure 6-5 Summing amplifier DAC.

If all four input resistors were the same resistance value, each would be able to contribute either zero current (logic zero input) or ¼ of the total current i_T (logic-HIGH input). Therefore, the DAC output could have the following states:

DAC Output   0V   (all logic inputs at 0V)
              -1V   (one logic input +5V, others 0V)
              -2V   (two logic inputs +5V, others 0V)
              -3V   (three logic inputs +5V, other 0V)
              -4V   (all logic inputs +5V)

Having all four resistors with identical resistance will only give us five output states with 1V steps. Knowing that four unequally weighted logic bits can form sixteen unique numbers, we change the resistance values accordingly to achieve sixteen unique current levels of i_T. Input resistors R0, R1, R2 and R3 are chosen such that their input resistance values differ by a power of two from one resistor to the next.

Resistor R0 contributes the equivalent of the binary number 2^0, R1 is equivalent to 2^1 etc., as shown in Figure 6-5. The current derived from the 2^0 logic input will be ⅛ th that of the 2^3 logic input, the current from the 2^1 logic input will be ¼ that of the 2^3 logic input and the current from the 2^2 logic input will be ½ that of the 2^3 logic input. By using the various combinations of logic input states, sixteen different combined current levels can be generated, resulting in sixteen different output voltages.

This DAC method has a drawback; to implement a DAC with finer voltage resolution (a larger number of logic input bits), requires more accurate resistor values. For example, an 8-bit DAC (having 256 states) would need to have resistance tolerances of much less than 1/256 in order to avoid over-stepping between states within the whole DAC output range. Also, the actual values of these resistors will be non-standard – more difficult to achieve. The following circuit known as a *R-2R ladder* avoids these problems.

6.2.2.2 R-2R Ladder DAC

This circuit also uses current-to-voltage conversion like the Summing Amplifier DAC. The difference between these two methods is the configuration of the input resistor arrays.

The input voltage to the resistor array is a very stable and precise voltage reference, marked as V_{REF} in Figure 6-6. This is needed to allow the generation of accurate current levels to flow through the circuit. If you examine the R-2R circuit, you will see four switches (actually semiconductor switches), each switch individually controlled by one of the four logic inputs. The switches connect with either the analog ground (shown by the ∇ symbol) at 0V, or they connect with the op-amp –ve input pin at *virtual ground*. The term virtual ground is used because this pin is nearly at 0V (as described previously in Section 6.2.1). Realising that the switches will connect with 0V in either switch position, the R-2R resistor array can be analysed as follows.

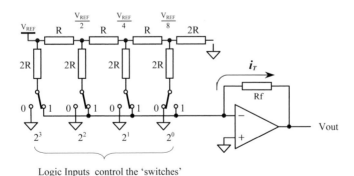

Figure 6-6 R-2R ladder DAC.

The current flowing through the switch controlled by the least significant bit, 2^0 is $1/8^{th}$ of the current flowing through the switch of the most significant bit, 2^3. Similarly, the current through the switch controlled by the 2^1 bit is ¼ of the current through the most significant bit switch, and the current through the switch controlled by the 2^2 bit is ½ of the current through the switch controlled by the most significant bit. When the switch is in the position marked '0', the current through the 2R resistor passes to analog ground and not towards the −ve input pin of the op-amp. In the other position marked '1', the current through the 2R resistor flows towards the op-amp −ve input pin, adding to the other currents from any other switches also in the '1' position. These currents combine to become i_T and generate the output voltage in the same way that any current-to-voltage amplifier circuit does. As mentioned previously, the positive current flowing through Rf produces a negative voltage output. DACs will often need to have a positive output voltage, and to achieve this, we add an inverting amplifier to reverse the polarity of the DAC output signal. A circuit performing such a function is found on the interface board and will be discussed shortly.

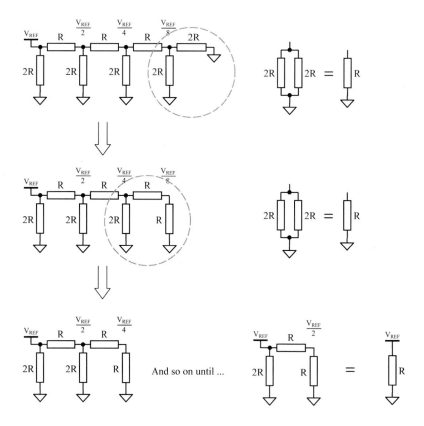

Figure 6-7 Reference voltage (VREF) loading.

There is one last point to be made concerning the R-2R ladder circuit. This circuit has a very nice feature in that the resistance the voltage reference 'sees' is always equal to a resistance value of R, no matter which position the switches are in. This excludes the short time when the switch is in 'mid air' and not connected to either contact. Voltage references or regulators have slight changes in their output voltage as their load varies. Because the R-2R ladder circuit presents a uniform load to the voltage reference, the output voltage of the voltage reference will remain quite steady. Figure 6-7 demonstrates the constant resistance, R, loading on the voltage reference – remembering that the switch ends of the 2R resistors always see zero volts, no matter which position the switches are in.

To summarise, the R-2R ladder array has the advantage of using a simple ratio of resistance for the entire array, the ratio of two. Also, the resistor array places a relatively constant load on the voltage reference, resulting in superior accuracy.

6.2.3 Operation of the DAC0800

The DAC used in the interface board is the DAC0800, an 8-bit device providing a programmable output current. We use this programmable output current with a current-to-voltage circuit to generate a voltage output. Figure 6-8 shows a block diagram of the DAC0800 circuit used on the interface board.

The DAC0800 produces an output current, where the current flowing through its output pin actually flows into the DAC (sink current). To generate an analog voltage, we need to use a current-to-voltage converter, in this case a single resistor, R, one end connected to the DAC output pin and the other end to a reference voltage, 0V, as shown in Figure 6-8. When current i flows through the resistor, a voltage ΔV ($\Delta V = i \times R$) is generated across the resistor. Knowing that current flows from a more positive voltage to a less positive voltage, the voltage at the end of the resistor, R, which connects to the DAC output, will be equal to $0V + -\Delta V = -\Delta V$. With a current i equal to zero, the output voltage generated across resistor R will be 0V.

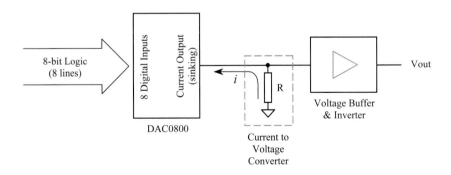

Figure 6-8 Block diagram - Interface board DAC0800 circuit.

So we have a programmable output voltage ranging from 0V to $-\Delta V$, where the actual value of ΔV for a given digital input will be dependent on the resistance value of resistor R. A DAC output range that always remains on the same side of zero volts (the negative side in this particular case) is termed a *unipolar output*. The other type of DAC output is one that which crosses between the positive and negative sides of zero volts, and is termed a *bipolar output*. Figure 6-9 shows the resistor configuration needed by the current-to-voltage converter for it to operate in a bipolar mode.

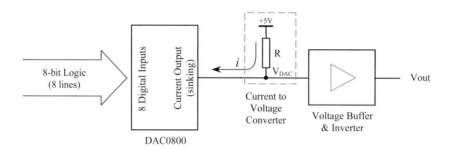

Figure 6-9 Bipolar current-to-voltage conversion.

The output voltage (V_{DAC}) of the bipolar current-to-voltage converter is evaluted from the relationship:

$$V_{DAC} = +5V - \text{Voltage drop across R}$$
$$= +5V - i \times R$$

The minus sign in the preceding equation comes from the fact that the current *i* flows into the DAC0800, and that current flows from a more positive voltage to a less positive voltage – therefore the voltage at the current-to-voltage converter will be equal to +5V minus the voltage drop across resistor R.

As current draw *i* into the DAC output increases, the voltage at the DAC output drops from +5V (all 8 logic inputs zero) to lower voltages, passing through zero volts and into the negative voltage region, eventually reaching –5V (all logic inputs high, and using the appropriate value for resistor R). The current-to-voltage converter on the interface board can be configured in unipolar or bipolar mode by fitting a link in either of two positions respectively.

Table 6-1 Current-to-Voltage Converter Output.

DAC Logic Input	Unipolar mode	Bipolar mode
255	-5V	-5V
0	0V	+5V

Table 6-1 summarises the output voltage of the current-to-voltage converter for both unipolar and bipolar configurations of the resistor R (chosen to have the appropriate value).

Usually, we expect the output voltage from a DAC to be lowest when all logic input bits are zero and highest when all logic input bits are high (255 for an 8-bit DAC). To meet this convention, the current-to-voltage converter output needs to be inverted to produce +5V for all logic input bits high and 0V for all logic input bits low when in unipolar mode. Bipolar mode should produce +5V when all logic input bits are high and –5V for all logic input bits at zero. The Buffer and Inverter circuit block shown ahead in Figure 6-11 and Figure 6-12 perform this voltage inversion.

A typical inverter circuit will need to draw some current i' through its input as shown in Figure 6-10. If this current draw is significant, it will adversely affect the voltage generated across R. This occurs because we now have two currents i and i' flowing through R with i' generating an error voltage $\Delta V_{ERROR} = i' \times R$. The total voltage across resistor R is $\Delta V = (i + i') \times R$, equal to the correct voltage ($i \times R$), plus ΔV_{ERROR}.

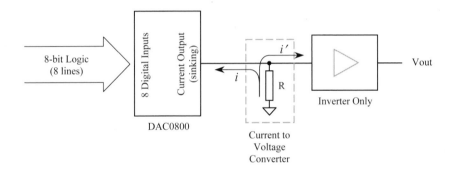

Figure 6-10 Inverter circuit affecting voltage ΔV.

To ensure that we do not draw a significant current into the inverter circuit, we precede it with a Voltage Buffer circuit. The Voltage Buffer draws a minute amount of current (~200 nA), not enough to interfere with the accuracy of the DAC current-to-voltage converter.

6.2.3.1 The Voltage Buffer Circuit

This circuit performs the function of buffering the output voltage of the DAC current-to-voltage converter as mentioned previously. The Buffer uses a special op-amp configuration that draws almost zero current. This circuit is shown in Figure 6-11.

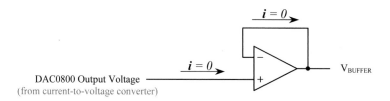

Figure 6-11 Voltage Buffer circuit.

Remember that the current flowing into an op-amp input pin is effectively zero, therefore the current flowing from the DAC0800 current-to-voltage converter circuit and into the op-amp +ve input pin is zero. This op-amp is configured using negative feedback. The output of the op-amp (V_{BUFFER}) will be driven to ensure there is effectively zero voltage between its –ve and +ve input pins. Therefore, the output voltage of the op-amp will follow or *buffer* the input voltage (from the DAC) present at its +ve input.

6.2.3.2 The Voltage Inverter Circuit

The DAC output circuitry produces a falling voltage as the input bit number value increases (see Table 6-1). The Voltage Inverter circuit inverts the voltage signal generated by the DAC current-to-voltage converter circuit, to give an increasing output voltage as the DAC input increases in value. We have already examined a circuit that produces an output voltage with opposite polarity to that presented to its input: the current-to-voltage op-amp circuit. That circuit performs voltage inversion and also amplification, where the amplification or gain is equal to the ratio of the resistor Rf to Ra. If we make Rf equal to Ra then we will have a gain of unity – now we have an inverter as shown in Figure 6-12. The circuit analysis of the inverter is identical to that carried out in Section 6.2.1.

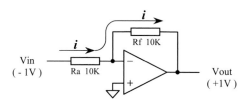

Figure 6-12 Op-amp Voltage Inverter circuit.

6.2.4 DAC Characteristics and Specifications

The DAC has some basic characteristics that specify its performance. These include *settling time*, *non-linearity* and *full-scale error*. The settling time is a measure of how fast the output of the DAC can change and settle to within half of a least significant bit (LSB - the smallest change in output voltage, caused by the 2^0 logic input). The DAC0800 has a typical settling time of 100 ns.

The linearity refers to the maximum deviation from the converter's ideal input/output relationship – being the relationship between the DAC output value and the DAC input value, over the whole input range of the converter. Ideally, this relationship should be a straight line, the output increasing in value as the input increases. Figure 6-13 shows an ideal input/output relationship along with two independent errors, the *offset error* and the *gain error*.

The offset error is the voltage (or current) present at the output of the DAC when the digital input is zero. Ideally the offset error should be zero. Gain error occurs when the output of the DAC doesn't increase by the correct amount for an increase in the digital input.

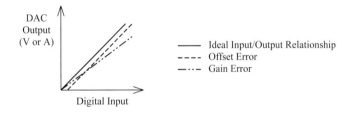

Figure 6-13 DAC Input/Output Relationship.

The DAC0800 used in the interface board has the following relevant specifications. Its non-linearity is ±0.1% over its rated temperature range (meaning that it has a linearity of 99.9%) and its full-scale error is ±1 LSB (least significant bit), meaning that its maximum output value will be within ±1 LSB of the true value corresponding to the ideal transfer function (input/output relationship).

6.3 Programming the Digital-to-Analog Converter

At the end of the previous chapter, we developed an object class named `ParallelPort`. This object has the capability to input and output data using all three addresses of the PC's parallel port. It can easily be used to drive an 8-bit Digital-to-Analog Converter. Only partial functionality of the `ParallelPort` class is needed to drive the DAC on the interface board. The DAC just needs to receive an 8-bit number from the PC. This can be done by sending an 8-bit number from the PC to the interface board using the port at address BASE. This is an output port with eight parallel signals we can connect to the interface board. Since we already have ParallelPort as a completely packaged object class, we can use it to drive the DAC.

The rest of the chapter will progress as follows. First, we will develop a program using the ParallelPort object to drive the DAC. When this program is executed, the DAC system will generate an analog voltage proportional to the 8-bit number sent to it. We will then proceed to learn about inheritance using an exercise in which a new class is derived to represent the DAC. This new object may require extra functionality that is specific to the DAC. We will then turn our attention to restrictions imposed by various access attributes. The proper use of access attributes will be described in detail. We will proceed through several different versions of the same program, each time strengthening code reuse. The final program presented will then be used in future chapters as the most appropriate object-oriented program to drive the DAC.

The first program to drive the DAC is given in Listing 6-1. The class definition and the member function definitions of the ParallelPort object class are exactly the same as given in the previous chapter. In the main() function, an object is instantiated using the ParallelPort class and it is given the name D_to_A. Then the WritePort0() function of the object is used repeatedly to send different data to the parallel port each time. Following each WritePort0() function is a getch() function. These getch() functions force the program to wait for a key press before executing the next statement. This will allow you time to carry out measurements on the interface board to verify whether or not the correct analog voltage has been generated by the DAC system.

Table 6-2 Connections for the DAC.

BASE Address (Buffer IC, U13)	DAC0800 (U8)
D0	D0 (12)
D1	D1 (11)
D2	D2 (10)
D3	D3 (9)
D4	D4 (8)
D5	D5 (7)
D6	D6 (6)
D7	D7 (5)

Before running your DAC programs, configure the interface board as follows. Ensure that an operational 9V battery is connected to the terminal block (J14). Fit the jumper on the interface board across the two-pin header position marked LINK1, to select unipolar mode (0V to +5V). When the connections are completed according to Table 6-2, the 8 bits of the port at address BASE of the parallel port will be connected to the 8-bit input of the DAC.

Listing 6-1 Digital-to-Analog Conversion using the `ParallelPort` object class.

```
/******************************************************
This program uses the ParallelPort object developed in the
previous chapter to write a byte of data to the Digital to
Analog Convertor (DAC). The DAC generates an analog voltage
proportional to the value of the data byte it receives.
******************************************************/
#include <iostream.h>
#include <conio.h>

class ParallelPort
{
    private:
        unsigned int BaseAddress;
        unsigned char InDataPort1;
    public:
        ParallelPort(); // default constructor
        ParallelPort(int baseaddress); // constructor
        void WritePort0(unsigned char data);
        void WritePort2(unsigned char data);
        unsigned char ReadPort1();
};

ParallelPort::ParallelPort()
{
    BaseAddress = 0x378;
    InDataPort1 = 0;
}

ParallelPort::ParallelPort(int baseaddress)
{
    BaseAddress = baseaddress;
    InDataPort1 = 0;
}

void ParallelPort::WritePort0(unsigned char data)
{
    outportb(BaseAddress,data);
```

```
}

void ParallelPort::WritePort2(unsigned char data)
{
    outportb(BaseAddress+2,data ^ 0x0B);
}

unsigned char ParallelPort::ReadPort1()
{
    InDataPort1 = inportb(BaseAddress+1);
// Invert most significant bit to compensate
// for internal inversion by printer port hardware.
    InDataPort1 ^= 0x80;
// Filter to clear unused data bits D0, D1 and D2 to zero.
    InDataPort1 &= 0xF8;

    return InDataPort1;
}

void main()
{
    ParallelPort D_to_A;

    cout << "Press a key ... " << endl;
    getch();
    D_to_A.WritePort0(0);

    cout << "Press a key ... " << endl;
    getch();
    D_to_A.WritePort0(32);

    cout << "Press a key ... " << endl;
    getch();
    D_to_A.WritePort0(64);

    cout << "Press a key ... " << endl;
    getch();
    D_to_A.WritePort0(128);

    cout << "Press a key ... " << endl;
    getch();
    D_to_A.WritePort0(255);
}
```

The analog output voltage can be measured at the pin on the interface board labelled VDAC, which is connected to pin 7 of the operational amplifier LM358 (U10B). Each time you press a key, the program will output a slightly higher 8-bit value and the DAC will produce a corresponding higher analog voltage. Read the analog voltage by connecting a voltmeter between the pin marked VDAC on the interface board and a ground pin.

As described earlier, additional circuitry has been provided on the interface board to facilitate both unipolar and bipolar output from the DAC. Move the jumper to the position marked LINK2 on the interface board and re-execute the program to check the bipolar operation of the DAC system.

6.4 Derivation of Object Classes

In the last example, we used the `ParallelPort` object class to operate the DAC. This object is designed for general-purpose use of the parallel port. The `ParallelPort` object class is more than capable of handling the simple requirements of the DAC. However, using the `ParallelPort` object class for digital-to-analog conversion is not particularly appropriate. Instead, it is desirable to create an object class, which at least has a name that suits digital-to-analog conversion. This is an ideal situation to derive a class. We are for now merely applying a change in name which will reduce the complexity associated with the derivation of the new class. As we progress through the book we will confront more involved class derivations.

One of the main strengths of object-oriented programming is the re-useability of the program segments developed in the past. The C++ language has excellent mechanisms in place to expand the capabilities of an existing class and thereby to form a new super class. As described in Chapter 4, these super classes are known as *derived* classes. To be able to derive a class, a *base* class must exist. The derived classes are meant for more specific purposes than the base class. In our case, the general-purpose object class `ParallelPort` can be considered as the base class. The new class to be created needs to be more specific to suit the Digital-to-Analog Converter. Listing 6-2 shows the simplest way to create the new class.

Listing 6-2 Derivation of DAC class.

```
class ParallelPort
{
    private:
        unsigned int BaseAddress;
        unsigned char InDataPort1;

    public:
        ParallelPort();
```

```
        ParallelPort(int baseaddress);
        void WritePort0(unsigned char data);
        void WritePort2(unsigned char data);
        unsigned char ReadPort1();
};


class DAC : public ParallelPort
{
};
```

The class derivation is kept simple so that we can direct our attention to the principles of derivation of classes rather than the actual functionality of the DAC object class. The class definition for the `ParallelPort` object class is identical to that given in Chapter 5. The only new part in the above listing is the class definition for the object type DAC shown in bold typeface. The first line of the new object class definition is:

```
class DAC : public ParallelPort
```

Here, we derive a new class named DAC, using the existing class `ParallelPort` as the base class. The keyword `public` before `ParallelPort`, is referred to as an *access specifier*. We will return to access specifiers after considering the member data and member functions we inherited from the base class.

Inherited members of the DAC class

Here, the word inheritance is used to describe the fact that the derived class inherits all the member data and member functions of the base class. If you now take a closer look at the DAC object class, you will see that the body of the DAC class definition is *empty*. It was explained in Chapter 5 that if we do not provide our own constructor and destructor for the object class, the compiler will provide them. Therefore, although not visible, the DAC class has a default constructor and a default destructor provided by the compiler. Also, because the DAC class has been derived from the `ParallelPort` class, it has inherited all the members (both data and functions including constructors) of the `ParallelPort` class.

If we list the data members of the DAC class, they are as follows:

```
        unsigned int BaseAddress;
        unsigned char InDataPort1;
```

The member functions of the DAC class are as follows:

```
        DAC(); // Compiler-generated constructor (hidden).
        ParallelPort();
        ParallelPort(int baseaddress);
        void WritePort0(unsigned char data);
        void WritePort2(unsigned char data);
        unsigned char ReadPort1();
```

```
~ParallelPort(); // Inherited compiler-generated
                 // destructor (hidden).
~DAC(); // Compiler-generated destructor (hidden).
```

> **NOTE**
>
> Note that we did not include destructors in the class definition for the ParallelPort class and also in the class definition for the DAC class. Therefore,the compiler will generate a hidden default destructor for each of these classes as listed above. All compiler-generated constructors and destructors are described as hidden to the programmer because they are invisible in the source code. See Chapter 8 for a more detailed description of destructors.

The DAC class is equivalent to the ParallelPort class except that it has a default constructor and destructor provided by the compiler. It is clear that this particular class derivation is just a name changing exercise since we did not expand the capability of the DAC class. It can only do as much as the ParallelPort class and no more. However, if we need to enhance the power of the DAC class, it is an easy matter to add member data and member functions to the new DAC class in its class definition and then provide the definitions of the added functions.

Instantiating DAC objects (calling constructors)

The question to consider is "Can we use this class?" To instantiate an object of type DAC, we must call the constructor of the DAC class. It would be called in the declaration statement for the new object named D_to_A as shown below:

```
DAC D_to_A;
```

When the constructor of a derived class is first called, it will call the constructor of its inherited base class that takes whatever parameters it is being passed from the derived class constructor (this will be further explained ahead). The base class constructor instantiates its members and completes whatever tasks it has been coded to do. When it is finished it returns program execution back to the start of the body of the derived class constructor. The derived class constructor then instantiates its members and completes whatever tasks it has been coded to do.

In the case of this particular program, the DAC class does not have a programmer-generated constructor, so the compiler adds its own default constructor (takes no parameters). This is the constructor that is called when the above statement is executed. *Before* it performs its own tasks, it will call its inherited base class constructor. Because the derived class constructor is not passing an argument to the constructor of the base class, the constructor ParallelPort() and not the constructor ParallelPort(int baseaddress) will be called to allocate memory for the inherited base class members. If you look at the body of this constructor in Listing 6-1 you will see that it also initialises BaseAddress to the value 0x378, (and `InDataPort1` to zero). Then the `DAC` class's default constructor instantiates memory for whatever members are added to it (none) and executes its empty body.

The public function `WritePort0()` has been inherited from the `ParallelPort` class. Therefore, it can be used to send the data (in this case, 255) to the parallel port at the initialised base address value (0x378) using the following statement:

```
D_to_A.WritePort0(255);
```

There is a problem with this program – the `DAC` class does not allow the user to specify a value for the `BASE` address of the parallel port. This is inadequate for those users that have their parallel ports at a different `BASE` address than 0x378. Therefore, we need some mechanism to provide the option to initialise the data member variable `BaseAddress` to a suitable value. Normally the best way to do this is to use the constructor. Similar to the early stages when developing the `ParallelPort` object, the compiler-generated constructor for the `DAC` class is not adequate for our application. We must provide our own constructors and code them in a manner that allows us to specify the value of `BaseAddress`. An improved `DAC` class is given in Listing 6-3.

Listing 6-3 An improved DAC class.

```
class ParallelPort
{
    private:

        unsigned int BaseAddress;
        unsigned char InDataPort1;

    public:
        ParallelPort();
        ParallelPort(int baseaddress);
        void WritePort0(unsigned char data);
        void WritePort2(unsigned char data);
        unsigned char ReadPort1();
};

class DAC : public ParallelPort
{
    public:
        DAC();  // default constructor.
        DAC(int baseaddress);  // constructor.
};
```

This time we have used a similar approach as for the `ParallelPort` class and declared two constructors for the `DAC` class. Note that the compiler will not provide a default constructor because we have provided the constructor `DAC(int baseaddress)` shown in Listing 6-3. Therefore, we must provide our own default constructor `DAC()`.

We can now define these constructors by providing their statements which will set the parallel port's `BASE` address by initialising inherited private data member `BaseAddress`. The first attempt to define the constructors is given in Listing 6-4.

Listing 6-4 A failed attempt to define the constructor.

```
DAC::DAC()  // Does work.
{
}


DAC::DAC(int baseaddress)
{
    BaseAddress = baseaddress; // Fails to work!
}
```

The following statement will use the `DAC` class's programmer-generated default constructor `DAC()` to instantiate an object of type `DAC` named `D_to_A`:

```
DAC D_to_A;
```

This default constructor of the `DAC` class will operate in the same way that its compiler-generated default compiler did. It will first call the inherited base class constructor that also takes no arguments - the base class's default constructor `ParallelPort()`. This constructor will instantiate the inherited base class members, initialise the private data member `InDataPort1` to 0, and initialise the private data member `BaseAddress` to the value 0x378. Then the `DAC` class's default constructor will be executed to instantiate whatever members have been added to it (none), followed by executing its body – in this case with nothing in it.

The derived class `DAC` inherits the members of its base class `ParallelPort` but does not have access to the *private* members of its inherited base class (regardless of the acess specifier used – public in this case). This means that the constructor `DAC(int baseaddress)` shown in Listing 6-4 will not be able to access the inherited base class data member `BaseAddress` (declared as private). Therefore, it cannot be compiled and so cannot work!

The solution in this situation is as follows. Instead of a member function from the derived `DAC` class trying to make an illegal attempt to directly access a private data member inherited from its base class `ParallelPort`, a call can be made to a public function inherited from the base class that can change the private data member `BaseAddress` of its class as shown in Figure 6-14. A proper constructor definition for the `DAC` class is given in Listing 6-5.

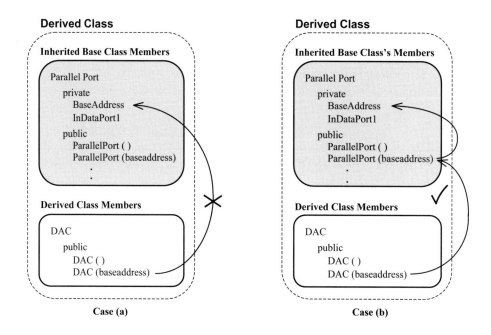

Figure 6-14 Accessing the inherited private data member `BaseAddress`.

Listing 6-5 Corrected definition of the constructor of the `DAC` class.

```
DAC::DAC()
{
}


DAC::DAC(int baseaddress)  : ParallelPort(baseaddress)
{
}
```

The part-line of bold font code shown in Listing 6-5 is new and is the mechanism that allows the value of private data member `BaseAddress` to be set to the value of the argument passed at the time of instantiating the `DAC` object named `D_to_A`:

```
DAC D_to_A(0x3BC);
```

When the program encounters this statement, it calls the appropriate constructor from the `DAC` class. Because the parallel port's BASE address is given as an argument when instantiating the `D_to_A` object, the constructor `DAC(int` baseaddress) will be called, and *not* the default constructor `DAC()`. When first called, the constructor calls the appropriate constructor of the inherited base class that takes the same arguments being passed to it. The bolded part-line:

`DAC::DAC(int baseaddress) :` **`ParallelPort(baseaddress)`**

informs the compiler that the derived class constructor `DAC(int baseaddress)` is to pass the parameter `baseaddress` to the inherited base class constructor `ParallelPort(baseaddress)`. This base class constructor will instantiate its members and then initialise its private data member `BaseAddress` to be equal to the argument passed to the parameter `baseaddress` (and initialise `InDataPort1` to zero). Immediately following this action, the constructor `DAC(int baseadress)` will be executed to instantiate its added members (none) and then complete its tasks contained within its empty body.

This is how to initialise the inaccessible private data member `BaseAddress` inherited from the `ParallelPort` class - with minimum code thanks to inheritance.

Now we can turn our attention to using this object class in a program to carry out digital-to-analog conversion. The complete program is shown in Listing 6-6. Instead of instantiating an object of type `ParallelPort`, this program instantiates an object of type `DAC`. The operation of the program is identical to the one shown in Listing 6-1.

Listing 6-6 The use of the DAC class for Digital-to-Analog Conversion.

```
/********************************************************
The new class DAC is used in the main() function to
sequentially write several bytes of data to the Digital
to Analog convertor.
********************************************************/
#include <iostream.h>
#include <conio.h>

class ParallelPort
{
    private:
        unsigned int BaseAddress;
        unsigned char InDataPort1;

    public:
        ParallelPort();
        ParallelPort(int baseaddress);
        void WritePort0(unsigned char data);
        void WritePort2(unsigned char data);
```

```cpp
        unsigned char ReadPort1();
};

ParallelPort::ParallelPort()
{
    BaseAddress = 0x378;
    InDataPort1 = 0;
}

ParallelPort::ParallelPort(int baseaddress)
{
    BaseAddress = baseaddress;
    InDataPort1 = 0;
}

void ParallelPort::WritePort0(unsigned char data)
{
    outportb(BaseAddress,data);
}

void ParallelPort::WritePort2(unsigned char data)
{
    outportb(BaseAddress+2,data ^ 0x0B);
}

unsigned char ParallelPort::ReadPort1()
{
    InDataPort1 = inportb(BaseAddress+1);
// Invert most significant bit to compensate
// for internal inversion by printer port hardware.
    InDataPort1 ^= 0x80;
// Filter to clear unused data bits D0, D1 and D2 to zero.
    InDataPort1 &= 0xF8;
    return InDataPort1;
}

class DAC : public ParallelPort
{
    public:
        DAC();
        DAC(int baseaddress);
};

DAC::DAC()
{
```

```
}

DAC::DAC(int baseaddress) : ParallelPort(baseaddress)
{
}

void main()
{
    DAC D_to_A;

    cout << "Press a key ... " << endl;
    getch();
    D_to_A.WritePort0(0);

    cout << "Press a key ... " << endl;
    getch();
    D_to_A.WritePort0(32);

    cout << "Press a key ... " << endl;
    getch();
    D_to_A.WritePort0(64);

    cout << "Press a key ... " << endl;
    getch();
    D_to_A.WritePort0(128);

    cout << "Press a key ... " << endl;
    getch();
    D_to_A.WritePort0(255);
}
```

6.5 Adding Members to Derived Classes

Having considered a simple class definition to understand the principles of inheritance and the derivation of new classes, we can now proceed to add extra functionality to the derived class. In the next example program, the DAC class has had a data member and a member function of its own added. The data member will remember the data last output to the DAC. The member function provides an interface to the outside world, allowing any function to query the last value output to the DAC. The DAC class will have a modified version of the inherited function WritePort0() to enable it to store the last output value. We will now see how to carry out the following:

1.   Add new members to a derived class.

2.   Modify an inherited function.

The *new* DAC class definition and its member function definitions are given in Listing 6-7. Note that in this listing we have re-declared the function WritePort0() as a member function of the DAC class. The C++ language requires us to do this so the body of the inherited function WritePort0() can be changed to suit our requirements. See Section 6.5.2 for additional details.

Listing 6-7 The DAC class with new members added.

```
class DAC : public ParallelPort
{
    private:
        unsigned char LastOutput;

    public:
        DAC();
        DAC(int baseaddress);
        void WritePort0(unsigned char data);
        unsigned char GetLastOutput();
};

DAC::DAC()
{
    LastOutput = 0;
}

DAC::DAC(int baseaddress) : ParallelPort(baseaddress)
{
    LastOutput = 0;
}

void DAC::WritePort0(unsigned char data)
{
    outportb(BaseAddress,data); // Will not work!
    LastOutput = data;
}

unsigned char DAC::GetLastOutput()
{
    return LastOutput;
}
```

The private data member LastOutput is added to the derived class DAC for the purpose of storing the value output by the WritePort0() function. A member function named `GetLastOutput()` is also added to the derived class. It returns the value stored in `LastOutput`. Therefore, the only statement within the body of the `GetLastOutput()` function is:

```
return LastOutput;
```

Any function in the program that requires the value of the last data number output to the DAC (from the port at address `BASE`) must call the `DAC` class's public function `GetLastOutput()`. This is the only way to access the value stored in the private data member `LastOutput` and ensures that functions outside the `DAC` class have no direct access to it. Once again, this demonstrates the controlled access to private data of a class by an object-oriented program.

Both constructors have been modified to initialise `LastOutput` to 0 by including the statement:

```
LastOutput = 0;
```

Therefore, at the time of instantiating a `DAC` object, the constructor will initialise the new data member `LastOutput` to 0.

We have modified the `WritePort0()` function in order to store the latest value output to the DAC by adding the line:

```
LastOutput = data;
```

However, the definition of the `WritePort0()` function given in Listing 6-7 will *fail* to compile. This is because the `WritePort0()` function of the derived class `DAC` is trying to use the inherited data member `BaseAddress` which is *private* to the base class. Although the private data of the base class is inherited, the derived classes cannot access this private data as shown in Figure 6-15.

One means to allow access to `BaseAddress` is by relaxing its access attributes. This can be done by declaring `BaseAddress` in the base class with `protected` access. Then `BaseAddress` can be accessed by all functions of all derived classes provided the classes are derived using a `public` or `protected` base class *access specifier*. Access specifiers are described in more detail ahead in section 6.5.1. The modified class definition of the base class is given in Listing 6-8.

NOTE

Declare variables as **private** unless you plan to derive other classes using the current class as a base class. When you want to use the current class as a base class to derive new classes, carefully determine the variables of the current class you would want the derived class to have access to. Declare *only* these variables as **protected** in the current class.

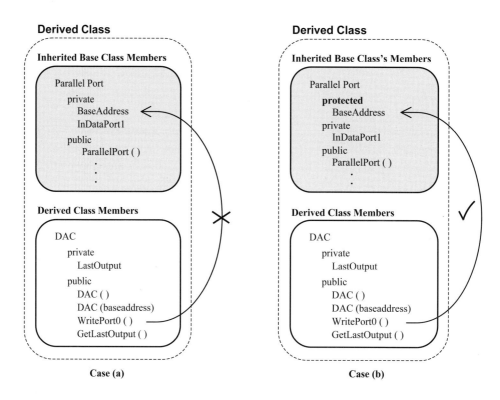

Figure 6-15 Private and protected access specifiers.

Listing 6-8 Base class `ParallelPort` – `BaseAddress` as protected data member.

```
class ParallelPort
{
    protected:
        unsigned int BaseAddress;

    private:
        unsigned char InDataPort1;

    public:
        ParallelPort();
        ParallelPort(int baseaddress);
        void WritePort0(unsigned char data);
        void WritePort2(unsigned char data);
        unsigned char ReadPort1();
};
```

```
class DAC : public ParallelPort
{
    private:
        unsigned char LastOutput;

    public:
        DAC();
        DAC(int baseaddress);
        void WritePort0(unsigned char data);
        unsigned char GetLastOutput();
};
```

A program that uses the class definition shown above is given in Listing 6-9.

6.5.1 Access specifiers

Consider the line:

```
class DAC : public ParallelPort
```

The keyword `public` in this line is an *access specifier*. Access specifiers change the access attributes as follows. The protected variables of the `ParallelPort` class can be accessed by the member functions of the `DAC` class, if and only if, the `DAC` class is derived from `ParallelPort` using a `public` or `protected` base class access specifier. Access specifiers can also be `private`. Figure 6-16 shows how the access specifiers determine access attributes of inherited members.

Access Specifier `public`

When deriving a class using a `public` base class access specifier, all inherited `public` members of the base class will become `public` members of the derived class, all inherited `protected` members of the base class will become `protected` members of the derived class. All inherited `private` members will remain `private` to the base class, and so the derived class *cannot* access them.

Access Specifier `protected`

When deriving a class using a `protected` base class access specifier, all inherited `public` and `protected` members of the base class will become `protected` members of the derived class. All inherited `private` members will remain `private` to the base class, and so the derived class *cannot* access them.

Access Specifier `private`

When deriving a class using a `private` base class access specifier, all inherited `public` and `protected` members of the base class will become `private` members of the derived class. All inherited `private` members will remain `private` to the base class, and so the derived class *cannot* access them.

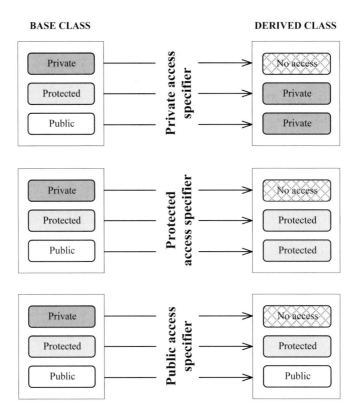

Figure 6-16 Access specifiers determine access attributes of derived class members.

6.5.2 Polymorph Functions

The attempt to redefine the member function `WritePort0()` in Listing 6-9 is quite legitimate. However, `WritePort0()` is a function the `DAC` class inherited from the base class `ParallelPort`. To allow derived classes to redefine inherited functions, the inherited functions must be explicitly included in the derived class definition as done in Listing 6-9 (and Listing 6-7). In this example, there are two `WritePort0()` functions: one of them belonging to the `ParallelPort` class; and the other belonging to the `DAC` class. The existence of functions of the same name throughout a class hierarchy is termed *polymorphism*. These functions not only have the same name, but also the same number of parameters, same types of parameters, and the same sequence of parameters.

The declaration of the `DAC` class is given in Listing 6-8. Despite the `DAC` class inheriting the function `WritePort0()` from the base class `ParallelPort`, it is explicitly coded again in the `DAC` class. This allows us to redefine the body of the `WritePort0()` function to suit the needs of the `DAC` class.

> **NOTE**
>
> There is a clear difference between the term polymorphism and overloading. Overloaded functions also have the same function name. They differ in the number or type of parameters passed to the functions. In addition, overloaded functions do not need to be member functions.

The complete program, including the class hierarchy that can be compiled without errors is given in Listing 6-9.

Listing 6-9 Digital-to-analog conversion with the expanded DAC object.

```
/*****************************************************
In this program, the compilation error has been
eliminated by  changing the access attribute of
BaseAddress in  the base class (ParallelPort) from
private to protected.  Now the functions of the publicly
derived class can access the inherited BaseAddress.
This accessibility is only available to the derived
classes of the base class and to the base class
itself.  The function WritePort0(), which is re-declared
in the derived class, can now be modified without any
compilation errors.
*****************************************************/
#include <iostream.h>
#include <stdio.h>
#include <conio.h>
#include <dos.h>

class ParallelPort
{
    protected:
        unsigned int BaseAddress;

    private:
        unsigned char InDataPort1;

    public:
        ParallelPort();
        ParallelPort(int baseaddress);
        void WritePort0(unsigned char data);
        void WritePort2(unsigned char data);
```

```cpp
        unsigned char ReadPort1();
};

ParallelPort::ParallelPort()
{
    BaseAddress = 0x378;
    InDataPort1 = 0;
}

ParallelPort::ParallelPort(int baseaddress)
{
    BaseAddress = baseaddress;
    InDataPort1 = 0;
}

void ParallelPort::WritePort0(unsigned char data)
{
    outportb(BaseAddress,data);
}

void ParallelPort::WritePort2(unsigned char data)
{
    outportb(BaseAddress+2,data ^ 0x0B);
}

unsigned char ParallelPort::ReadPort1()
{
    InDataPort1 = inportb(BaseAddress+1);
// Invert most significant bit to compensate
// for internal inversion by printer port hardware.
    InDataPort1 ^= 0x80;
// Filter to clear unused data bits D0, D1 and D2 to zero.
    InDataPort1 &= 0xF8;
    return InDataPort1;
}

class DAC : public ParallelPort
{
    private:
        unsigned char LastOutput;

    public:
        DAC();
        DAC(int baseaddress);
        void WritePort0(unsigned char data);
```

```
        unsigned char GetLastOutput();
};

DAC::DAC()
{
    LastOutput = 0;
}

DAC::DAC(int baseaddress) : ParallelPort(baseaddress)
{
    LastOutput = 0;
}

void DAC::WritePort0(unsigned char data)
{
    outportb(BaseAddress,data);
    LastOutput = data;
}

unsigned char DAC::GetLastOutput()
{
    return LastOutput;
}

void main()
{
    DAC D_to_A;


    D_to_A.WritePort0(0);
//  printf("\nDAC byte:%3d   ", D_to_A.LastOutput); // Does
                                    // not work, why?
    printf("\nDAC byte:%3d   ", D_to_A.GetLastOutput());
    cout << "   Measure voltage and press a key" << endl;
    getch();

    D_to_A.WritePort0(32);
    printf("\nDAC byte:%3d   ", D_to_A.GetLastOutput());
    cout << "   Measure voltage & then press a key" << endl;
    getch();

    D_to_A.WritePort0(64);
    printf("\nDAC byte:%3d   ", D_to_A.GetLastOutput());
    cout << "   Measure voltage and press a key" << endl;
    getch();
```

```
    D_to_A.WritePort0(128);
    printf("\nDAC byte:%3d    ", D_to_A.GetLastOutput());
    cout << "   Measure voltage and press a key" << endl;
    getch();

    D_to_A.WritePort0(255);
    printf("\nDAC byte:%3d    ", D_to_A.GetLastOutput());
    cout << "   Measure voltage and press a key" << endl;
    getch();
}
```

In the above program, the constructor `DAC()` is called at the time of instantiating the DAC class object `D_to_A`. Referring to the function definition of the default `DAC()` constructor; it makes a call to the default constructor of the class `ParallelPort` before entering the body of the `DAC()` constructor. The default constructor of the `ParallelPort` class will initialise `BaseAddress` to 0x378 (and set `InDataPort1` to 0). Then execution of the body of the constructor `DAC()` begins. It will initialise the value of the DAC class's private data member `LastOutput` to 0.

Note that the member function `GetLastOutput()` is called when the value of `LastOutput` needs to be printed onscreen. This needs to be done because the `printf()` function does not have direct access to the private data member `LastOutput`.

The definition of the `WritePort0()` function can be modified slightly to revert the access attribute of `BaseAddress` back to `private` for the following reasons. Consider the function `WritePort0()` from Listing 6-9 reproduced in Listing 6-10.

Listing 6-10 `WritePort0()` function of the DAC class.

```
void DAC::WritePort0(unsigned char data)
{
    outportb(BaseAddress,data);
    LastOutput = data;
}
```

The only time `BaseAddress` is accessed is when the data is sent out the port. The polymorphic function `WritePort0()` of the `ParallelPort` class can do this. It has no problem in accessing `BaseAddress` since the function and the data are in the same class. It is possible to call the polymorphic function `WritePort0()` of the `ParallelPort` class from inside the polymorphic function `WritePort0()` of the DAC class by using the scope resolution operator;

the double colon (::). This process is shown in Figure 6-17 and is implemented by modifying the fragment of code from Listing 6-10 to become that given in Listing 6-11.

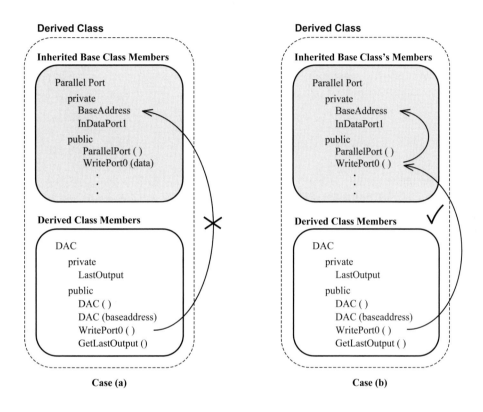

Figure 6-17 Use of inherited polymorphic functions (`BaseAddress` private again).

Listing 6-11 Calling a polymorphic function of a base class.

```
void DAC::WritePort0(unsigned char data)
{
    ParallelPort::WritePort0(data);
    LastOutput = data;
}
```

In the `ParallelPort` class definition, the access attribute of the data member `BaseAddress` can now be set back to `private` as shown in Figure 6-17. The new and preferred program is given in Listing 6-12.

Listing 6-12 Use of polymorphic functions.

```
/*************************************************************
In this program, the access attribute of the data
member BaseAddress has been changed back to private.
BaseAddress is accessed via the polymorphic WritePort0()
function of the base class, which can access BaseAddress.
*************************************************************/
#include <iostream.h>
#include <stdio.h>
#include <conio.h>

class ParallelPort
{
    private:
        unsigned int BaseAddress;
        unsigned char InDataPort1;

    public:
        ParallelPort();
        ParallelPort(int baseaddress);
        void WritePort0(unsigned char data);
        void WritePort2(unsigned char data);
        unsigned char ReadPort1();
};

ParallelPort::ParallelPort()
{
    BaseAddress = 0x378;
    InDataPort1 = 0;
}

ParallelPort::ParallelPort(int baseaddress)
{
    BaseAddress = baseaddress;
    InDataPort1 = 0;
}

void ParallelPort::WritePort0(unsigned char data)
{
    outportb(BaseAddress,data);
}

void ParallelPort::WritePort2(unsigned char data)
{
```

```
        outportb(BaseAddress+2,data ^ 0x0B);
    }

    unsigned char ParallelPort::ReadPort1()
    {
        InDataPort1 = inportb(BaseAddress+1);
    // Invert most significant bit to compensate
    // for internal inversion by printer port hardware.
        InDataPort1 ^= 0x80;
    // Filter to clear unused data bits D0, D1 and D2 to zero.
        InDataPort1 &= 0xF8;
        return InDataPort1;
    }

    class DAC : public ParallelPort
    {
        private:
            unsigned char LastOutput;

        public:
            DAC();
            DAC(int baseaddress);
            void WritePort0(unsigned char data);
            unsigned char GetLastOutput();
    };

    DAC::DAC()
    {
        LastOutput = 0;
    }

    DAC::DAC(int baseaddress) : ParallelPort(baseaddress)
    {
        LastOutput = 0;
    }

    void DAC::WritePort0(unsigned char data)
    {
        ParallelPort::WritePort0(data);
        LastOutput = data;
    }

    unsigned char DAC::GetLastOutput()
    {
        return LastOutput;
```

```
}

void main()
{
    DAC D_to_A;

    D_to_A.WritePort0(0);
    printf("\nDAC byte:%3d    ", D_to_A.GetLastOutput());
    cout << "   Measure voltage and press a key" << endl;
    getch();

    D_to_A.WritePort0(32);
    printf("\nDAC byte:%3d    ", D_to_A.GetLastOutput());
    cout << "   Measure voltage and press a key" << endl;
    getch();

    D_to_A.WritePort0(64);
    printf("\nDAC byte:%3d    ", D_to_A.GetLastOutput());
    cout << "   Measure voltage and press a key" << endl;
    getch();

    D_to_A.WritePort0(128);
    printf("\nDAC byte:%3d    ", D_to_A.GetLastOutput());
    cout << "   Measure voltage and press a key" << endl;
    getch();

    D_to_A.WritePort0(255);
    printf("\nDAC byte:%3d    ", D_to_A.GetLastOutput());
    cout << "   Measure voltage and press a key" << endl;
    getch();
}
```

Having learnt this elegant means of manipulating private data of a base class from inside a derived class, we can complete our improvements to the DAC class by changing the name of the WritePort0() function of the DAC class to something more appropriate. Let us choose the name SendData() as a replacement name for the function WritePort0() of the DAC class.

The class definition and the complete program to carry out the exact same tasks as the program in Listing 6-12, is given in Listing 6-13. We will be using this final version of the DAC class when we need to use the DAC system on the interface board in future chapters.

Listing 6-13 Replacing `WritePort0()` of DAC class by `SendData()`.

```
/******************************************************
In this program, the Function WritePort0() of the DAC
class is given the new name SendData() which is more
appropriate for the DAC class.
******************************************************/
#include <iostream.h>
#include <stdio.h>
#include <conio.h>
#include <dos.h>

class ParallelPort
{
    private:
        unsigned int BaseAddress;
        unsigned char InDataPort1;

    public:
        ParallelPort();
        ParallelPort(int baseaddress);
        void WritePort0(unsigned char data);
        void WritePort2(unsigned char data);
        unsigned char ReadPort1();
};

ParallelPort::ParallelPort()
{
    BaseAddress = 0x378;
    InDataPort1 = 0;
}

ParallelPort::ParallelPort(int baseaddress)
{
    BaseAddress = baseaddress;
    InDataPort1 = 0;
}

void ParallelPort::WritePort0(unsigned char data)
{
    outportb(BaseAddress,data);
}

void ParallelPort::WritePort2(unsigned char data)
{
```

```
        outportb(BaseAddress+2,data ^ 0x0B);
}

unsigned char ParallelPort::ReadPort1()
{
    InDataPort1 = inportb(BaseAddress+1);
// Invert most significant bit to compensate
// for internal inversion by printer port hardware.
    InDataPort1 ^= 0x80;
// Filter to clear unused data bits D0, D1 and D2 to zero.
    InDataPort1 &= 0xF8;
    return InDataPort1;
}

class DAC : public ParallelPort
{
    private:
        unsigned char LastOutput;

    public:
        DAC();
        DAC(int baseaddress);
        void SendData(unsigned char data);
        unsigned char GetLastOutput();
};

DAC::DAC()
{
    LastOutput = 0;
}

DAC::DAC(int baseaddress) : ParallelPort(baseaddress)
{
    LastOutput = 0;
}

void DAC::SendData(unsigned char data)
{
    ParallelPort::WritePort0(data);
    LastOutput = data;
}

unsigned char DAC::GetLastOutput()
{
    return LastOutput;
```

```
}

void main()
{
    DAC D_to_A;

    clrscr(); // clear screen

    D_to_A.SendData(0);
    printf("\nDAC byte:%3d    ", D_to_A.GetLastOutput());
    cout << "   Measure voltage and press a key" << endl;
    getch();

    D_to_A.SendData(32);
    printf("\nDAC byte:%3d    ", D_to_A.GetLastOutput());
    cout << "   Measure voltage and press a key" << endl;
    getch();

    D_to_A.SendData(64);
    printf("\nDAC byte:%3d    ", D_to_A.GetLastOutput());
    cout << "   Measure voltage and press a key" << endl;
    getch();

    D_to_A.SendData(128);
    printf("\nDAC byte:%3d    ", D_to_A.GetLastOutput());
    cout << "   Measure voltage and press a key" << endl;
    getch();

    D_to_A.SendData(255);
    printf("\nDAC byte:%3d    ", D_to_A.GetLastOutput());
    cout << "   Measure voltage and press a key" << endl;
    getch();
}
```

6.6 Summary

The operational amplifier, discussed in this chapter, is the building block for many analog electronic systems. This device is used in conjunction with the interface board DAC0800 IC to form a complete digital-to-analog voltage converter system. Basic principles of two types of DAC circuits have been discussed including DAC characteristics and specifications.

In this chapter the important concepts of inheritance and polymorphism have been explained. How various access attributes interact with each other, and how various access specifiers affect the access attributes has also been described. We also learned how to use the scope resolution operator to call a polymorphic function from a base class. The DAC object created at the end of the chapter has all the functionality to drive the Digital-to-Analog Converter, and protects the member data of both the base class and the derived class at private level.

6.7 Bibliography

NS *DATA CONVERSION/ACQUISITION Databook*, National Semiconductor Corporation, 1984.

Bentley, J., *Principles of Measurement Systems*, Second edition, Longman Scientific & Technical, Essex, 1988.

Horowitz, P. and Hill, W., *The Art of Electronics*, Cambridge University Press, Cambridge, 1989.

Loveday, G., *Microprocessor Sourcebook*, Pitman Publishing Limited, London, 1986.

Savant, C.J., et al., *Electronic Design Circuits and Systems*, Second Edition, Benjamin-Cummings, Redwood City, 1987.

Webb, R.E., *Electronics for Scientists*, Ellis Horwood, New York, 1990.

Wobschall, D., *Circuit Design for Electronic Instrumentation*, McGraw-Hill, 1987.

Lafore, R. *Object Oriented Programming in MICROSOFT C++*, Waite Group Press, 1992.

Wang, P.S., *C++ with Object Oriented Programming*, PWS Publishing, 1994.

Pohl, I., *Object Oriented Programming Using C++*, Benjamin Cummins, 1993.

Johnsonbaugh, R. and M. Kalin, *Object Oriented Programming in C++*, Prentice Hall, 1995.

Barton, J.J. and L.R. Nackman, *Scientific and Engineering C++ - An Introduction with Advanced Techniques and Examples*, Addison Wesley, 1994.

7

Driving LEDs

Inside this Chapter

- Iterative loops.

- Conditional Branching.

- Object classes for Driving LEDs.

- Arrays.

- Default actual arguments to functions.

- Pointers.

- Dynamic memory allocation.

7.1 Introduction

In this chapter we will first explain how to apply the widely used C/C++ constructs such as iterative loops and conditional branching. We will then discuss the use of pointers that are employed extensively in many C++ programs. Knowledge of pointers is essential when using dynamic memory allocation and virtual functions as discussed in the next chapter. You will gain a familiarity with pointers when they are used to scan an array of numbers. These numbers will then be used to light LEDs on the interface board to visualise the array scanning operation.

7.2 Iterative Loops

7.2.1 The `for` Loop

The `for` loop is an iterative loop. It executes one or more statements repeatedly. In general, a `for` statement takes the form shown in Figure 7-1. The braces in the `for` statement are essential only if the body has a *compound statement*. If the body is a single statement, the braces may be used but are not essential.

C++   **Compound statement**

A Compound statement or *block* is a number of single statements grouped together between matching braces ({ }).

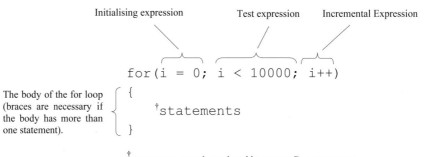

Figure 7-1 An example of a `for` loop.

Three expressions are enclosed within the pair of parentheses belonging to the `for` loop. The first of these statements is:

```
i = 0;
```

This expression is executed only once at the start of the `for` statement and is known as the *initialising expression*. The initialising expression can be quite complex. It may be used to initialise a number of variables. In general, these variables are known as *loop counters*. In the preceding example, the value of `i` is used to keep a count of the number of times the `for` loop is executed; hence the name loop counter. In C++, the initialising expression may even include variable declarations (e.g. `int i = 0;`). Note that if the initialising expression is omitted the semicolon must still be used.

The second expression:

```
i < 10000;
```

is known as the *test expression*. This expression is evaluated just before the body of the `for` loop is executed. The result of evaluating this expression is considered in a logical sense. That is, it will be tested to determine whether the expression evaluates to *true* (one) or *false* (zero).

C++	***true* or *false***
A program that is given any values that are *zero* are considered to be false; *non-zero* values are considered to be true.	
When a program *evaluates* a logical expression, if the condition is *true* the result will be 1. If the condition is *false* the result will be zero.	

In this particular case, the test expression tests whether the value of `i` is less than `10000`. If the value of `i` is less than `10000` the expression evaluates to true, otherwise false. The body of the `for` statement will be executed immediately after the test expression, if and only if the test expression evaluates to true. If the test expression evaluates to false, the `for` statement terminates without executing the statements in its body.

The left angle bracket ($<$) is known as the *less than* operator. These operators belong to a class of operators named *relational operators*. Due to the presence of the relational operator, the expression (`i<10000`) is known as a relational expression.

C++	**Relational Operators**
$<$	less than
$>$	greater than
$<=$	less than or equal to
$>=$	greater than or equal to

In addition to relational expressions, we can also use equality expressions. These expressions contain *equality operators*.

C++	Equality Operators
==	equal to
!=	not equal to

The third expression in the `for` statement is:

```
i++;
```

This statement is known as the *incremental expression*. It will be evaluated immediately after executing the body of the `for` statement. Usually, it increments one or more loop counters. In this particular case it increments the value of i by 1.

The ++ operator can be used in two different ways. Using it before the identifier will cause a *pre-increment*, e.g. ++i. Using it after the identifier will cause a *post-increment,* e.g. i++. In the case of 'pre' operations, the operation (operation meaning increment or decrement) is carried out *before* using the identifier in the test expression. In the case of 'post' operations, the operation is carried out *after* using the identifier in the test expression. The -- operator is used exactly the same way; the only difference being that it will cause a decrement. These operators fall into a category known as *unary operators*. They are referred to as unary operators because they operate on just *one* argument.

C++	Unary Operators
+	unary plus
−	unary minus
++	pre-increment (prefix) or post-increment (postfix)
--	pre-decrement (prefix) or post-decrement (postfix)
~	bitwise complement. Toggles bit by bit.
!	logical negation. Change true to false and vice-versa.

The code fragment shown below demonstrates the operation of the `for` loop. It also shows how a `for` loop operates inside another (nested `for` loops):

```
int i, j;

for (i = 0; i < 5; i++)
{
    for(j = 0; j < i; j++)
        cout << '*';
```

```
    cout << endl;
}
```

Implementing this code in a program will produce the output shown in Figure 7-2:

```
*
* *
* * *
* * * *
* * * * *
```

Figure 7-2 The output of the nested `for` loop operation.

The body of the outer `for` loop starts at the open brace and ends at the close brace. Between these two braces is the inner `for` loop. The inner `for` loop has no braces because it only has the one statement as its body:

```
cout << '*';
```

The second statement of the outer `for` loop is:

```
cout << endl;
```

and will be executed after the inner `for` loop has completed all its iterations. Each iteration of the inner `for` loop prints the character '*' on the screen and increments the loop counter `j`. Printing for that line ceases when the value of `j` reaches that of the outer loop counter `i`. Therefore, each iteration of the outer `for` loop consists of `j` iterations of the inner `for` loop.

7.2.2 The `while` Loop and the `do-while` Loop

Repetitive iterations as performed with the `for` loop can be carried out using the `while` loop. The `while` loop is similar to the `for` loop without initialising and incremental expressions. It simply has a test expression enclosed within a pair of parentheses that is evaluated at the *start* of the loop. This expression must evaluate to true for the body of the `while` loop to be executed. If it evaluates to false (zero) the loop will be terminated. It is possible that the body of the `while` loop is not executed at all - if in the first entry of the loop, the test expression evaluates to false or zero.

Because initialising and incremental expressions are omitted, `while` loops do not generally deploy a loop counter. However, the `while` loop can be used to implement the behaviour of a `for` loop and vice-versa. In general, `while` loops are implemented when the exact number of iterations are not known.

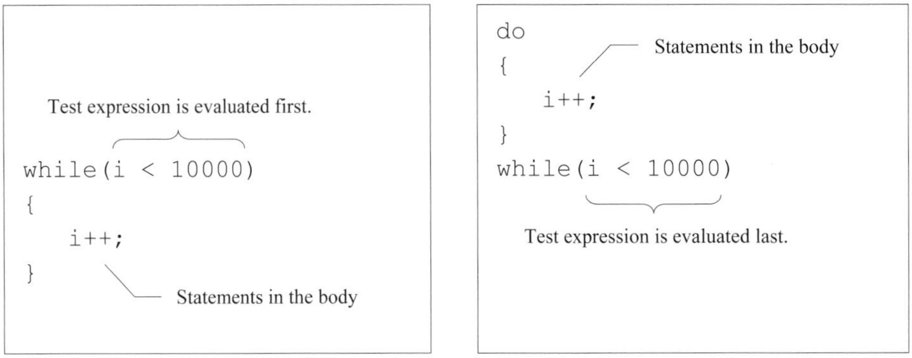

Figure 7-3 The `while` and `do-while` loops.

The `do-while` statement is very similar to the `while` statement. The difference being that the test expression is evaluated at the *end* of the loop resulting in at least one execution of the body of the loop. The statements in the body of the loop are between the keyword `do` and the keyword `while`. Braces must be used if the body is a compound statement. Figure 7-3 shows the anatomy of the two types of loop.

7.3 Branching

7.3.1 The `if` Statement

The `if` statement has a *conditional expression* enclosed within a pair of parentheses placed immediately after the keyword `if`. The most general form of the `if` statement has a *true clause* and a *false clause* separated by the keyword `else`. The true clause consists of the statements before `else` and the false clause consists of the statements after `else`. The conditional expression will evaluate to true or false. If it evaluates to true, the true clause will be executed and the false clause will be ignored, otherwise the false clause will be executed and the true clause will be ignored.

If there are multiple statements in any of the clauses, they must be placed within braces (`{` and `}`) to form compound statements. Within these compound statements there may be other `if` statements. If this is the case they are known as *nested* `if` statements. In nested `if` statements, the `else` keyword will bind to the last opened `if` statement without an `else`.

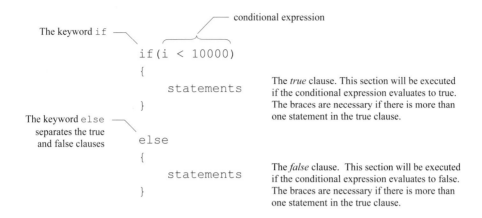

Figure 7-4 The if statement - the most general case.

It is also possible to have if statements with only one clause as shown in Figure 7-5. This clause must be the true clause.

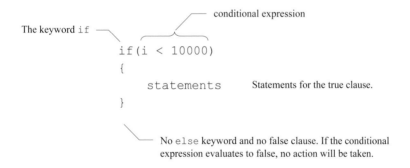

Figure 7-5 The if statement with only a true clause.

The if statements can be nested as shown in Figure 7-6. In the two examples shown, the else keyword binds to two different if statements. Use of proper indentation helps the programmer to see the correct association for each clause in nested if statements as shown in Figure 7-7.

```
if (<cond. exp. 1>)
{
if (<cond. exp. 2>)
{
    <true clause>
}
else
{
    <false clause>
}
}
```

The else binds to the second if statement. The first if statement only has a true clause which contains the second if statement. The second if statement has a true clause and also a false clause.

```
if (<cond. exp. 1>)
{
if (<cond. exp. 2>)
{
    <true clause>
}
}
else
{
    <false clause>
}
```

The else binds to the first if statement. The first if statement has a true clause and a false clause. The second if statement has only a true clause and is within the true clause of the first if statement.

Figure 7-6 Nested if statements without indentation.

```
if (<cond. exp. 1>)
{
    if (<cond. exp. 2>)
    {
        <true clause>
    }
    else
    {
        <false clause>
    }
}
```

```
if (<cond. exp. 1>)
{
    if (<cond. exp. 2>)
    {
        <true clause>
    }
}
else
{
    <false clause>
}
```

Figure 7-7 Indented if statements.

The use of an `if` statement is shown in the following fragment of code:

```
int Number;
cout << "Enter an integer Number ";
cin >> Number;

if(Number > 50)
     cout << "Number is greater than 50" << endl;
else
     cout << "Number is less than or equal to 50" << endl;
```

The number entered by the user will be tested by the `if` statement and a message will be printed on the screen displaying the result of the test.

A compact version of the `if` statement can be implemented using the so-called conditional operator (`?:`). For example, by checking a variable named `Switch`, ON/OFF status of the switch can be printed on the screen using:

```
Switch == 1 ? cout << "on" : cout "off";
```

This statement is equivalent to:

```
if(Switch == 1)
    cout << "on";
else
    cout << "off";
```

7.3.2 The `break` and `continue` Statements

The `break` and `continue` statements are two important statements that can be used efficiently to enhance the functionality of our programs. Of the two statements, the `break` statement is more widely used. Their syntax is very simple and always used as follows:

```
break;
continue;
```

The `break` statement is used to terminate the execution of a loop (such as `while`, `do-while`, or `for`) or a `switch` statement. The `continue` statement is used to skip and continue the execution of loops. Figure 7-8 shows the two cases.

As mentioned previously, iterative loops can be nested; i.e. one loop within another. Similarly, a `switch` statement can be placed within a loop. In such situations, the `break` statement will be associated with the nearest loop or `switch` statement. The `continue` statement will be associated with the nearest loop and cannot be used with the `switch` statement.

The C++ language also supports the use of `goto` to jump to a label. Use of `goto` can severely damage the structure of a program. Its use is discouraged and is not explained in this text - see references listed in Section 7.11.

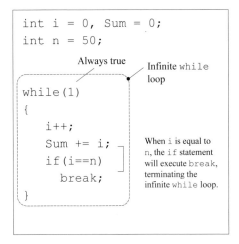

Figure 7-8 The break and continue statements.

7.3.3 The switch – case Statement

A switch-case statement is used to select and then execute one of several cases. Selection is carried out by a *switch expression* located after the keyword *switch* and enclosed between parenthesis.

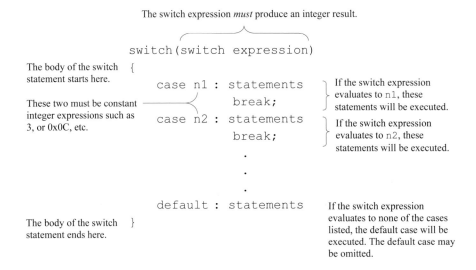

Figure 7-9 The switch statement.

The `switch` expression must be of integral type such as `char`, `unsigned char`, `int`, `unsigned int`, etc. Program control will be transferred to a `case` statement that matches the value of the `switch` expression.

The cases are listed within the body of the `switch` statement. Immediately after the keyword `case`, there must be a *constant integer expression*, which must have a unique value. Each `case` may have any statement including an empty statement. All statements under a `case` will be executed sequentially.

The `break` statement must be used to exit a `switch` statement at the end of a particular `case`. If `break` is not used, program execution will flow on to the next `case`. Optionally, a special case named `default` may be used to take necessary action if no matching `case` is found.

C++   **Constant Integer Expression**

A constant integer expression must produce an integer result and cannot contain any variables.

`#define TWIN 2`

Here the symbolic constant `TWIN` is defined to be a substitute for the number 2. Then `TWIN+1` is a constant integer expression. Note that `TWIN` is not a variable.

However, if `int a=0;`

Then `a+1` is *not* a constant integer expression, because `a` is a variable.

7.4 Arrays

An *array* is a collection of objects of the same type. The objects could be fundamental data types or user-defined data types. Each individual object of the collection is referred to as an *element*. Towards the end of the chapter, we will be using arrays to store LED lighting patterns for the program.

Arrays can be represented in different configurations or number of dimensions as shown in Figure 7-10. Each cell can store one object of the designated type. There is practically no limit to the number of dimensions an array can have.

The size of the array is given by the number of elements (cells) for each dimension. Using Figure 7-10 as an example, the sizes of the arrays are:

1-dimensional array: 5 elements.

2-dimensional array: 3 rows, 5 elements/row. Size = 15 elements.

3-dimensional array: 3 rows, 4 elements/row, 2 elements/row. Size = 24 elements.

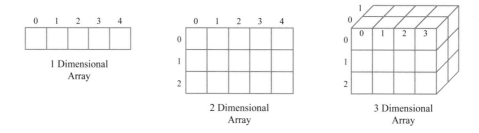

Figure 7-10 Diagramatic representation of arrays.

Although the arrays can be represented as shown in Figure 7-10, in your computers memory the elements of the array are stored sequentially row after row, termed *row-major* fashion. In storing a 3-D array, the first layer is stored first in a row-major fashion and then the second layer in row-major fashion and so on.

One-Dimensional Arrays

When declaring one-dimensional arrays, the size of the array is specified within a pair of square brackets immediately after the array identifier. The array subscripts always start with 0 and range to the array size minus one. An example of a declaration is:

```
int a[10];
```

This declares an array of 10 `int` type objects. They are stored in adjacent memory locations starting from the element `a[0]` ranging up to `a[9]`, making available a set of 10 elements as shown in Figure 7-11. Note that there is no element named `a[10]`. Attempting to access such an element will be illegal, since this memory location is not part of the array `a`.

a[0]	a[1]	a[2]	a[3]	a[4]	a[5]	a[6]	a[7]	a[8]	a[9]

Figure 7-11 Schematic of a one-dimensional array.

If a variable subscript is used to access array elements, as in `a[i]`, then `i` must be an integer expression and as just mentioned must not evaluate to a value outside the permitted range of array subscript values, in this case 0 to 9.

The array elements can be initialised individually during program execution by assigning each element a value. For example, the following code fragment sets the value of all elements to zero:

```
for(int i = 0; i < 10; i++)
    a[i] = 0;
```

Alternatively, the array elements can be initialised when the array is declared. The values for each element to be initialised need to be listed within braces and separated by commas as shown in the following line:

```
int a[8] = {2,3,7,4};  // 8 elements
```

In this example the array `a` is partially filled with the values listed between the braces; element `a[0]=2`, `a[1]=3`, `a[2]=7`, and `a[3]=4`. The remaining elements (`a[4]` to `a[7]`) that have not been explicitly initialised are initialised by default with values of zero. Therefore, to declare and initialise an array with all zero values can be done as follows:

```
int a[8] = {};
```

Accessing array elements

Individual elements of the array can be accessed by using a subscript. In the following example the array subscripts range from 0 to 7 (8 elements).

```
int a[8];       // declare a to be 8 elements
int Result;     // declare a variable named Result

a[0] = 2;       // access and set a[0] to 2
a[1] = 3;       // access and set a[1] to 3

Result = a[0]*a[1];    // 2*3 = 6
```

Two-Dimensional Arrays

A two dimensional array can be viewed as an array of one-dimensional arrays. Two-dimensional arrays have two sizes specified. The total number of elements is the product of the two sizes. For example, a two-dimensional array can be declared as follows:

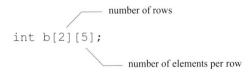

Figure 7-12 Declaring a 2-D array.

Two subscripts are used to access each element of the array. In array `b`, one of the array dimensions ranges from 0 to 1 and the other dimension from 0 to 4. The array can be thought of as the arrangement shown in Figure 7-13. Note that elements are stored in consecutive memory locations in row-major fashion. That is, the first row is stored first, immediately followed by the second row, and so forth. Thus, the element b[0][4] is immediately followed by the element b[1][0].

b[0][0]	b[0][1]	b[0][2]	b[0][3]	b[0][4]
b[1][0]	b[1][1]	b[1][2]	b[1][3]	b[1][4]

Figure 7-13 2-D array schematic representation.

The array elements can be individually initialised during program execution by assigning each element a value. For example, the following code fragment sets the values of all elements to zero.

```
int i, j;

for(i = 0; i < 2; i++)
    for(j = 0; j < 5; j++)
        b[i][j] = 0;
```

Alternatively the array elements can be initialised when the array is declared. This is shown in the following line.

```
int b[2][5] = {{0,0,0,0,0};{0,0,0,0,0}};
```

Each row of initialised elements is enclosed by inner braces, and separated from adjacent rows by a semicolon.

7.5 Pointers

A pointer is an address of an entity that resides in memory. Examples of entities that reside in memory are; class objects, fundamental data type objects such as int, float, char, long, etc., arrays of objects (a group of objects of the same type), and functions. A pointer in C++ will hold *where* the object is and most of the time the pointer will know the *type* of object. For example, a pointer to an integer knows that the data type is integer and it will also know where the integer is, but it does not know the value of the integer.

Pointers play an important role in helping to make C++ programs very efficient. There are three major uses for pointers that offer distinct advantages: passing large objects to functions, dynamic memory allocation, and using virtual functions. In the most common case when passing a parameter to a function, we replace the parameter by a copy of the actual argument. If the actual argument is very large, the program will need to consume a large amount of memory when it creates a copy of the argument. It is more efficient to make a copy of *where* the large object is than copy the entire object. This is done by using a pointer which occupies a small amount of memory in order to store the address of the object.

Dynamic memory allocation involves the provision of storage space at run-time. Normally, dynamic memory allocation is a need-based process – i.e. if during program operation there is a need for more memory it can be requested and will be granted depending on availability. The dynamic memory allocation process returns a pointer indicating the location in memory where the allocation has been made. This pointer can then be used to manipulate the data in the allocated memory area.

Perhaps the most obscure use of pointers is in association with virtual functions which will be described in detail in Chapter 8. The following sections describe general use of pointers in the C++ language.

Two unary operators are used closely with pointers. As mentioned before, a unary operator takes only one argument. These operators are given in Table 7-1.

Table 7-1 Unary operators used with pointers.

Operator	Name
&	address of operator
*	indirection operator

The *address of* operator can be used to find the address of an object in memory.

The *indirection* operator can be used to obtain the contents of a location given its memory address. This is also known as *de-referencing*.

7.5.1 Declaration of Pointer Variables

As you know, there is a dedicated data type named `int` to represent integers and many other fundamental data types. Programmers can also create their own data types such as DAC created in Chapter 6. However, there is no unique data type named *pointer*. Since all memory addresses are integers, all pointer data types carry integer values. The locations pointed to by these addresses can contain all types of data or functions.

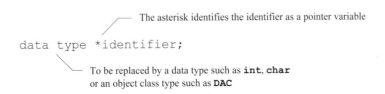

Figure 7-14 Syntax of a pointer declaration.

When declaring a pointer variable, the C++ language requires us to specify the type of data or function pointed to by the pointer variable. We will see the significance of knowing the data type pointed to by the pointer variables when pointer arithmetic is explained in section 7.5.6. In the simplest of cases, the syntax for declaring pointer variables takes the form shown in Figure 7-14. Pointers to different entities are each declared differently as described ahead.

7.5.2 Pointers to Scalar Quantities

A single item is referred to as a scalar quantity. If a pointer variable is declared to point to one solitary integer, then that pointer is said to point to a scalar quantity. This is in contrast to pointers that point to arrays. Examples of declarations of ordinary variables and declaration of pointers to scalar quantities are shown in Table 7-2.

Table 7-2 Declaration of scalar identifiers and pointers to scalar identifiers.

Declaration – scalar identifiers	Declaration – pointers to scalar identifiers
`int a;`	`int a;`
	`int *IntPtr = &a;`
`int b = 0;`	`int b = 0;`
	`*IntPtr = b;`
`float p = 0.0;`	`float p = 0.0;`
	`float *FltPtr = &p`

The following line is a combined declaration and initialisation of a pointer variable to an `int`:

```
int *IntPtr = &a;
```

The same effect can be achieved with the following two lines:

```
int *IntPtr;
IntPtr = &a;
```

The first statement declares a pointer to an `int`. The second statement uses the 'address of' operator '&' to obtain the address of the integer variable a which is assigned to the pointer variable `IntPtr`. Note that the `int` type variable a must be declared *before* assigning its address to `IntPtr`.

The statement:

```
*IntPtr = b;
```

carries out a de-referencing and an assignment operation. The expression `*IntPtr` reads as 'the contents of the location pointed to by `IntPtr`'. This is known as de-referencing. Therefore, the entire expression reads as 'the contents of the location pointed to by `IntPtr` is assigned the value of b'. Since `IntPtr` already points to the location of a, the effect is same as:

```
a = b;
```

An example of declaring a pointer to a float type variable and assigning it a value is given in Table 7-2.

NOTE

Given the following two declarations;

```
int a=0;
float* FltPtr;
```

An assignment of the form;

```
FltPtr = &a; // Illegal!
```

is illegal. The pointer `FltPtr` is expected to carry an address of a `float` type object. However, `&a` is an address of an integer object. These two do not match and therefore it is an illegal assignment.

7.5.3 Pointers to Class Objects

Pointers to class objects are declared in a similar manner to pointers to scalar quantities. An example is given below:

```
ParallelPort *PortPtr;
```

Here the data type is `ParallelPort` and the pointer variable is `PortPtr`. A pointer to an object of the `DAC` class can be declared as follows:

```
DAC *DACPtr;
```

An object of type `DAC` can be declared as follows:

```
DAC Dac;
```

Then the following assignment is valid:

```
DACPtr = &Dac;
```

Membership Access Operators

If we use the object Dac, we can call the SendData() function as follows using the dot operator (.), also known as the *membership access* operator:

```
Dac.SendData(255);
```

We can also use a pointer variable such as DACPtr to call the SendData() function, although the syntax is different. In this case the membership pointer operator is used (->), formed by combining the minus sign (-) and the right angle bracket (>):

```
DACPtr->SendData(255);
```

Pointers to Base Class Objects can point to Objects of Derived Classes

This is one of the most useful and important concepts in object-oriented programming. In earlier sections it was explained that a pointer pointing to a float type variable cannot point to a location containing an int. This rule does not apply to base classes and derived classes. Although the two objects are different, a pointer to a base class object *can* point to an object of a derived class:

```
ParallelPort *PortPtr;
DAC Dac;
PortPtr = &Dac; // is allowed!
```

We are yet to discuss the advantages of using this type of pointer assignment. Its major use is associated with virtual functions and will be explained in Sections 8.5 and 8.6.

7.5.4 Pointers to Arrays

Pointers to One-Dimensional Arrays

When an array is declared to be equivalent to that shown in Figure 7-15, the address of the array will be a (no subscripts) which points to the first element of the array. Therefore, a and &a[0] are equivalent and both point to the first element. The important thing to note is that a is a *pointer constant*. It cannot be incremented, decremented or assigned any other values. Since the array has been stored in a specific memory space, the address value is fixed.

a[0]	a[1]	a[2]	a[3]	a[4]	a[5]	a[6]	a[7]	a[8]	a[9]

Figure 7-15 Schematic of a one-dimensional array.

The following statements are all valid:

```
int a[10];
int *ElementPtr;
int b = 0;
ElementPtr = a;        // same as ElementPtr = &a[0];
*a = b;                // the value of b is deposited
                       // in a[0]
```

Some of the statements shown below are illegal:

```
int a[10];
float *FltPtr;
int b = 0;
FltPtr = a;    // illegal - type mismatch
a = &b;        // illegal - a is constant
```

Pointers to Two-Dimensional Arrays

As mentioned earlier, a two-dimensional array can be viewed as an array of one-dimensional arrays. An example two-dimensional array can be declared as:

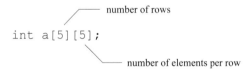

Recall that elements are stored in consecutive memory locations. For example, the element a[1][0] is stored next to a[0][4].

Unlike the case for one-dimensional arrays, the array name a is a pointer to the *entire row* starting at a[0][0] and ending at a[0][4]. The pointer a is still a constant.

a[0][0]	a[0][1]	a[0][2]	a[0][3]	a[0][4]
a[1][0]	a[1][1]	a[1][2]	a[1][3]	a[1][4]

.

.

.

| a[4][0] | a[4][1] | a[4][2] | a[4][3] | a[4][4] |

Figure 7-16 Schematic of a two-dimensional array.

A pointer to a row of five elements can be declared as follows:

```
int (*RowPtr)[5];
```

Note the subtle difference between the presence and absence of the parentheses in the above declaration. Compare this declaration to the declaration of an array of pointers discussed earlier.

If a is de-referenced, the result will be a pointer to the first element of the first row, i.e. &a[0][0]. This resulting pointer is still a constant. To access the value of a[0][0], the pointer a must be de-referenced twice. The following statements illustrate this:

```
int *ElementPtr;
int b;
int a[5][5];
int (*RowPtr)[5];
RowPtr = a;          // pointer to the first row
ElementPtr = *a;     // pointer to the first element
                     // of the first row
b = **a;             // same as b = *ElementPtr;
ElementPtr = a;      // Illegal - type mismatch
RowPtr = *a;         // Illegal - type mismatch
*a = &b;             // Illegal - *a is constant
```

An important observation is that when an array name is de-referenced, it points to the next lower level entity. For example, if the name of a two-dimensional array is de-referenced, it will point to a one-dimensional array. If the name of a one-dimensional array is de-referenced it will evaluate to be the contents of the first element of the array. Any further de-referencing is illegal.

7.5.5 Arrays of Pointers

It is also possible to declare arrays of pointers. In such an array, each element itself is also a pointer. An example of a pointer array declaration is given as follows:

```
int *IntPointers[20];
```

In this declaration, IntPointers is a constant pointer. It points to the first element of the array of pointers. If we use the de-referencing operator as shown below we will obtain the contents of the first element, which itself is a pointer to an int. Therefore it must be assigned to a compatible pointer variable. Consider the following declaration:

```
int a;
int *IntPtr;
int *IntPointers[20];
```

IntPointers is the start address of the array of pointers to int. In other words, it holds a memory address. This location contains a pointer to an int. Thus, the contents of any element of the array can be assigned to a pointer to an `int`, such as `IntPtr`. To obtain the contents of the first element of the `IntPointers` array, we can de-reference `IntPointers`. Once de-referenced, it can be assigned to `IntPtr` as shown in the line below:

```
IntPtr = *IntPointers; // contents of 1st element
```

A pointer to an `int` is stored at the address obtained by de-referencing `IntPointers` (i.e. `*IntPointers`). If we need the contents of the location pointed to by this pointer to `int`, we must de-reference `*IntPointers` once more to obtain the integer value. Such an integer value can be assigned to an `int` type variable such as `a`. Thus:

```
a = **IntPointers;         // Same as a = *IntPtr;
```

7.5.6 Pointer Arithmetic

As already seen, a pointer variable can be incremented or decremented. Likewise an integer value can be added or subtracted. However, the results produced by pointer arithmetic are different to the results produced by normal arithmetic. Before explaining this further, we should understand where pointer arithmetic is useful.

One-dimensional Arrays

Pointer arithmetic is especially useful for accessing array elements. Consider the example:

```
int a[5];
```

Here we have declared an array of 5 integers. The array name `a` is a constant pointer and it points to the first element of the array. Figure 7-17 shows an example of such an array in memory.

Memory Address	Array Element and its value
600000	a[0] is 4
600002	a[1] is 9
600004	a[2] is 8
600006	a[3] is 3
600008	a[4] is 5

Figure 7-17 An example of 5 integers in memory.

In the example shown in Figure 7-17, the memory address values change by 2 when we move from one address to the next address. This is because we have assumed each integer occupies two bytes. Thus, the element a[0] occupies the addresses 600000 and 600001. The next element a[1] begins at 600002 and so on.

The value of a is 600000. The pointer a is not a variable, it is a constant. It points to the first element of the array and therefore cannot be assigned another value. It can be de-referenced like other pointers as follows:

*a ⟶ a[0] which is 4

Although a cannot be changed, we can add an integer value to a to obtain a new value. Thus:

a+1 is a valid expression.

If we interpret this result using normal arithmetic, it evaluates to 600001. However, in pointer arithmetic it evaluates to 600002. Since a is a pointer to an integer, '+1' really means plus one int type data, which is *two bytes* in our example. The compiler takes into account the size of the data type pointed to by the pointer when evaluating pointer arithmetic expressions. The result a+1 still remains a pointer and points to the next element of the array:

a+1 ⟶ 600002
a+2 ⟶ 600004
a+3 ⟶ 600006

Just like the pointer a, the following three pointers can also be de-referenced:

*(a+1) ⟶ a[1] which is 9
*(a+2) ⟶ a[2] which is 8
*(a+3) ⟶ a[3] which is 3

Use of parentheses is very important in the cases shown above. If the parentheses had not been used, the results would be as follows:

*a+1 ⟶ a[0]+1 which is 5
*a+2 ⟶ a[0]+2 which is 6
*a+3 ⟶ a[0]+3 which is 7

As can be seen from this discussion, pointer arithmetic can be used to access an array element of a one-dimensional array. For the array a, the ith element can be accessed by:

*(a+i)

Two-dimensional arrays

Pointer arithmetic is applied slightly differently to two-dimensional arrays. Consider the example:

```
int a[3][2];
```

This declares an array of 6 elements stored in sequential memory as shown in Figure 7-18.

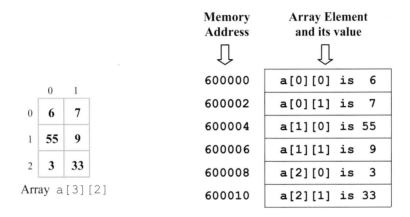

Figure 7-18 Two-dimensional array in memory.

As for the case of one-dimensional arrays, the array name a is still a pointer. Its value is 600000. It is a constant and cannot be changed. The difference for two-dimensional arrays is that a points to the *first row* of elements, not the first element of the array. Therefore, it represents a data size of 2 integers as specified by the second subscript of the array declaration. As a result the pointer a represents 2 integers, i.e. points to 4 bytes of memory. With this in mind, pointer arithmetic works as follows.

```
a+1    ⟶    600004  points to the second row
a+2    ⟶    600008  points to the third row
```

Using this notation, a pointer to the ith row can be obtained by adding **i** to **a**:

```
a+i    ⟶    points to the i[th] row
```

Like any other pointer, these pointers can also be de-referenced. When these pointers are de-referenced, the result is still a pointer:

$$*(a+1) \longrightarrow 600004 \text{ points to the first element of the second row}$$

$$*(a+2) \longrightarrow 600008 \text{ points to the first element of the third row}$$

$$*(a+i) \longrightarrow \text{points to the first element of the } (i+1)^{th} \text{ row}$$

The size of data pointed to by these de-referenced pointers is no longer an entire row. They now point to elements, which are single integers. Now pointer arithmetic will be based on single integers or two bytes. Thus:

$$*(a+1) + 1 \longrightarrow 600006 \text{ points to } a[1][1]$$

$$*(a+2) + 1 \longrightarrow 600010 \text{ points to } a[2][1]$$

Therefore, the pointer that points to the jth element of the ith row is:

$$*(a+i) + j \longrightarrow \text{points to } a[i][j]$$

By de-referencing the above pointers, the values of the elements can be accessed:

$$*(*(a+1) + 1) \longrightarrow a[1][1] \text{ which is } 9$$

$$*(*(a+i) + j) \longrightarrow a[i][j]$$

The same arguments we presented for two-dimensional arrays can be extended in the same logical manner to higher-dimensional arrays.

7.5.7 Pointers to Functions

Pointers can be declared to point to functions. Just like all the pointers discussed previously, function pointers will also have an address in memory. These addresses point to the first instruction to be executed.

An example of a function pointer declaration is:

```
int (*CalcFunctionPtr)(int,int);
```

In the above example, the name of the function pointer is `CalcFunctionPtr`. This pointer can only point to a particular category of function as specified by the pointer declaration. The declaration specifies that the function to be pointed to by `CalcFunctionPtr` must receive two integer parameters and must return an integer value. An example of a function that can be pointed to by `CalcFunctionPtr` is:

```
int Add(int a, int b)
{
    return (a + b);
}
```

Another function that can be pointed to by the `CalcFunctionPtr` is:

```
int Sub(int a, int b)
{
```

```
    return (a - b);
}
```

In a similar manner to arrays, where the array name is a constant pointer, function names are also constant pointers for the simple reason that when a program is running the location of a function in memory is fixed.

In the above two functions, Add and Sub are two constant function pointers. Each of them point to the first instruction to be executed in their respective functions. These constant pointer values can be assigned to a declared pointer variable. An example program is given in Listing 7-1.

Listing 7-1 Use of pointers to functions.

```cpp
#include <iostream.h>
#include <conio.h>

int Add(int a, int b)
{
    return (a + b);
}

int Sub(int a, int b)
{
    return (a - b);
}

void main()
{
    int a, b, Result;
    char key;
    int (*CalcFunctionPtr)(int,int);

    cout << "Enter two integer values " << endl;
    cin >> a >> b;
    cout << "Press '+' or '-' key" << endl;

    key = getch(); // getch() reads the key pressed

    switch(key)
    {
    case '+' : CalcFunctionPtr = Add;
        break;
    case '-' : CalcFunctionPtr = Sub;
    }
```

```
        Result = CalcFunctionPtr(a,b);
        cout << "The result is " << Result << endl;
}
```

If the '+' key is pressed, the `switch` statement will set the `CalcFunctionPtr` to point to the `Add()` function. If the '-' key is pressed, `CalcFunctionPtr` will be set to point to the `Sub()` function. The second-last line of the code fragment will execute either the `Add()` function or the `Sub()` function depending on the key pressed. This is an example where a function pointer is used to carry out different tasks using the same statement for different cases (`Result = CalcFunctionPtr(a,b)`).

In a more complex program you may have a large portion of the program written using the function pointer variable. If we need to change the cases, we do not need to change the part of the program that calculates the result. The program in Listing 7-2 shows how we can add another case for multiplication and yet the result will be calculated using the same statement as for Listing 7-1; 'Result = CalcFunctionPtr(a,b);'.

Listing 7-2 Adding more functionality to the program in Listing 7-1.

```
#include <iostream.h>
#include <conio.h>

int Add(int a, int b)
{
    return (a + b);
}

int Sub(int a, int b)
{
    return (a - b);
}

int Mult(int a, int b)
{
    return (a * b);
}

void main()
{
    int a, b, Result;
    char key;
    int (*CalcFunctionPtr)(int,int);
```

```
    cout << "Enter two integer values " << endl;
    cin >> a >> b;
    cout << "Press '+', '-' or '*' key" << endl;

    key = getch(); // getch() reads the key pressed

    switch(key)
    {
    case '+' : CalcFunctionPtr = Add;
        break;
    case '-' : CalcFunctionPtr = Sub;
        break;
    case '*' : CalcFunctionPtr = Mult;
    }

    Result = CalcFunctionPtr(a,b);
    cout << "The result is " << Result << endl;
}
```

Functions Returning Pointers

Functions returning pointers are discussed here since their declarations are similar. A function with the name AnyFunction that receives two int type parameters and returns a pointer to an int is declared as follows:

```
int *AnyFunction(int,int);
```

Compare this declaration with the declaration of FunctionPtr, which is a pointer to a function taking two int type parameters and returning an int type value:

```
int (*FunctionPtr)(int,int);
```

In one case a pair of parentheses is present and in the other case the parentheses are omitted. Although only slightly different, these two declarations are completely different. AnyFunction is a function name and therefore is a constant pointer. FunctionPtr is a pointer variable.

7.5.8 Pointers to void

Pointers can also be declared to be of type 'pointer to void'. These pointers do not have any restrictions as to the type of data or functions they can point to. The following example outlines their use.

```
    int a;       // declaration of an int
    float b;     // declaration of a float
    void *VoidPtr;    // declaration of a void pointer
    int Add(int,int); // declaration of a function
```

.
.
.

```
VoidPtr = &a;   // int address assigned to void
                // pointer
VoidPtr = &b;   // float address assigned to void
                // pointer
VoidPtr = Add;  // function address assigned to
                // void pointer
```

The advantage when using pointers to void is that the same pointer can be used to point to many different types of entities without needing to create specific pointers to specific objects.

7.5.9 The Pointer this

In object-oriented programming, each object maintains an invisible pointer named this which points to itself. Although invisible, if need be, the this pointer can be used exactly like any other pointer. The member functions of a particular class can determine exactly which object the function should operate on by using the this pointer. To understand how the pointer this operates, consider the function GetLastOutput() of the DAC class described in Chapter 6:

```
unsigned char DAC::GetLastOut()
{
    return LastOutput;
}
```

If we make the this pointer visible, the function would appear as follows:

```
unsigned char DAC::GetLastOut()
{
    return this->LastOutput;
}
```

Now suppose we had created two DAC objects:

```
DAC Dac1, Dac2;
```

We will call the GetLastOutput() function once each for each of the two objects as shown below:

```
Dac1.GetLastOutput();
Dac2.GetLastOutput();
```

When the first of these functions is executed, the this pointer will point to the address of the object Dac1 and thus this->LastOutput will select the LastOutput member of the Dac1 object. When the second GetLastOutput() function prefixed with Dac2 is called, the this pointer will point to the address of the object Dac2, and this->LastOutput will select the LastOutput member of the object Dac2.

To demonstrate situations where the this pointer is used explicitly, consider the constructor of the ParallelPort class:

```
ParallelPort::parallelPort(int baseaddress)
{
    BaseAddress = baseaddress;
}
```

We have deliberately named the parameter baseaddress. However, if instead we named it BaseAddress, the constructor would be:

```
ParallelPort::ParallelPort(int BaseAddress)
{
    BaseAddress = BaseAddress; // confusion!
}
```

As can be seen, the parameter cannot be differentiated from the member data. The solution to this is to modify the function as follows:

```
ParallelPort::parallelPort(int BaseAddress)
{
    this->BaseAddress = BaseAddress;
}
```

Within the body, the part-statement this->BaseAddress definitely refers to the member data.

7.6 Using Pointers

To demonstrate the use of pointers, we will create an array of numbers that can be sent out to the interface board to light up the LEDs in a specific pattern. This array will then be scanned using a pointer to fetch consecutive values from the array. The port at address BASE will be used to output the numbers to the interface board.

7.6.1 Number arrays for the LEDs

Firstly we will develop a program which 'walks a LED' along the bank of eight LEDs. This program uses a fixed pattern.

Walking LEDs – Fixed Array Defined Within the Class

We will be using an array and scanning it cyclically to light up the LEDs using the port at address BASE. The effect of cyclic scanning will be the appearance of a "walking LED" across the bank of 8 LEDs. The array will contain eight elements, each element used to light just one LED of the group. When we move from one element to the next element in the array, the LED that is currently lit will turn off and the adjacent LED in the direction of the 'walk' will light up.

Table 7-3 shows each hexadecimal number and corresponding binary bit pattern for each array element that in-turn must be output to the port.

Table 7-3 LED pattern values stored in `Pattern` array.

Array Element	Binary Number								Hex value
	D7	D6	D5	D4	D3	D2	D1	D0	
Pattern[0]	0	0	0	0	0	0	0	1	0x01
Pattern[1]	0	0	0	0	0	0	1	0	0x02
Pattern[2]	0	0	0	0	0	1	0	0	0x04
Pattern[3]	0	0	0	0	1	0	0	0	0x08
Pattern[4]	0	0	0	1	0	0	0	0	0x10
Pattern[5]	0	0	1	0	0	0	0	0	0x20
Pattern[6]	0	1	0	0	0	0	0	0	0x40
Pattern[7]	1	0	0	0	0	0	0	0	0x80

A new object class named `LEDs` will be created which will have `Pattern` as a data member and also have the functionality to initialise the `Pattern` array to the desired values. The contents of the array `Pattern` will be *fixed* for the class and cannot be changed by the user within the `main()` function. The class must also have a function to sequentially output the appropriate values in `Pattern` to the port at address `BASE`. The `LEDs` class can be derived from the `ParallelPort` class to inherit the required interface functionality. Listing 7-3 shows the class definition.

Listing 7-3 `LEDs` class definition.

```
class LEDs : public ParallelPort
{
    private:
        unsigned char Pattern[8];
        int PatternIndex;

    public:
        LEDs();
        LEDs(int baseaddress);
        void LightLEDs();
};
```

The private data member `PatternIndex` is required to store the number of the LED that was previously lit so cycling can be controlled as we move through the array to produce the 'LED walk'. Note that `Pattern` is an array of eight `unsigned char` elements. The elements need to be `unsigned char` to provide just 8 bits in each element, and to avoid the added complications that would be involved if signed numbers were used instead.

The member functions for the class can now be defined as shown in Listing 7-4.

Listing 7-4 Member functions for the LEDs class.

```
LEDs::LEDs()
{
    // Fill in the Pattern array
    for(int i = 0; i < 8; i++)
        *(Pattern + i) = 1 << i;

    PatternIndex = 0;    // initialise to 0
}


LEDs::LEDs(int baseaddress) : ParallelPort(baseaddress)
{
    // Fill in the Pattern array
    for(int i = 0; i < 8; i++)
        *(Pattern +i) = 1 << i;

    PatternIndex = 0;    // initialise to 0
}

void LEDs::LightLEDs()
{
    while(!kbhit())    // key press terminates function
    {
        WritePort0(*(Pattern + PatternIndex++));

        // Reset PatternIndex when it gets to 8
        if(PatternIndex == 8) PatternIndex = 0;

        delay(500);
    }
}
```

The array name `Pattern` (without the subscripts) is a pointer and points to the first element of the array. Therefore:

```
Pattern + i
```

points to the ith element of the array. To refer to the value pointed to by `Pattern + i`, we must de-reference it as follows:

```
*(Pattern + i)
```

The constructors of the `LEDs` class initialise the array by left-shifting the number 1 by `i` bit places using:

```
1 << i
```

The constructors will initialise the `Pattern` array with the values shown in Table 7-3 as their respective `for` loops complete each iteration of the statement:

```
*(Pattern + i) = 1 << i;
```

The `while` loop in the `LightLEDs()` function is conditioned on `!kbhit()` and will continue to execute provided a key is not hit. Inside the `while` loop the inherited `WritePort0()` function is used to write the array `Pattern` to the port at address `BASE` – one element at a time. Each element is accessed using `PatternIndex` as an offset with respect to the starting memory address of the array `Pattern`. The constant integer pointer `Pattern` is added the value of `PatternIndex` each time. This allows the `Pattern` array to be scanned from beginning to end. When the end is reached, `PatternIndex` is reset to 0 to allow a new cycle of scanning the `Pattern` array to repeat. Note that `PatternIndex` is post-incremented within the expression:

```
*(Pattern + PatternIndex++)
```

In this expression, the current value of `PatternIndex` is used to evaluate the current address of the element to access. Following this activity, the value of `PatternIndex` is incremented. A delay of 500 ms is included to provide sufficient time to see the LED walk. Connect the `BASE` address signals on the interface board (U13) to the LED Driver IC (U3) to test the complete program shown below.

Listing 7-5 Complete program to 'Walk a LED'.

```
// Complete Program to 'walk' a LED
#include <iostream.h>
#include <conio.h>
#include <dos.h>

class ParallelPort
{
    private:
        unsigned int BaseAddress;
```

```
        unsigned char InDataPort1;

    public:
        ParallelPort();
        ParallelPort(int baseaddress);
        void WritePort0(unsigned char data);
        void WritePort2(unsigned char data);
        unsigned char ReadPort1();
};

ParallelPort::ParallelPort()
{
    BaseAddress = 0x378;
    InDataPort1 = 0;
}

ParallelPort::ParallelPort(int baseaddress)
{
    BaseAddress = baseaddress;
    InDataPort1 = 0;
}

void ParallelPort::WritePort0(unsigned char data)
{
    outportb(BaseAddress,data);
}

void ParallelPort::WritePort2(unsigned char data)
{
    outportb(BaseAddress+2,data ^ 0x0B);
}

unsigned char ParallelPort::ReadPort1()
{
    InDataPort1 = inportb(BaseAddress+1);
// Invert most significant bit to compensate
// for internal inversion by printer port hardware.
    InDataPort1 ^= 0x80;
// Filter to clear unused data bits D0, D1 and D2 to zero.
    InDataPort1 &= 0xF8;
    return InDataPort1;
}

class LEDs : public ParallelPort
{
```

```
     private:
         unsigned char Pattern[8];
         int PatternIndex;

     public:
         LEDs();
         LEDs(int baseaddress);
         void LightLEDs();
};

LEDs::LEDs()
{
  // Fill in the Pattern array
    for(int i = 0; i < 8; i++)
       *(Pattern + i) = 1 << i;

    PatternIndex = 0;    // initialise to 0
}

LEDs::LEDs(int baseaddress) : ParallelPort(baseaddress)
{
  // Fill in the Pattern array
    for(int i = 0; i < 8; i++)
       *(Pattern + i) = 1 << i; // Shift '1' left 'i' places
                                // and fill Pattern array.
    PatternIndex = 0;  // initialise to 0
}

void LEDs::LightLEDs()
{
    while(!kbhit())     // keypress terminates function
    {
       WritePort0(*(Pattern + PatternIndex++));

       // Reset PatternIndex when it reaches 8
       if(PatternIndex == 8) PatternIndex = 0;

       delay(500);
    }
}

void main()
{
    LEDs Leds;
```

```
    Leds.LightLEDs(); // Displays a 'walking' LED.
    getch();

    cout << endl << "Halted !" << endl;
    cout << "Press a key to continue" << endl;
    getch();

    Leds.LightLEDs(); // 'Walking' restarts with the LED
                      // alight in the next position.
}
```

Walking LEDs – User Definable Contents with Fixed Array Size

The program shown in Listing 7-5 is rather inflexible. The contents of the array Pattern are fixed within the class. It is more appropriate to give the user the ability to define the contents of the array, therefore, the LEDs class must be modified.

The user will define the contents of the array used to light the LEDs. Therefore, the constructors of the LEDs class are not needed to initialise this array. Instead, the class can maintain a pointer that points to the array the user will define. The user-defined array can be scanned by the LightLEDs() function if the starting address of the array and its size are known. Therefore, the LEDs class only needs to have a data member to store the address of the array and another data member to store its size.

To facilitate these changes we will replace the member data Pattern (unsigned char array) with PatternPtr, a pointer type (pointer to unsigned char). A member function must also be included to extract the address of the array and to assign it to the pointer maintained within the class. Since the size of the array can now be arbritrarily set, a new data member must be included to store the maximum number of elements in the array. This new member is used to determine when the final element of the array has been scanned, whereupon PatternIndex can then be reset to 0. The modifications to the LEDs class are shown in Listing 7-6.

Listing 7-6 Modified LEDs class.

```
class LEDs : public ParallelPort
{
    private:
        unsigned char* PatternPtr;
        int PatternIndex;
        int MaxIndex;

    public:
```

```
        LEDs();
        LEDs(int baseaddress);
        void SetPatternAddress(unsigned char* pattern,
                                              int maxidx);
        void LightLEDs();
};
```

The definitions of the member functions of this class are given in Listing 7-7.

Listing 7-7 Member function definitions for the modified class.

```
LEDs::LEDs()
{
    MaxIndex = 0;
    PatternIndex = 0;
}


LEDs::LEDs(int baseaddress) : ParallelPort(baseaddress)
{
    MaxIndex = 0;
    PatternIndex = 0;
}


void LEDs::SetPatternAddress(unsigned char* pattern, int maxidx)
{
    PatternPtr = pattern;   // Pointer PatternPtr assigned
                            // address of pattern.
    MaxIndex = maxidx;
}


void LEDs::LightLEDs()
{
    if(MaxIndex <= 0)
    {
        cout << "No Patterns to display " << endl;
        return;
    }

    while(!kbhit())
    {
        WritePort0(*(PatternPtr + PatternIndex++));

        // Reset PatternIndex when it gets to MaxIndex.
        if(PatternIndex == MaxIndex) PatternIndex = 0;
```

```
        delay(500);
    }
}
```

Listing 7-8 showns a main() function which asks a user to enter patterns into the LED pattern array during program execution.

Listing 7-8 Main function – user fills in the array (LED pattern).

```
void main()
{
    unsigned char LightPattern[8];
    int UserPattern;

    LEDs Leds;
    int i;

    cout << "Enter 8 user patterns in the range 0x00-0xFF ";
    cout << endl;
    for(i = 0; i < 8; i++)   // fill 8 element Array
    {
        cin >> UserPattern;
        *(LightPattern + i) = UserPattern;
    }

    Leds.SetPatternAddress(LightPattern, 8);
    Leds.LightLEDs();
    getch();
    Leds.LightLEDs();
}
```

Note that in the main() function shown in Listing 7-8, the programmer is required to code the array name (LightPattern[8]) with the number of elements of the array (each element will store one of the patterns sent out to the LEDs). The user enters the actual values for each sequential pattern at run-time. The local variable UserPattern, of type int, is used to read integer data. The data is then assigned to the array LightPattern of type unsigned char through the use of the array name LightPattern as a pointer.

The statement in Listing 7-8:

```
Leds.SetPatternAddress(LightPattern,8);
```

can be replaced by:

```
Leds.SetPatternAddress(LightPattern,
                        sizeof(LightPattern));
```

Programming with functions can often be made more efficient by defining macros and then calling the functions through the macro. As can be seen in the above statement, the word `LightPattern` occurs twice in the function call. In order to minimise the possibility of coding an error, a macro can be written that requires the parameter `LightPattern` be specified only once. The following section describes the process of defining a macro.

7.7 Macros

Macros can be viewed as placeholders that the preprocessor replaces with an expression. For example, consider when the compiler encounters the following statement:

```
y = x*x*x;
```

We can define a macro to facilitate programming as follows:

```
#define CUBE(x)    ((x)*(x)*(x))
```

The preprocessor will replace all occurrences of `CUBE(x)` with `((x)*(x)*(x))`. Thus, `CUBE(x)` can be used freely in the program. The extra pair of parenthesis is necessary to adhere strictly with the intended precedence of operations. For example, if `CUBE(x)` was defined as:

```
#define CUBE(x)    x*x*x;            then,
```

`y = CUBE(3);`   will be replaced by, `y = 3*3*3`   evaluating to 27.

On the other hand, the line:

`y = CUBE(2+1);` will be expanded to, `y = 2+1*2+1*2+1`   incorrectly evaluating to 7.

If we now use the definition with every x placed in a pair of parentheses; (x)*(x)*(x), the expression will be expanded to:

`y = (2+1)*(2+1)*(2+1)`   which correctly evaluates to 27.

If the preceding definition for `CUBE` was used for the following expression, an incorrect result will be generated.

```
y = 81/CUBE(3);
```

This will expand to:

```
y = 81/(3)*(3)*(3);
```

which evaluates to 243 instead of the intended result 3. Using an additional outer pair of parenthesis ensures a correct result.

To improve the program given in Listing 7-8, we can include a macro as follows:

```
#define SetArray(x)    SetPatternAddress((x), sizeof(x))
```

The `main()` function is shown in Listing 7-9.

Listing 7-9 The `main()` function – user fills in the LED pattern.

```
#define SetArray(x) SetPatternAddress((x), sizeof((x)))

void main()
{
    unsigned char UserPattern, LightPattern[4];
    LEDs Leds;
    int i;

    cout << "Enter " << sizeof(LightPattern);
    cout << " user patterns in the range 0x00-0xFF ";
    cout << endl;
    for(i = 0; i < sizeof(LightPattern); i++)
    {
        cin >> UserPattern;
        *(LightPattern + i) = UserPattern;
    }
    Leds.SetArray(LightPattern);
    Leds.LightLEDs();
    getch();
    Leds.LightLEDs();
}
```

7.8 Dynamic Memory Allocation

It often happens that a C++ program will use dynamic memory allocation to request and have memory allocated at run-time. Dynamic memory allocation is used very widely in C++ programs and can greatly reduce the size of the executable program. It is especially useful when the actual storage requirements of the program are not known at the time of programming. For example, a program written to process the marks of a class of students will operate on various sized classes, their size unknown at the time of programming.

Memory allocation can be static or dynamic. In the static case, the compiler allocates the required memory at compile time. Programs with statically allocated memory tend to be bigger than programs with dynamically allocated memory. Furthermore, the statically allocated memory is obtained from the *data area* - an area specially set aside to store data. The memory region used to store program instructions is known as the *code area* and is a different region to the data area. In general, the program instructions or code do not change during the life of the program. However, data will change as the executing program operates on it.

Temporary data is created and destroyed during program execution in another area named the *stack*. The stack is a last-in-first-out (LIFO) type queue using a specially allocated region of computer memory. Some examples of temporary data are; parameters passed to a function, local variables declared within a function, and values returned by functions.

In the case of dynamic memory allocation, the requested memory is granted from yet another area named the *heap* (also known as the *free store*). Note: memory that has been allocated is not automatically returned back to the heap by the system. It is the programmer's responsibility to include instructions to return the dynamically allocated memory for other use.

The two operators that manage dynamic memory allocation are shown in Table 7-4. The operator `new` allocates the requested memory and returns a pointer that points to the beginning of the allocated area. The simplest use of the `new` operator is shown in the following example:

```
int *IntPtr;
IntPtr = new int;
```

The same effect can be achieved in one statement as follows:

```
int *IntPtr = new int;
```

Table 7-4 Operators used with dynamic memory allocation.

Operator	Function
new	To request memory
delete	To free up memory

In this example, we have requested the dynamic allocation of space from the heap for one `int` type data. We do not have a name for the allocated space, however, the pointer `IntPtr` knows *where* it is. Since we know how to manipulate data using pointers, we can use the allocated space as we need. At the end of its use, the memory space must be returned using the `delete` operator as follows:

```
delete IntPtr;
```

This operation does not remove the pointer variable `IntPtr`. Instead it releases or returns the portion of memory pointed to by `IntPtr`, thereby making the dynamically allocated integer no longer available. Now the *memory* previously occupied by that integer is available for any future dynamic memory allocation operations. If the memory was not released using the `delete` operator, then that memory will not be able to be used during remaining program operation. In this situation we have created what is known as a *memory leak* and the computer will have a reduced amount of memory it can use.

A slightly more complex example is now given that requests space for an array of ten integers:

```
int *IntPtr;
IntPtr = new int[10];
```

When the allocated memory space is no longer needed, it must be relinquished using the delete operator as follows:

```
delete IntPtr;
```

The following example requests space for a two-dimensional array of 100 int type data:

```
int(*RowPtr[10]);
RowPtr = new int[10][10];
```

RowPtr is a pointer to a row of 10 elements. The new operator asks for memory to store 10 sets of 10 element arrays of type int. The allocated storage can be freed by using:

```
delete RowPtr;
```

Space can be dynamically allocated to class objects as well. To dynamically create a DAC type object, we can use the following statements:

```
DAC *DACPtr;
DACPtr = new DAC;
```

We do not have a name for the DAC object that has been dynamically created on the heap. However, the pointer DACPtr knows where the object is. The pointer can be used just as efficiently as an object name to manipulate the object.

The allocated space must be deleted by:

```
delete DACPtr;
```

In all these memory allocation operations the new operator calls the constructor of the object. Although we did not create a constructor for int type objects, the data type int has its own constructor. For example, to create space for one int type data which is initialised to 0, the following statements can be used:

```
int *IntPtr;
IntPtr = new int(0);
```

or they can be combined into one line as follows:

```
int *IntPtr = new int(0);
```

Similarly, if we want to create a new DAC type object that communicates with the PC using the BASE address 0x3BC instead of 0x378, we use the following statement:

```
DAC* DACPtr = new DAC(0x3BC);
```

When the new operator calls the constructor of the DAC class, the BASE address that is specified (in this case 0x3BC) will be passed.

When the object belongs to a class hierarchy, the pointer does not necessarily need to be of the same type – it can be a pointer to one of its base classes. In the case of the DAC class, the following is still valid:

```
ParallelPort *Ptr;
Ptr = new DAC;
```

The same pointer can even be made to point to a new, yet different object of the same hierarchy. Suppose we had an object type named DAC16Bit further down in the hierarchy. Then, the following statement is allowed:

```
Ptr = new DAC16Bit;
```

In Section 8.6 we will be using this concept together with virtual functions to improve our programming.

It is also possible for the dynamic memory allocation operation to fail. This can occur if there is insufficient memory available to allocate. If memory allocation fails, then the pointer returned by the new operator will be set to NULL, a predefined constant. A simple test such as that shown below can be carried out to check if memory allocation has been unsuccessful:

```
if(Ptr == NULL)
{
    cout << "Memory allocation failed ";
    exit(1);
}
```

The function exit() is a library routine that can be called to terminate the program (if you decide insufficient memory on the heap justifies termination). Another approach is to write the program to cause an exception as described in Section 7.9. Any pointer returned by the new operator can be tested this way before proceeding. The usual practice is to pass an actual argument of 1 to the exit() function. This indicates to the system that runs your application that the program has terminated prematurely.

Typecasting

Wherever permitted, typecasting can be used to convert an existing type to match another type. The application of typecasting to fundamental data types is demonstrated by the following example:

```
int a;
```

```
float b = 8.73;
a = (int) b; // a will be 8.
```

The float type variable b is *typecast* to match that of a (data type int) and hence its value rounds down to 8. Similarly, pointers can also be typecast as shown in the following example:

```
int *IntPtr;
IntPtr = (*int) new int[10][10];
```

The pointer returned by the new operator is type casted to match the pointer type of IntPtr by using (*int), which can be interpreted as 'pointer to int'. Note that the new operator returns a pointer to an array of 10 elements (because the row size of the array being created is 10). However, the pointer IntPtr is a pointer to just one int type data.

7.9 Exception Handling

Exception handling allows a program to take appropriate actions in the event of exceptional conditions ocurring. These situations usually happen when a program cannot continue exececution as expected due to events that occur outside the scope of normal program control. For example, the program may not be able to continue its normal operation if there is insufficient memory to fulfil a memory allocation request.

There are many situations that can cause a program to terminate abnormally, such as not having enough disk space to write to a file, attempting to write to a file that is already opened for reading, etc. Such situations often arise due to the circumstances under which the program is running and not necessarily due to programming errors. To manage such situations, C++ uses *exception handling*. Exception handling can only manage routine events that arise when executing a program. It is not used to handle events such as user-driven abortion of program execution by pressing 'Control-C'.

The three keywords associated with exception handling are *try*, *throw* and *catch*. The keyword try is used to form a try *block*. A try block consists of the keyword try followed by the try block contained within a pair of matching braces:

```
try
{
    . . .
}
```

All statements that are likely to cause exceptional situations are executed within the try block. Each situation that may lead to an exception must be identified and a throw statement must then be executed. In the example shown in Listing 7-10 we are attempting to allocate memory from the free store; being n unsigned char locations. This attempt may fail; (a) if the value of n is less than 1, (b) there is no space available in the free store.

Listing 7-10 An example try block for dynamic memory allocation.

```
unsigned char* LightPattern;
try
{
   if(n < 1)
      throw(n);
   LightPattern = new unsigned char[n];
   if(LightPattern == NULL)
      throw("Memory error");
}
```

The most significant observation to be made here is the signature of the throw statements. The first throw, throws just one integer, being n. The second throw throws a string. Immediately after the try block, there must be matching catches. In our case there must be a catch that matches the throw of one integer and another catch that matches the throw of one string. When these two catches are included, Listing 7-10 will become the code shown in Listing 7-11.

Listing 7-11 Try block with the catches.

```
unsigned char* LightPattern;

try
{
   if((n < 1) || (n > 4))   // Note: || is logical OR
      throw(n);

   LightPattern = new unsigned char[n];
   if(LightPattern == NULL)
      throw("Memory error");
}

catch(int n) // catches the throw of integer
{
   cout << "Illegal number of elements requested" << endl;
   cout << "Array size defaults to 4" << endl;
   n = 4;
   LightPattern = new unsigned char[n];
}
```

```
    catch(char* memerror)
    {
       cout << "Memory allocation failed " << endl;
       cout << "Terminating program " << endl;
       exit(1);
    }
```

Walking LEDs – User Definable Array Size and Contents

We can improve the program presented in Listing 7-9 so the user has the flexibility to define the *size* as well as the *contents* of the LED pattern array. Each number entered into the array by the user will be output to the bank of LEDs in sequence, followed by a short delay. Therefore, the main() function shown in Listing 7-9 can be re-written using dynamic memory allocation and exception handling as shown in Listing 7-12.

Listing 7-12 Dynamic memory allocation and exception handling for the LED walk .

```
#include <iostream.h>
#include <conio.h>
#include <stdlib.h>
#include <dos.h>

#define SetArray(x) SetPatternAddress((x), sizeof((x)))

class ParallelPort
{
   private:
      unsigned int BaseAddress;
      unsigned char InDataPort1;

   public:
      ParallelPort();
      ParallelPort(int baseaddress);
      void WritePort0(unsigned char data);
      void WritePort2(unsigned char data);
      unsigned char ReadPort1();
};

ParallelPort::ParallelPort()
{
   BaseAddress = 0x378;
   InDataPort1 = 0;
}
```

```
ParallelPort::ParallelPort(int baseaddress)
{
   BaseAddress = baseaddress;
   InDataPort1 = 0;
}

void ParallelPort::WritePort0(unsigned char data)
{
   outportb(BaseAddress,data);
}

void ParallelPort::WritePort2(unsigned char data)
{
   outportb(BaseAddress+2,data ^ 0x0B);
}

unsigned char ParallelPort::ReadPort1()
{
    InDataPort1 = inportb(BaseAddress+1);
// Invert most significant bit to compensate
// for internal inversion by printer port hardware.
   InDataPort1 ^= 0x80;
// Filter to clear unused data bits D0, D1 and D2 to zero.
   InDataPort1 &= 0xF8;
   return InDataPort1;
}

class LEDs : public ParallelPort
{
   private:
      unsigned char* PatternPtr;
      int PatternIndex;
      int MaxIndex;

   public:
      LEDs();
      LEDs(int baseaddress);
      void SetPatternAddress(unsigned char* pattern,
                                          int maxidx);
      void LightLEDs();
};


LEDs::LEDs()
{
```

```
      MaxIndex = 0;
      PatternIndex = 0;
}


LEDs::LEDs(int baseaddress) : ParallelPort(baseaddress)
{
      MaxIndex = 0;
      PatternIndex = 0;
}


void LEDs::SetPatternAddress(unsigned char* pattern, int maxidx)
{
      PatternPtr = pattern;    // pointer Pattern assigned address
                               // of pattern
      MaxIndex = maxidx;
}


void LEDs::LightLEDs()
{
      if(MaxIndex == 0)
      {
          cout << "No Patterns to display " << endl;
          return;
      }

      while(!kbhit())
      {
          WritePort0(*(PatternPtr + PatternIndex++));

          // Reset PatternIndex when it gets to MaxIndex.
          if(PatternIndex == MaxIndex) PatternIndex = 0;
          delay(500);
      }
      getch(); // absorb the key that was hit
}



void main()
{
      LEDs Leds;
      unsigned char* LightPattern;
      int TempPattern;
      int n, i;

      cout << "Pass in the desired size of LightPattern => ";
```

```
cin >> n;

try
{
    if(n < 1)
        throw(n);
    LightPattern = new unsigned char[n];
    if(LightPattern == NULL)
        throw("Memory error");
}

catch(int n) // catches the throw of integer
{
    cout << "Illegal number of elements requested" << endl;
    cout << "Array size defaults to 4" << endl;
    n = 4;
    LightPattern = new unsigned char[n];
}

catch(char*) // catches the throw of the string
{
    cout << "Memory allocation failed " << endl;
    cout << "Terminating program " << endl;
    return;
}

cout << "Enter " << n ;
cout << " numbers in the range (0x00 - 0xFF)" << endl;

for(i = 0; i < n; i++)
{
    cin >> TempPattern;
    *(LightPattern + i) = TempPattern;
}

Leds.SetPatternAddress(LightPattern,n);
Leds.LightLEDs();
}
```

7.10 Summary

This chapter explained the operation and use of various types of iterative loops
such as `for`, `while` and `do-while`. The `for` loop is used primarily when the

number of iterations is known at programming time. When this is not the case, either a `while` loop or a `do-while` loop can be used. In some situations the body of a `while` loop may not execute at all, whereas the body of `do-while` loops will execute at least once. Various control mechanisms such as `if`, `switch-case`, `break`, and `continue` can be used in conjunction with loops to enhance program flow. The `switch` statement can be used when one of several cases needs to be selected for execution.

Pointers are an important and powerful feature of the C++ language and have been explained in this chapter. They contain the memory addresses that "point" to locations in memory storing various objects or functions. Arithmetic used with pointers automatically takes into account the size of the object the pointer is pointing to. Pointers can be used very efficiently to scan arrays as demonstrated by the programs developed in this chapter. Most importantly, a pointer pointing to a base class object can also point to a derived class object.

We have used dynamic memory allocation to allow memory to be made available for new objects and arrays of objects during program execution. When memory is dynamically allocated, the `new` operator returns a pointer to the allocated memory. To free the allocated memory, the `delete` operator must be used.

Exception handling was introduced in this chapter to contain predictable run-time errors. Programming statements, which have the potential to cause run-time errors can be contained within a `try` block. Depending on what is thrown from within a `try` block, a `catch` statement can be executed to indicate the cause of the run-time error. We used exception handling to manage any erroneous situations occurring from out-of-range array sizing or insufficient memory when attempting to allocate memory.

7.11 Bibliography

Kelley, A. and I. Pohl, *A Book on C – programming in C*, Benjamin Cummins, 1995.

House, R., *Beginning with C – An Introduction to Professional Programming*, International Thompson Publishing, 1994.

Deitel H.M. and P.J. Deitel *C: How to Program*, Prentice Hall, 1994.

L Miller and A Quilici, *C programming Language – an applied perspective*, John Wiley Publishing, 1987.

Hanly, J.R., E.B. Koffman and J.C Horvath, *C Program Design for Engineers*, Addison Wesley, 1995.

Rudd, A., *Mastering C*, John Wiley, 1994.

Lafore, R, *Object Oriented Programming in C++*, Waite Group Press, 1995.

8

Driving Motors - DC & Stepper

Inside this Chapter

- DC motors – types, performance and control.

- Stepper motors and stepper motor drive techniques.

- Object classes for DC and stepper motors.

- Virtual functions and class hierarchies.

- Using Pointers and Dynamic Memory Allocation.

- Motor control using the keyboard.

8.1 Introduction

Almost every industry and household has motors used in their equipment or appliances. Motors that are often controlled by computers have also become an essential part of many motion control systems. This chapter describes the basic construction of DC motors, their performance, control techniques, and the effect of applying loads to them. The technique known as Pulse Width Modulation (PWM) is explained and implemented to control DC motors. This is followed by descriptions of stepper motor types, their construction, their different modes of operation, and control methods.

The software segment of the chapter begins with the development of an abstract class to represent motors in general. Using this abstract class and the `ParallelPort` class as base classes, a new `Motor` class is derived using multiple inheritance. This new class has the capability to communicate with the PC via the parallel port. With this new class as a base class, further classes are derived for DC motors and various forms of stepper motors, demonstrating the development of object class hierarchies.

Dynamic memory allocation is used in the programs that drive the various types of motors. Towards the end of the chapter, the power of virtual functions and the concept of late binding will be demonstrated by a motor control program using keyboard control.

8.2 DC Motors

Most motors employed for control purposes are DC permanent magnet types. They are characterised by having a linear torque-speed curve and come in several popular configurations.

8.2.1 DC Motor Construction, Performance

The basic construction of a DC motor is shown in Figure 8-1. Current flows from the power source (shown as a battery) and through the armature coil winding via the brushes in contact with the commutator. This current flow induces a magnetic field around the armature coil that opposes the magnetic field produced by the magnets, and generates a motor torque that is capable of doing work. To produce a smoother output torque, many groups of coil windings are used in a typical motor armature. Each group of windings connects to opposing contacts on the commutator with each commutator contact being insulated from its neighbouring contacts.

Today, permanent magnets are widely used in motors. These magnets are made from Alnico, with high-performance motors using high-strength rare-earth magnets (samarium cobalt).

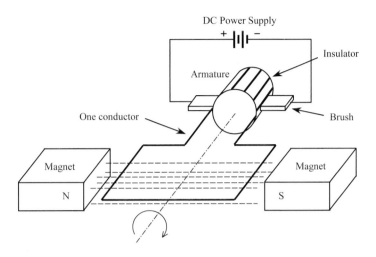

Figure 8-1 Basic DC motor construction.

A different type of DC motor, known as the *brushless motor*, does not have commutator brushes; instead current is switched through the coil windings by electronic means. These motors use permanent magnets for the rotor and have the field windings on the stator. This arrangement gives a greatly improved path for heat flow out of the motor and results in high reliability, low noise, high speed, and high peak torque characteristics.

Motor performance depends upon the load connected to the motor and the applied voltage driving the motor. Friction in the motor bearings, brush losses, iron losses, and short-cut circuit losses (from each brush contact overlapping an adjacent contact) limit the upper speed during no-load situations, and the torque load at the motor shaft determines the lower speed limit. As the load applied to the motor increases, motor speed drops in a linear manner, proportional to the load applied. The steady power produced by the motor can be increased if air-cooling is used or if the motor is used for short periods of operation. When selecting a suitable motor for a given task, the load inertia, load torque, and load power should all be considered.

8.2.2 DC Motor Control

To control DC motors, a variable voltage must be generated and applied to power the motor. There are two main sources of voltage supply for a motor: the linear power supply, and the switching power supply. The linear power supply technique uses power transistors acting in their linear mode to provide a smooth and continuously adjustable output voltage depending upon load drive requirements. In this mode, the transistors are used as variable resistors, conducting according to the level of the input signal applied. This method of control wastes power when the transistors are used to drop voltage and in doing so dissipate power.

Switching power supply techniques use transistors to switch the voltage through to the motor in a series of pulses that applies an 'average' voltage to the motor. Two switching methods are commonly used, *pulse-width modulation* (PWM), and *pulse-frequency modulation* (PFM) as shown in Figure 8-2.

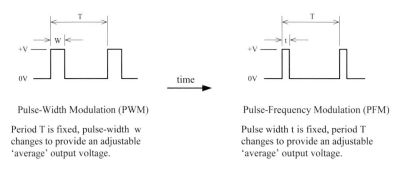

Pulse-Width Modulation (PWM)

Period T is fixed, pulse-width w changes to provide an adjustable 'average' output voltage.

Pulse-Frequency Modulation (PFM)

Pulse width t is fixed, period T changes to provide an adjustable 'average' output voltage.

Figure 8-2 Voltage-control using switching methods.

Pulse-width modulation is the more common method used for motor control. This method uses a constant period (T) between sequential pulses, while the width (w) of the pulses is altered (duty cycle) to change the effective or average voltage applied to the motor.

Pulse-frequency modulation varies the frequency of pulses having a constant pulse width (t) to control the average voltage applied to the motor.

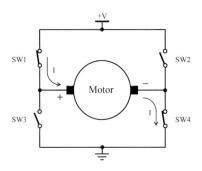

Figure 8-3 Motor control using a H-bridge.

The most common means of controlling the voltage applied to DC motors is through the use of a H-bridge circuit as shown in Figure 8-3. Two switches are closed at a time to switch the DC voltage through to the motor. When switches designated as SW1 and SW4 are closed and switches SW2 and SW3 are open, as shown in the diagram, the motor will rotate in one direction. Swapping the positions of the open and close switches to make SW2 and SW3 closed and SW1 and SW4 open will reverse the flow of current through the bridge and reverse the direction of motor rotation.

The average voltage delivered to the motor can be controlled using pulse-width modulation applied to the switches. In practice, the switch function is performed using power transistors in place of each switch. As the switch is opened, a large voltage is generated across the motor winding, known as *back-emf*. This voltage must be limited to avoid damage to the active switching transistors. Damage is prevented by fitting diodes across the transistors that clamp the voltage to much lower levels that will not cause damage.

Motors can be controlled using both open- and closed-loop control techniques. Open-loop control does not use any form of feedback from the output of the motor shaft to indicate motor speed or motor shaft rotation. This mode of operation relies upon estimating motor speed or rotation by knowing the motor load and motor operating characteristics. If accurate motor control is needed and motor load is not known or is not constant, closed-loop control is typically used. When controlling motor speed, the feedback taken from the motor output will be motor shaft speed. This is usually measured using a tachometer that generates an output voltage proportional to the motor speed. Alternatively, the tachometer can output a digital pulse-train.

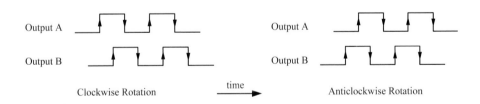

Figure 8-4 Motor encoder output – quadrature.

When position control is needed, resolvers or digital encoders are connected to the motor output to provide position feedback. Resolvers generate two out-of-phase sinusoidal waves whose amplitudes provide position information. Optical encoders output two digital waveforms that are separated in phase by approximately 90 degrees to provide directional information in *quadrature* format as shown in Figure 8-4. It is possible to determine the direction of motor rotation by knowing the order in which the rising or falling edges of each waveform occur.

8.3 Stepper Motors

Stepper motors are often used for motor control applications requiring positive positioning and good levels of torque at low speed. Provided they are not overloaded, their output shaft speed and rotational position are inherently known and controlled through simple digital switching of their winding currents – often without the need for expensive shaft encoders.

8.3.1 Stepper Motor Construction

Three main types of stepper motor construction are in use; these being *permanent magnet*, *variable reluctance*, and *hybrid* construction.

The **permanent magnet** type has a permanent magnet rotor with many poles. This motor has a residual holding torque when the windings are not energised and requires less power to operate than other types. It is low in cost, produces low torque and runs at low speed. However, it suffers from resonance effects, relatively long settling times and rough performance at low speed. It is mainly used in non-industrial applications.

Variable reluctance motors use a soft iron core rotor with a different number of teeth than the number of poles on the stator. As the stator poles are energised in-turn, the rotor aligns itself with the magnetic field of the energised stator pole, to create a rotor position having minimum magnetic reluctance. This type of motor has a good ratio of torque to inertia but lacks a residual torque when the stator field windings are de-energised. Variable reluctance motors are seldom used in industrial applications and require a different driving arrangement from the other stepper motor types.

Hybrid motors use a combination of permanent magnet and variable reluctance features in their construction. This is the most widely used type of stepper motor for industrial applications due to its high torque and residual holding torque.

8.3.2 Stepper Motor Configuration

To understand how a stepper motor works, consider the simplified motor shown in Figure 8-5. This motor has a permanent magnet rotor with one north-south pole pair and uses a stator with four teeth. The motor drive sequence shown uses what is known as *full-stepping* – in this case each full-step corresponds to a 90° rotation. There are two independent coil windings used on the stator, meaning this is a *two-phase* motor.

The motor rotates in a clockwise direction as shown in Figure 8-5 when using the coil energising sequence in Table 8-1. Both of the motor coils are energised for every step position during each drive sequence.

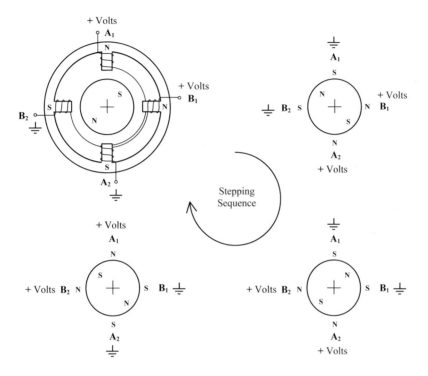

Figure 8-5 Simplified 2-phase stepper motor – full-stepping sequence.

Table 8-1 Full-stepping sequence for a 2-phase motor.

Step Position	Coil Contact Voltage			
	A_1	A_2	B_1	B_2
1	+	⏚	+	⏚
2	⏚	+	+	⏚
3	⏚	+	⏚	+
4	+	⏚	⏚	+

Smaller step angles can be achieved when a *half-stepping* drive sequence is used. This involves energising only one of the coil windings for every second step angle as shown in Figure 8-6 and Table 8-2.

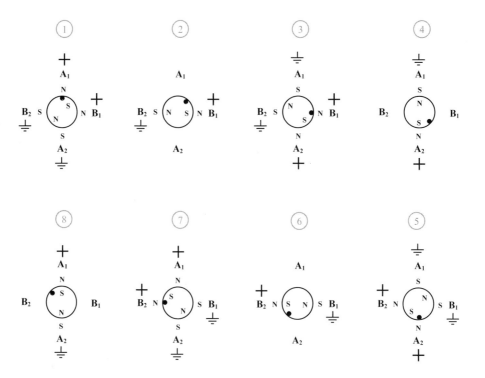

Figure 8-6 Half-stepping sequence for a 2-phase motor.

Table 8-2 Half-stepping sequence for 2-phase motor.

	Coil Contact Voltage			
Step Position	A_1	A_2	B_1	B_2
1	+	⏚	+	⏚
2			+	⏚
3	⏚	+	+	⏚
4	⏚	+		
5	⏚	+	⏚	+
6			⏚	+
7	+	⏚	⏚	+
8	+	⏚		

Stepper motors are categorised depending upon the way current is driven through their windings as will now be explained.

Bipolar Motor

This type of motor uses coil currents that reverse in direction throughout the stepping sequence, as shown in the preceding text. To achieve this reversal of current, *bipolar* voltages are applied to the coil windings. A bipolar power source with +ve, −ve and ground potentials can be used with four switching transistors to drive this type of motor. However, the usual means of implementing this form of drive is to use a single polarity power supply with eight switching transistors, configured using two separate H-bridge circuits as shown in Figure 8-7. Bipolar type motors are easily recognized since they have only four connection leads.

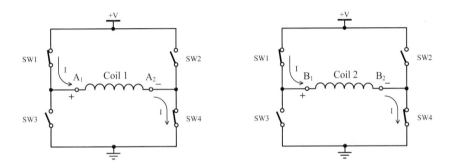

Figure 8-7 Bipolar drive using two H-bridge circuits.

Unipolar Motor

This type of motor uses coil winding currents which flow in one direction only. In order to obtain a reversal of magnetic field at each stator tooth, the coil is wound in two halves as shown in Figure 8-8. One half of the coil is wound clockwise around the stator tooth and the other half of the coil is wound anticlockwise around the stator tooth – known as a *bifilar* winding.

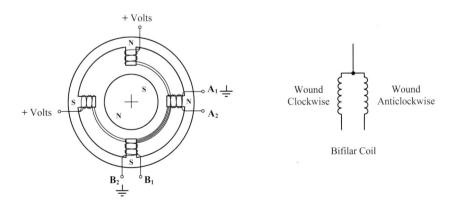

Figure 8-8 Unipolar motor – bifilar coil.

The wire connection halfway through the coil is connected to the positive power supply. Only one half of the coil is energised at a time. A north pole is produced when one half of the coil is energised by connecting its end to ground potential; a south pole is produced by energising just the other half of the coil by grounding its end. This is shown in Figure 8-9, Figure 8-10 and Table 8-3. Note that a north pole is produced when coil winding A_1 or B_1 are grounded; conversely grounding A_2 or B_2 results in a south pole.

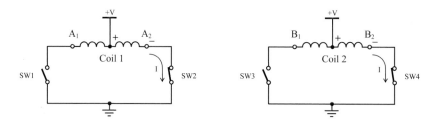

Figure 8-9 Unipolar coil drive.

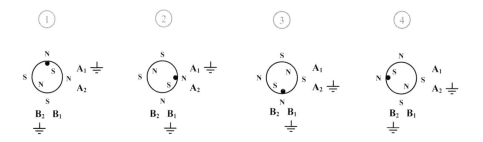

Figure 8-10 Full-stepping sequence – unipolar motor.

Table 8-3 Full-stepping sequence for unipolar motor.

Step Position	Coil Contact Voltage			
	A_1	A_2	B_1	B_2
1	⏚			⏚
2	⏚		⏚	
3		⏚	⏚	
4		⏚		⏚

Half-stepping is achieved by energising two half coils together followed by energising only one coil in a repetitive sequence, generating a similar sequence of magnetic fields when half-stepping a bipolar motor. Figure 8-11 and Table 8-4 show the half-stepping sequence for a unipolar motor.

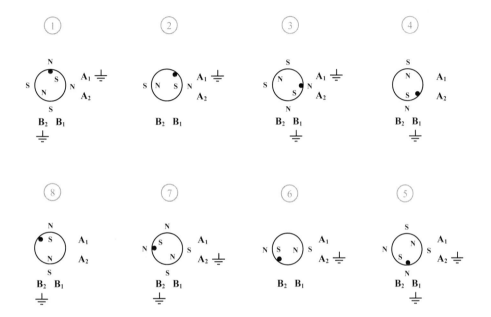

Figure 8-11 Half-stepping sequence – unipolar motor.

Table 8-4 Half-stepping sequence for unipolar motor.

	Coil Contact Voltage			
Step Position	**A_1**	**A_2**	**B_1**	**B_2**
1	⏚			⏚
2	⏚			
3	⏚		⏚	
4			⏚	
5		⏚	⏚	
6		⏚		
7		⏚		⏚
8				⏚

For a unipolar motor to have the same number of turns per winding as a bipolar motor, the wire diameter must be decreased due to its bifilar winding scheme. This reduction in wire diameter leads to an increase in coil resistance and lowers the motor torque by approximately 30% at low step rates. At higher step rates, unipolar motor performance exceeds that of bipolar wound motors.

Stepper motors come in a variety of wire configurations as shown in Table 8-5. The five-wire unipolar motor is identical to the six-wire unipolar motor except the two power supply wires are connected together internally and only one wire is brought outside the motor case.

Table 8-5 Stepper motor wire configuration.

Wire Arrangement	Motor Type
4-wire	Bipolar – 2 phase
4-wire	Variable Reluctance – 3 phase
5-wire	Unipolar – 2 phase
6-wire	Unipolar – 2 phase
8-wire	Bipolar – 4 phase

The eight-wire bipolar motor has two pairs of two phases (independently wound coils). This arrangement allows a pair of phases to be connected in series or in parallel. A series-connected configuration effectively doubles the amp-turns producing twice the torque at lower speeds. The inductance of the effective coil is proportional to the number of turns squared, meaning that the inductance is now raised four-fold. Winding resistance is now double, which lowers the maximum value of current through the winding in order to not exceed the motor power rating.

Connecting the eight-wire motor in a parallel configuration does not change the effective number of turns and therefore does not increase the winding inductance. The effective winding resistance is now halved, meaning that the motor can be driven at higher levels of winding current for the same power dissipation. This will give improved torque at this higher current.

A series-connected configuration of phases leads to a rapid drop in torque as speed increases. This occurs because the time constant of the winding, being equal to inductance divided by resistance, is now twice that of a single connected phase. The parallel-connected windings perform much better at high speed than the series configuration since their winding time constant is half that of a single winding and therefore ¼ that of the series configuration. Additionally, the torque curve is flatter for a motor connected in parallel, producing greater shaft power.

8.3.3 Stepper Motor Control

Stepper motor performance is compromised when driven by a simple H-bridge circuit as shown previously in this chapter. The coil current requires finite time to increase in level once the controlling switch (transistor) is 'closed'. Since the winding has inductance (L) and resistance (R), the current (I) will increase exponentially with increasing time according to the value of the winding time constant (τ) as shown in Figure 8-12.

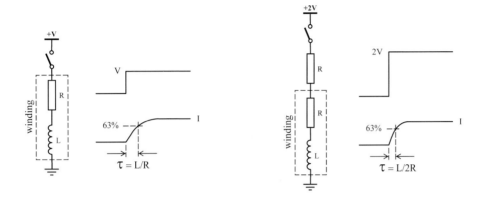

Figure 8-12 Regular drive and Series resistance drive.

One means of reducing this winding time constant is by using *series resistance drive*. In this scheme, an external resistor is added in series to the winding circuit. If this resistor was say equal in value to the resistance of the wound coil and the applied voltage doubled, the peak current (I) through the winding would remain the same, but the winding time constant is now halved. Although this form of motor control increases the high-speed torque, it is inefficient due to the loss of power generated by the current flowing through the added external resistor.

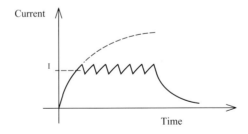

Figure 8-13 Coil winding current under chopper drive.

A better approach to stepper motor control is known as *chopper drive*. The voltage applied to the winding is raised similar to that for the series resistance drive scheme but an added resistor is not used. This improves the rise-time of the current through the winding. Without using an added resistor, the current would eventually increase and exceed the motor's rated winding current if some form of voltage control was not used. To prevent this excessive current build-up, a current sense resistor (with small resistance) is used to measure the winding current. When the level of rising current reaches the rated value of the winding, the applied voltage to the motor control circuit is turned off, preventing any further current rise. Naturally, the winding current will now decay while the winding is not being powered. The sense resistor is used to monitor this decay so that voltage can be re-applied to the motor control circuit when the level of current has dropped a small amount below the rated value. This cycle of current build-up and decay continues until it is time to turn off the winding current for the next step sequence as shown in Figure 8-13.

There is a more refined technique used to drive stepper motors known as *microstepping*. This drive scheme proportions the level of individual coil winding current to produce intermediate stepping positions within a normal full-step. Accurate control of coil current is needed and high-resolution microstepping in excess of several thousand steps per revolution is possible.

Motor controllers that use integrated circuits are available to simplify the task of stepper motor control. These circuits contain waveform-generating logic, power transistor switches, and associated protection diodes to control the damaging effect of back-emf (electromagnetic force) produced when coil currents are switched off.

8.3.4 Stepper Motor Specification

Several terms such as *holding torque, dynamic torque, pull-in torque, pull-out torque*, and *ramped step rate* are used to specify stepper motor performance. These terms are briefly explained in Table 8-6.

Table 8-6 Stepper motor terminology.

Holding torque	That torque which the motor generates when stationary.
Dynamic torque	The torque generated by the motor when rotating. This torque drops as motor speed increases due to the effect of the motor's time constant and the reduced time to build-up current.
Pull-in torque (*start without error torque*)	The pull-in torque is the maximum torque available to be applied to the motor load when starting from rest for a particular step rate (or when coming to a stop without losing steps). It does not include that portion of motor torque needed to accelerate the inertia of the motor itself.
Pull-out torque (*running torque*)	The maximum torque that can be applied to the motor load during steady speed without losing steps. This torque is higher than the pull-in torque because the motor is not being accelerated and therefore no torque is consumed for this purpose.
Ramped step rate	This is the step rate which avoids any loss of steps during periods of acceleration or deceleration.

So far in this chapter we have discussed the principles of operation of DC motors and stepper motors. Let us now turn our attention to writing object-oriented programs to drive these motors using the motor drive circuits of the interface board. In the coming sections we will encounter the most powerful feature of object-oriented programming – virtual functions.

The approach taken in the following sections is to first develop a class hierarchy to represent DC motors and all forms of stepper motors discussed earlier in this chapter. Then we will use the classes in this hierarchy to develop a generic program that will drive any type of motor in the hierarchy.

8.4 A Class Hierarchy for Motors

All the motors to be included in our class hierarchy will be driven using the interface board. Therefore, while the class hierarchy is in principle for *motors*, *interfacing* also plays an important role. We can identify two major categories for all motors described in this chapter. They are; i) *DC motors* and ii) *Stepper Motors*. Stepper motors fall into two types; *Unipolar Stepper Motors* and *Bipolar Stepper Motors*. These two types of stepper motors can be controlled in two different ways; *full-step control* and *half-step control*. DC motors have not been further subdivided.

To start with, we can think of motors 'in general' as abstract objects. A 'motor' will remain an abstract concept until we can describe all its relevant details. Therefore, a good starting point for our class hierarchy is an abstract motor class that encompasses the most common features of all motors in the hierarchy. Interfacing these motors to the interface board is also a very basic requirement for all the motors. In our case, all motors will be controlled via the parallel port and so the `ParallelPort` object we developed earlier can be used for this purpose. Since interfacing is necessary for all motors of the hierarchy, the `ParallelPort` class must join the hierarchy at a very early stage. The proposed class hierarchy is shown in Figure 8-14.

At the root of the class hierarchy is the abstract class `AbstractMotor` that represents all motors. The `ParallelPort` class is also at the same level as the `AbstractMotor` class. However, the `ParallelPort` class developed earlier is not an abstract class. It is a real class since objects can be instantiated from it.

The `Motor` class is derived by multiple inheritance from the two base classes `AbstractMotor` and `ParallelPort`. The `Motor` class is also an abstract class since it is not yet a fully described object class of the motor hierarchy. This means the class lacks the finer details of the specific motor types needed to complete the member function definitions. However, the objects of the `Motor` class have more capabilities than the objects of the `AbstractMotor` class, namely they can communicate with devices via the parallel port.

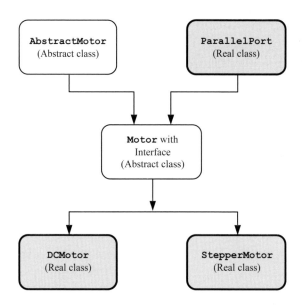

Figure 8-14 Motor class hierarchy.

The Motor class is then used to derive the two classes DCMotor and StepperMotor. At this level DCMotor and StepperMotor are completely described and must be real classes. The class hierarchy ends here.

The question arises; why not derive further classes to represent, for example, dual-phase bipolar stepper motors in half-step control? The answer is as follows. The functional characteristics of a dual-phase bipolar stepper motor in half-step control are different from a dual-phase unipolar stepper motor in full-step control. However, from a software point of view, this is analogous to two cars having different colour. The different motor drive schemes are certainly not analogous to a normal car and a luxury car. When two different coloured cars are needed, the choice of colour can be handled by changing the value of a parameter. A new class is not necessary for each colour. Our approach in treating different kinds of stepper motors will be along similar lines.

8.5 Virtual Functions – An Introduction

A virtual function is a function that has the same function signature throughout a class hierarchy although behaves according to its definition in each class. Virtual functions are polymorphic in nature but with a subtle difference. As discussed in Chapter 4, polymorphic functions are functions throughout a class hierarchy with the same name, the same number of parameters, same sequence of parameters and same types of parameters. However, their bodies are programmed differently to suit the requirements of each class. Virtual functions have all the features of polymorphic functions, except the keyword virtual is added right at the front of function declarations within their classes. Although virtual and non-virtual functions appear to be quite similar, there are significant advantages when using virtual functions. They allow the implementation of a very powerful feature known as *late binding* which will be explained in more detail later in this chapter.

The benefits of late binding are strictly linked to virtual functions within a class hierarchy. Virtual functions are added at first to a base class and then later implemented throughout the derived classes of the hierarchy. This arrangement provides the primary link needed between a virtual function of the base class and a virtual function of a derived class somewhere down the class hierarchy. Any virtual functions of *any* derived class can be called using a pointer to the base class. Despite using a base-class pointer, rather than a pointer to the derived class itself, at run-time the base-class pointer will select the correct function from the derived class, thanks to the mechanism that links virtual function within a class hierarchy.

If virtual functions are not used to select the correct function to match the object chosen by the user at run-time, the developer needs to provide additional code to carry out these tasks. This extra code will include all the necessary program statements placed within a framework of "if-then-else" logic or switch statements. When class hierarchies are large and complex, this extra programming can be an immense burden and produce programs that are difficult to debug and maintain. If a new class is added to the hierarchy, the entire program needs to be modified. As can be imagined, this is not a very efficient approach to programming.

Virtual functions and associated late binding alleviates the program developer from needing to generate this extra code. The developer writes a generic program using virtual functions. The compiler and linker are now managing this task. As a result, programming and debugging times are markedly reduced. Furthermore, minimal change to code is required to incorporate a new addition to the class hierarchy. These advantages will become evident as we work through the example programs ahead.

AbstractMotor *Class*

As mentioned previously, the principal motive underlying the development of the AbstractMotor class as a base class is to form a foundation for the network of virtual functions. All classes derived from the AbstractMotor class will inherit all its functions. Only some of the functions can be completely defined at this early stage, while the remaining functions will be skeletal due to a lack of exact details for the 'motors' involved. A thoughtful selection of data and functions to be included in a base class for motors is as follows:

1. A data member to store the set speed of the motor.
2. A mechanism for functions outside the class to obtain the set speed.
3. A function to drive 'a motor' forward.
4. A function to drive 'a motor' in reverse.
5. A function to brake 'a motor'.
6. A function to turn 'a motor' off (no applied power).

A class definition for the `AbstractMotor` class is given in Listing 8-1.

Listing 8-1 `AbstractMotor` class.

```
class AbstractMotor
{
    private:
        int Speed;

     public:
        AbstractMotor();
        void SetSpeed(int speed);
        int GetSpeed();
        virtual void Off()=0;
        virtual void Forward()=0;
        virtual void Reverse()=0;
        virtual void Brake()=0;
};
```

We haven't seen the bold font code shown in Listing 8-1. It will be discussed in the sections ahead.

The `AbstractMotor` class has `Speed` as a private data member, which is of type `int`. The member function `SetSpeed()` receives `speed` as a parameter and then sets the value of data member `Speed` to equal `speed`. The public member function `GetSpeed()` can be called by any function in the program to obtain the value of private data member `Speed`. The `Off()` function turns off all power to the motor. The function `Forward()` is used to drive the motor forward at the speed specified in `Speed`. Similarly, the function `Reverse()` is used to drive the motor in reverse direction at the speed specified by `Speed`. The `Brake()` function can be used to short-circuit the motor windings to stop the motor rotation, in the minimum time.

We will now provide the definitions for all member functions of the `AbstractMotor` class. The `AbstractMotor` class does not drive specific types of motors. Rather it provides general functions that will be overridden in its derived classes to suit specific types of motor. Without knowing the exact details of the motor, we cannot define the functions `Off()`, `Forward()`, `Reverse()` and `Brake()`. If we cannot define the functions, why did we include them as member functions? One reason is that it is good to specify the form of the objects of the hierarchy at its start to maintain a high level of conformity throughout the hierarchy. This helps us write code that maintains the relationships with objects of the entire hierarchy. For example, all classes of the hierarchy will have a function named `Forward()`. The subsequent derived classes will then redefine the inherited `Forward()` function to apply specifically to their respective class. The other reason is associated with *virtual functions* and *late binding*.

The motor driving part of the program can be very complex. It may contain sophisticated operations to change the speed, change direction of rotation and to brake the motor. Using virtual functions, the motor control program in the `main` function can be written completely, without knowing exact specifics of the particular motor to be driven. When the user selects a particular control action for the chosen motor, the correct function from that class will be deployed automatically to operate on the associated object type (without the programmer needing to provide explicit code). When we use virtual functions, the compiler generates the actual code that allows this virtual function mechanism to operate in this way. This behaviour will be shown in the sections ahead. This part of the program will even work for objects that will be added to the class hierarchy in the future. This gives us the flexibility to expand the class hierarchy as desired without needing to rewrite the motor control portion.

When the program is running, the user will select an object type (`DCMotor` or `StepperMotor`) to be used from a choice of motor types. For example, the user may select a motor of type `DCMotor` to be driven forward. Then the program will automatically bind the `DCMotor` object to the `Forward()` function of the `DCMotor` class. This deferred decision-making is known as *late binding*. In other words, the program selects the correct function to drive the motor forward based on the object type selected by the user at run-time. Late binding is also known as *dynamic binding*. The word dynamic is used because the binding takes place while the program is running. Note that polymorph functions cannot be used in late binding - only virtual functions can use this feature.

If virtual functions were not used in the program, the programmer would need to provide the extra logic to select the correct function. In this case the end of each logic branch will be hard-coded to bind a specific object to its associated member function. Binding is no longer deferred to a later time. In such a situation the compiler can identify the correct function and bind it at programming time. This is known as *early binding* or *static binding*. The word static is used to signify that binding takes place before the program runs - the program is already coded for each possible combination of object/function the user might select.

At first glance, virtual functions and late binding appear to be overkill! However, without them it is very difficult to write sound generic programs that work with a variety of situations that are decided by the user at run time. Using virtual functions can significantly reduce the amount of programming required. The programmer no longer needs to write a number of code segments to control each object type. Instead, one code segment will be written to control all object types of a given hierarchy. Also, the user (or the program itself) has complete flexibility to choose the object type at run time.

8.5.1 Pure Virtual Functions

It was mentioned previously that without knowing the physical construction and interfacing of the motor, it is impossible to define the bodies of the functions `Off()`, `Forward()`, `Reverse()`, and `Brake()`. In order to inform the compiler of our inability to provide the bodies of these functions we must declare them as *pure* functions. Note that only virtual functions can be declared pure. A function is declared pure by appending '=0' at the end of the declaration.

A normal function is declared as:

```
void Forward();
```

A virtual function is declared as:

```
virtual void Forward();
```

A pure virtual function is declared as:

```
virtual void Forward()=0;
```

Pure Virtual Functions and Abstract Classes

An effect of declaring at least one pure virtual function (which has no executable code) in a class definition is that the class becomes an abstract class. Some abstract classes have useful functions. If the class has a function whose body cannot be defined, then the class cannot be used to instantiate completely usable objects. The reason for this is that a program may attempt to call a pure virtual function (with no executable code). Therefore, a rule is in place that prevents objects from abstract classes being instantiated. We will revisit virtual functions when we make use of them for controlling the DC and Stepper motors in the class hierarchy (shown in Figure 8-14).

Returning to the class definition given previously in Listing 8-1, only three of the member functions can be defined: `AbstractMotor()`, `SetSpeed()` and `GetSpeed()`. As explained previously, the other four functions; `Off()`, `Forward()`, `Reverse()`, and `Brake()` cannot be defined at this stage, which is why they are declared as pure virtual functions. The member function definitions are given in Listing 8-2.

Listing 8-2 `AbstractMotor` class member function definitions.

```
AbstractMotor::AbstractMotor()
{
    Speed =0;
}

void AbstractMotor::SetSpeed(int speed)
{
    Speed = speed;
    if(Speed > 255) Speed = 255; // Limit upper value
    if(Speed < 0) Speed = 0; // Limit lower value
}

int AbstractMotor::GetSpeed()
```

```
{
    return Speed;
}
```

The constructor of the `AbstractMotor` class initialises the data member `Speed` to 0.

The public function `Setspeed()` can be called by any function to change the speed of any of the motors in the hierarchy. This function assigns the actual argument passed in `speed` to the member data `Speed`. The two `if` statements in the `SetSpeed()` function ensure the actual argument passed in place of `speed` is restricted to within the acceptable range $0 - 255$ we have decided to use. If for example, a value such as 300 is passed, then the first `if` statement will limit and override that value to become 255. Therefore, the data member `Speed` will be set to 255. In other words, any value above 255 will be forced to be 255. The second `if` statement will force any value below 0 to be 0. Any value inside the acceptable range will be left as is.

The `GetSpeed()` function returns the current value of `Speed` to any function. This is the only mechanism provided for a function outside the class to obtain the value of `Speed`.

Motor *Class*

Next in the class hierarchy is the `Motor` class. This class inherits functions and data from the two classes `AbstractMotor` and `ParallelPort`. Its class definition is given in Listing 8-3.

Listing 8-3 Motor class.

```
class Motor : public AbstractMotor, public ParallelPort
{
    public:
        Motor(int baseaddress=0x378);
        void Off();
        virtual void Forward()=0;
        virtual void Reverse()=0;
        virtual void Brake()=0;
};
```

Note the program line:

```
class Motor : public AbstractMotor, public ParallelPort
```

The new class name is `Motor` and it is derived using the `public` access specifier with the two base classes `AbstractMotor` and `ParallelPort`. A comma separates the base classes. No new data members are added. The two functions that must be defined are the constructor of the `Motor` class and the `Off()` function. The other three functions remain as pure virtual functions. It was mentioned earlier that if at least one of the member functions is a pure virtual function, the class becomes an abstract class. Therefore the `Motor` class is an abstract class and so cannot be used to instantiate objects. The difference between the classes `AbstractMotor` and `Motor` is that the latter has the ability to communicate with devices via the parallel port. As such, any objects derived from the `Motor` class can communicate via the parallel port.

The definition of the `Motor` class constructor and the `Off()` function is given in Listing 8-4.

Listing 8-4 Member function definitions for the Motor class.

```
Motor::Motor(int baseaddress):ParallelPort(baseaddress)
{
    Off();
}

void Motor::Off()
{
    WritePort0(0x00);
}
```

All motors can be controlled using either one or both H-bridge circuits on the interface board. Each H-bridge circuit uses four 'switches' with a corresponding logic control signal for each switch. Therefore, to control both H-bridges requires eight control signals, and so our programs will use the port at address BASE for this purpose.

A motor can be turned off by opening switches in the H-bridge. For DC motors, open all four switches, and for bipolar stepper motors, open all eight switches of the two H-bridges. In general, we can turn off any type of motor by opening all eight switches of the two H-bridges if we set all eight bits of the port at address BASE to zero. The `Off()` function does just this. It is not a virtual function since none of the derived classes require further specialisation of this function.

Note that the constructor calls the `Off()` function. This turns all transistors off in both H-bridges to eliminate any risk of a short-circuit path being generated between the power supply positive output and its ground return. This means that the user must allow the program to execute *before* connecting power to the interface board and motor.

Default Parameter Values for Functions

The constructor for the `Motor` class is an improvement on the coding of the constructors for the classes in previous chapters. These earlier chapters used a default constructor that took no parameters, *and* a constructor that took an argument of type integer for the `baseaddress` parameter. The constructor `Motor(int baseaddress=0x378)` takes a parameter to be passed to the base class constructor `ParallelPort(int baseaddress)` and it initialises the value of this parameter to 0x378 if no argument is provided when instantiating the `Motor` object. This allows this single constructor to perform the role of the two different constructors used in earlier chapters.

The default parameter value for `baseaddress` is shown in bold typeface in Listing 8-3. As implied by the word 'default', if no actual argument is passed when declaring an instance of an object, the actual argument will be taken as 0x378. Note that the default argument value is mentioned within the pair of parentheses of the constructor in the class definition (Listing 8-3). This is one way to specify default actual argument values. Another method for specifying default arguments is to include them in the function definition as shown in Figure 8-15. In this case we do not specify the default actual argument values in the class definition as is shown in Listing 8-5. Note that we must not specify default values in both places!

```
Motor::Motor(int baseaddress=0x378):AbstractMotor(),
    ParallelPort(baseaddress)
{
    Off();
}
```

Default actual argument for the parameter **baseaddress**

Figure 8-15 An alternative way of specifying default actual argument values.

Listing 8-5 `Motor` class definition with no default arguments specified inside.

```
class Motor : public AbstractMotor, public ParallelPort
{
    public:
        Motor(int baseaddress);
        void Off();
        virtual void Forward(int speed)=0;
        virtual void Reverse(int speed)=0;
        virtual void Brake()=0;
};
```

The `Motor` class constructor can be called in one of the two following forms as shown in Figure 8-16 and Figure 8-17:

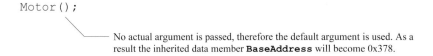

Figure 8-16 Calling the **Motor** class constructor without arguments.

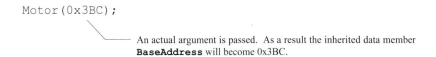

Figure 8-17 Calling the **Motor** class constructor with an actual argument.

The function Motor() being a constructor, can also be called in the manner shown in Figure 8-18.

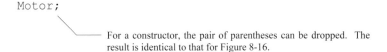

Figure 8-18 Calling a constructor with default actual arguments.

These examples show how one parameter can be given a default actual argument. If the function has more than one parameter, any of them can be assigned default actual arguments subject to the following condition. All parameters to the right of the parameter being considered must also have default actual argument values assigned. The two examples in Figure 8-19 show the valid and invalid declarations.

```
void AnyFunction(int x, int y=0, int z=1);    // correct
```

This situation is acceptable. Parameter **y** can be assigned a default actual argument value because parameter **z** is already assigned a default actual argument value.

```
void AnyFunction(int x, int y=0, int z);    // illegal
```

This situation is illegal. Parameter **y** cannot be assigned a default actual argument value because parameter **z** has not been assigned a default actual argument value.

Figure 8-19 Assigning multiple default actual arguments.

DCMotor *class*

The DCMotor class is at the bottom end of the class hierarchy shown in Figure 8-14. This class is developed to represent real DC motors. As such, the class is no longer an abstract class. The class definition for the DCMotor class is given in Listing 8-6.

Listing 8-6 DCMotor class.

```
class DCMotor : public Motor
{
    public:
        DCMotor(int baseaddress=0x378);
        virtual void Forward();
        virtual void Reverse();
        virtual void Brake();
};
```

The DCMotor class is derived from the Motor class. Although the DCMotor class looks small, it has inherited all member data and member functions from the Motor, AbstractMotor, and ParallelPort classes in the hierarchy. Note that there are no *pure* virtual functions in the class definition for the DCMotor class. Therefore, the class is not an abstract class, but a real class that can be used to instantiate objects once complete function definitions are provided.

To be able to define the member functions of the class, we must know how to drive the DC motor and connect it to the interface board. Figure 8-20 shows how a DC motor is connected using a H-bridge. We can use the data bits D0 to D3 of the port at BASE address to control the switches SW1 to SW4 as shown in Table 8-7.

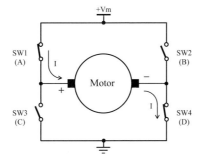

Figure 8-20 H-bridge connections to a DC Motor.

Table 8-7 DC Motor H-bridge to Parallel Port connections (Figure 8-20).

Switch No	H-bridge label on Interface Board	BASE Address Data Bit
SW1	A	D0
SW2	B	D1
SW3	C	D2
SW4	D	D3

A data-bit set to 1 closes the switch, whereas a 0 opens the switch. To drive the motor forward, the data bits D0 and D3 must be set to 1 and bits D1 and D2 must be set to 0. To drive the motor in reverse, we must do the opposite, i.e. data bits D1 and D2 must be set to 1 and data bits D0 and D3 must be set to 0.

To brake the motor, short-circuit the ends of the armature together. This can be done by either setting bits D0 and D1 to 1 (D2, D3 set to 0), or by setting bits D2 and D3 to 1 (D0, D1 set to 0). We will choose to short-circuit the armature to ground. For this, bits D2 and D3 will be set to 1 *and* bits D0 and D1 will be set to zero. Table 8-8 summarises the values to be written to the port to achieve forward motion, reverse motion and braking.

Table 8-8 Data to be written to the port to control the DC motor.

Operation	Data Bits[†]								Output to port at address BASE
	D7	D6	D5	D4	D3	D2	D1	D0	
Forward	x	x	x	x	1	0	0	1	0000 1001 = 0x09
Reverse	x	x	x	x	0	1	1	0	0000 0110 = 0x06
Brake	x	x	x	x	1	1	0	0	0000 1100 = 0x0C

† data bits represented as 'x' can take either 0 or 1. They are not connected to the H-bridge to drive a DC motor, and so do not have any effect. They have been taken to be 0 for bits D4 to D7.

WARNING

Do NOT at any time allow SW1 and SW3 to be turned on together at anytime. Likewise, do NOT at any time allow SW2 and SW4 to be turned on together. If these events were allowed to happen, the transistors and possibly the power supply will be damaged.

The DC motor is connected to the interface board as shown in Table 8-9 and Figure 8-21. The interface board is designed to drive +12V DC motors that require less than 1A of current. However, DC motors can be driven from external DC power supplies (up to a maximum 30V) as explained in the following note. Leave the interface board **unpowered** until the program is being executed since some bits of the port to be used may have previously been left in an on state.

> **NOTE**
>
> The two contacts on the 4-way terminal block labelled VM1 and VM1 GND are used to connect power to the H-bridge (VM2 and VM2 GND for the other H-bridge).
>
> The interface board is capable of providing +12V at up to 1A to drive motors. To use this power supply, connect Vm1 (or Vm2 as appropriate) via an insulated wire to the +12V of the power supply's 2-way terminal block (where the power-pack connects). A second insulated wire is used to connect the ground of the respective H-bridge Vm1 GND (or Vm2 GND) to the ground of the power supply. Most importantly, a wire needs to be connected across the two contacts of each 2-way terminal block. These 2-way terminal blocks have been provided to allow resistive drive schemes to be trialed by fitting resistors of suitable value inplace of the wires.
>
> Motors may need to be driven with voltage or current supply requirements differing from those of the power supply on the interface board (+12V, 1A). In this case, an external power supply up to a maximum of 30 V DC can be used to power the motors. **DO NOT** at any time attempt to connect **mains power** to the board.
>
> When using an external power supply, connect the positive contact of its output to the 4-way terminal block contact Vm1 (or Vm2 as appropriate). Also, connect the negative contact of the external power supply to the 4-way terminal block contact Vm1 GND (or Vm2 GND). The external power supply MUST NOT be connected to the 2-way terminal block of the interface board power supply.

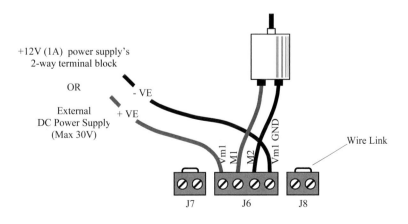

Figure 8-21 Connecting a DC power supply to power a DC motor.

Table 8-9 H-bridge to DC motor connections.

H Bridge 1	Connections
Vm1	Motor Power Supply +ve[*]
M1	Motor + terminal
M2	Motor - terminal
Vm1 GND	Motor Power Supply –ve[*]

Having determined all the connections, we can now proceed to implement a method to drive the motor, in particular to control its torque/speed. The most appropriate means to do this is by using Pulse Width Modulated (PWM) signals.

8.5.2 Generating Pulse Width Modulated Signals

As described earlier in Section 8.2.2, using pulse width modulation with wider pulse widths generates higher levels of motor speed/torque. In other words, the higher the *duty cycle*, the higher the motor speed. In our application the duty cycle is the proportion of time in a repetitive cycle that the motor will have power applied to it. This ratio is usually represented as a percentage. To control the speed we must control the duty cycle. Therefore, the design of the program should allow the duty cycle to be changed as desired. In general, exact timing is normally used to control duty cycle. In this chapter, we will not use exact time delays. Instead, duty cycle will be controlled by executing software loops. Different computers will have different execution speeds and therefore produce different PWM cycle times (frequencies), although the shape/duty cycle of the PWM signal will be the same.

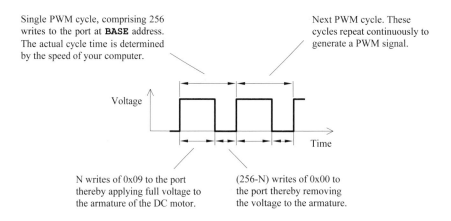

Single PWM cycle, comprising 256 writes to the port at **BASE** address. The actual cycle time is determined by the speed of your computer.

Next PWM cycle. These cycles repeat continuously to generate a PWM signal.

Voltage

Time

N writes of 0x09 to the port thereby applying full voltage to the armature of the DC motor.

(256-N) writes of 0x00 to the port thereby removing the voltage to the armature.

Figure 8-22 Generation of a PWM signal.

The PWM signal can be generated as follows. Suppose one cycle consists of 256 uninterrupted writes to the port. The cycle time is then equal to the time the

computer would take to complete 256 writes to the port. Depending on the desired
duty cycle, a portion of the 256 writes to the port can be used to send 0x09 to the
port (see Table 8-8) to turn the motor on in the forward direction. The remainder of
the writes can send 0x00 to the port to turn the motor off. For example, if we want
50% duty cycle, we will output 0x09 to the port for the first 128 writes and output
0x00 to the port for the remaining 128 writes. This is a means to control the speed
of the motor. A number between 0 and 255 (forming 256 different speed settings)
can be used to specify the speed. Figure 8-22 illustrates the PWM process.

The control of speed in the reverse direction can be achieved by using the same
procedure, except the data written to the port during the ON portion of the PWM
cycle is 0x06 instead of 0x09. In strict terms, the speed control our software uses is
known as *open-loop* speed control. When using open-loop speed control we do not
measure the speed of the motor nor apply corrective action for variations in speed.
In other words, we do not have a form of feedback to measure and correct actual
motor speed.

The member function definitions of the **DCMotor** class are shown in Listing 8-7.
Referring to the earlier Listing 8-6, it can be seen that a default actual argument is
specified for baseaddress. This is not evident in the function definitions in
Listing 8-7.

Listing 8-7 Member functions of the DCMotor class.

```
DCMotor::DCMotor(int baseaddress):Motor(baseaddress)
{
}


void DCMotor::Forward()
{
    int j;
    for(j = 0; j < GetSpeed(); j++)
        WritePort0(0x09);
    for(;j < 256; j++)
        WritePort0(0x00);
}


void DCMotor::Reverse()
{
    int j;
    for(j = 0; j < GetSpeed(); j++)
        WritePort0(0x06);
    for(;j < 256; j++)
        WritePort0(0x00);
}
```

```
void DCMotor::Brake()
{
    WritePort0(0x0C);
}
```

The `Forward()` function and the `Reverse()` functions are very similar. They differ in the actual argument passed to the `WritePort0()` function within the first `for` loop of each function. Therefore, we will only explain operation of the `Forward()` function.

The `Forward()` function operates using the functions shown in bold typeface below:

```
void DCMotor::Forward()
{
    for(int j = 0; j < GetSpeed(); j++)
        WritePort0(0x09);
    for(;j < 256; j++)
        WritePort0(0x00);
}
```

The highlighted functions are all member functions of various classes of the hierarchy. The function `GetSpeed()` is inherited from the abstract class `AbstractMotor` and the function `WritePort0()` is inherited from the `ParallelPort` class.

As described previously, the speed of the motor is determined by the duty cycle of the PWM signal. The duty cycle is controlled by the number of writes to the port at BASE address that allows the H-bridge to apply the power supply voltage to the motor. One cycle of the PWM signal comprises 256 writes to the port at address BASE. Counting from 0 to 255 (inclusive) will give us 256 numbers. Therefore, 255 specifies 100% duty cycle (full speed) and 0 specifies 0% duty cycle (zero speed).

When the program enters the first `for` loop, the inherited data member `Speed` must have a valid value, i.e. a value between 0 and 255. The first `for` loop initialises its loop count variable `j` to 0. It then uses the inherited `WritePort0()` function to write 0x09 to the port at BASE. This action will apply full voltage to the motor. The loop count variable `j` is then incremented using `j++`. Next, the test expression `j < GetSpeed()` is evaluated. The `Forward()` function of the DCMotor class has no *direct* access to the inherited data member `Speed` as it is a private member of the `AbstractMotor` class. Therefore, to obtain the value of `Speed`, the public function `GetSpeed()` of the `AbstractMotor` class must be used. Provided the test condition evaluates to true, the body of the first `for` loop will execute again. For example, if the value of `Speed` is 5, the function `WritePort0()` will be executed for `j` equal to 0, 1, 2, 3 and 4. When `j` is incremented to become 5, the test expression will evaluate to *false* and the first `for` loop will terminate. Control will then be transferred to the second `for` loop.

Note that the second `for` loop is unusual in that it does not have an initialising expression. The value of `j` is passed on to the next loop to complete the rest of the PWM cycle - which is why the loop count must not be re-initialised.

Using the values given in the example above, the value of `j` will be 5 upon entering the second `for` loop. The second `for` loop will run until the value of `j` exceeds 255. Each time the `for` loop executes 0x00 will be output to the BASE address part of the port, preventing voltage being supplied to the motor. When `j` becomes 256, the `for` loop will terminate, having completed one PWM cycle. This would produce a duty cycle of 5 ON writes out of 256 total writes in the cycle; i.e. 2% duty cycle.

Consecutive PWM cycles must be made to generate a continuous PWM signal. Therefore, the `Forward()` function must be called repeatedly until the user ceases to issue a forward command. This can be accomplished, for example, by implementing a `while` loop conditioned on `!kbhit()`. The `while` loop will continue provided the keyboard is not hit. The function `kbhit()` will return a non zero value when a key is pressed. When the logical negation operator (`!`) is placed in front, `!kbhit()` will return a non-zero value as long as a key is *not* pressed. Now the `while` loop will continue execution until a key is pressed. Each iteration of the `while` loop generates one PWM cycle and so the PWM signal will continue until a key is pressed. The `while` loop can be implemented outside the member functions, in say the `main()` function.

The `Reverse()` function operates in a similar manner to the `Forward()` function. The difference being that the first `for` loop writes a different value to the H-bridge to apply a voltage with reversed polarity to drive the motor in the opposite direction. The PWM signal is generated in the same manner as for the `Forward()` function.

The `Brake()` function calls the `WritePort0()` function once to close the switches of the H-bridge and short-circuit the DC motor armature to brake the motor.

In Section 8.6 we explain how the user can use this class to drive a DC motor.

StepperMotor *class*

The `StepperMotor` class must cater for several types of stepper motors and their two modes of operation. These modes and their acronyms are shown in Table 8-10. The class definition for the `StepperMotor` class is given in Listing 8-8.

Table 8-10 Acronyms for Stepper Motors.

Stepper Motor	Mode	Acronym
Unipolar	Full-Step	UPFS
Unipolar	Half-Step	UPHS
Bipolar	Full-Step	BPFS
Bipolar	Half-Step	BPHS

Listing 8-8 New data type MOTORTYPE and StepperMotor class.

```
enum MOTORTYPE {UPFS, UPHS, BPFS, BPHS};

class StepperMotor : public Motor
{
    private:
        MOTORTYPE MotorType;
        unsigned char Switching[8];
        int CycleIndex;
        int MaxIndex;

    public:
        StepperMotor(MOTORTYPE motortype = UPFS,
                            int baseaddress=0x378);
        virtual void Forward();
        virtual void Reverse();
        virtual void Brake();
};
```

A new programmer-defined data type named MOTORTYPE has been created at the top of this listing. This data type is termed an *enumerated data type* because all possible values the data type can be given are listed in the type declaration. These values are named enumeration constants and are always of type int. Use mnemonic identifiers that are meaningful to improve readability of the program.

Figure 8-23 Enumerated data types.

In the declaration shown in Figure 8-23, the new data type has the name
MOTORTYPE. The enumerated constants 0, 1, 2, and 3 are automatically assigned
to the mnemonic identifiers UPFS, UPHS, BPFS, and BPHS, respectively. We can
declare a variable of this type as follows:

```
MOTORTYPE NewMotor;
```

Note that if you use a C compiler (Not a C++ compiler) then the equivalent
declaration must be:

```
enum MOTORTYPE NewMotor;
```

NewMotor can take any value enumerated; i.e. the values listed between the two
braces. Had we specified that the mnemonic UPFS was equal to say 2 as shown
below, then UPHS, BPFS and BPHS would become 3, 4 and 5 respectively.

```
enum MOTORTYPE {UPFS = 2, UPHS, BPFS, BPHS};
```

Since all enumerated values are represented by integers, it is possible to carry out
integer arithmetic on the data type. For example:

```
MOTORTYPE NewBreed;
NewBreed = UPFS + BPFS; //valid but meaningless!
```

While this is possible, it has no meaning whatsoever for our purposes. However,
under certain circumstances integer arithmetic with enumerated types can be of
benefit when used with care.

Referring to Table 8-10, the enumerated data type shown in Listing 8-8 can
represent any of the stepper motor types we aim to control using our interface
board. We can now return to study the StepperMotor class definition in Listing
8-8. It has four private data members described as follows.

MOTORTYPE MotorType

The first member data is MotorType. Once initialised this data member will store
the type of stepper motor to be controlled.

unsigned char Switching[8]

This statement declares an array named Switching that has eight unsigned
char elements. The program will load this array with the unique switching
patterns for the type of stepper motor the user selects. These switching patterns are
specified ahead when the functions for the class are defined.

int CycleIndex

The member data CycleIndex will be used as a subscript to scan through the
array Switching and select the values of each of its single byte elements in the
proper sequence to step-wise drive a stepper motor. To drive a stepping motor
forward using full-steps, CycleIndex will have values 0, 1, 2, 3, 0, etc. To drive
the same motor in reverse direction CycleIndex will have values 0, 3, 2, 1, 0,
etc. Similarly, to drive a stepping motor forward using half-steps, CycleIndex

will have values 0, 1, 2, 3, 4, 5, 6, 7, 0, etc. When driving a stepping motor in reverse using half-steps, `CycleIndex` will have the values 0, 7, 6, 5, 4, 3, 2, 1, 0, etc.

int MaxIndex

This data member will be used to detect the position in the array `Switching` when a new cycle must recommence. As such `MaxIndex` stores the maximum value `CycleIndex` takes. For full-step control, only four elements of the `Switching` array are used, so the value of `MaxIndex` will be 4. However, all eight elements of the array `Switching` must be used for half-step control. In this case the value of `MaxIndex` will be 8.

The member function definitions of the `StepperMotor` class are given in Listing 8-9.

Listing 8-9 Member functions of the StepperMotor class.

```
StepperMotor::StepperMotor(MOTORTYPE motortype,
                int baseaddress): Motor(baseaddress)
{
    MotorType = motortype;
    CycleIndex = 0;

    switch(MotorType)
    {
        case UPFS: MaxIndex = 4;
                        Switching[0] = 0x11;
                        Switching[1] = 0x12;
                        Switching[2] = 0x22;
                        Switching[3] = 0x21;
                        break;
        case UPHS: MaxIndex = 8;
                        Switching[0] = 0x01;
                        Switching[1] = 0x11;
                        Switching[2] = 0x10;
                        Switching[3] = 0x12;
                        Switching[4] = 0x02;
                        Switching[5] = 0x22;
                        Switching[6] = 0x20;
                        Switching[7] = 0x21;
                        break;
        case BPFS: MaxIndex = 4;
                        Switching[0] = 0x99;
                        Switching[1] = 0x69;
                        Switching[2] = 0x66;
```

```
                                Switching[3] = 0x96;
                                break;
          case BPHS: MaxIndex = 8;
                                Switching[0] = 0x99;
                                Switching[1] = 0x09;
                                Switching[2] = 0x69;
                                Switching[3] = 0x60;
                                Switching[4] = 0x66;
                                Switching[5] = 0x06;
                                Switching[6] = 0x96;
                                Switching[7] = 0x90;

     }
}


void StepperMotor::Forward()
{
    if(++CycleIndex == MaxIndex) CycleIndex = 0;
    WritePort0(Switching[CycleIndex]);
    delay(259-GetSpeed());
}


void StepperMotor::Reverse()
{
    if(--CycleIndex == -1) CycleIndex = MaxIndex -1;
    WritePort0(Switching[CycleIndex]);
    delay(259-GetSpeed());
}


void StepperMotor::Brake()
{
    switch(MotorType)
    {
        case UPFS: case UPHS:
                        WritePort0(0x11);
                        break;
        case BPFS: case BPHS:
                        WritePort0(0x99);

     }
}
```

As usual, the constructor initialises the private data members of the class. If a motor type is specified in the actual argument for the parameter motortype, it will be assigned to the private data member MotorType. The CycleIndex is always initialised to 0. The MaxIndex is either set to 4 or to 8 depending on full-step control or half-step control. The array `Switching` is initialised depending on the motor type and the operating mode, explained as follows.

The `MotorType` is tested in a `switch` statement which is used to fill the array `Switching` with appropriate values for that combination of stepper motor and drive mode depending on the case value. The values that are written into the array `Switching` control the sequential switching of the H-bridges to drive a given stepper motor through its sequence of steps. Note that the stepper motors will use both H-bridges and have almost full supply voltage applied to their respective windings during each step. Their speed/position is controlled by the rate/number of steps. As such they do not use pulse width modulation for speed or torque control.

The `Forward()` function of the `StepperMotor` class operates in a similar manner as the `Reverse()` function, and so only the `Forward()` function is explained (see Figure 8-24).

```
void StepperMotor::Forward()
{
    if(++CycleIndex == MaxIndex)
        CycleIndex = 0;
    WritePort0(Switching[CycleIndex]);
    delay(257-GetSpeed());
}
```

CycleIndex is incremented and tested for exceeding its limit. If exceeded it will be reset.

Contents of the array **Switching** are written to the port, one element per step delay.

Speed is controlled by inserting a controlled delay between consecutive writes to the port.

Figure 8-24 Operation of the `Forward()` function.

The `Brake()` function implements a `switch` statement to apply braking appropriate to the motor type; a unipolar stepper motor or a bipolar stepper motor. Dynamic braking is applied uniquely for two of the four cases by closing and opening the required switches of the H-bridge. Note that in the configuration we have used for unipolar stepper motors, the armature cannot be short-circuited. Instead, voltage is not applied to the armature windings.

The H-bridge connections for Bipolar and Unipolar Stepper Motors are shown in Figure 8-25 and Figure 8-26 respectively. All the classes have been defined and the definitions of all member functions have been provided. The implementation of the class hierarchy is now complete. We now need to develop a `main()` function to make use of these classes.

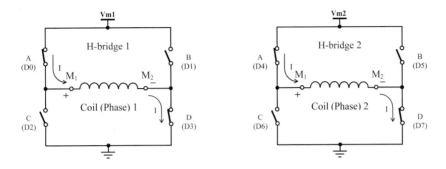

Figure 8-25 H-bridge connections for a Bipolar Stepper Motor.

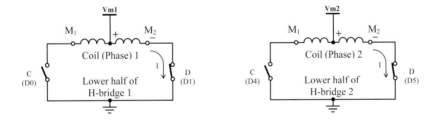

Figure 8-26 H-bridge connections for a Unipolar Stepper Motor.

8.6 Virtual Functions - Application

We are now ready to develop an application that makes use of virtual functions. This application will enable any of the types of motor accommodated in our class hierarchy developed earlier to be driven:

1. DC motors.
2. Unipolar stepper motors with dual-phase full-step control.
3. Unipolar stepper motors with half-step control.
4. Bipolar stepper motors with dual-phase full-step control.
5. Bipolar stepper motors with half-step control.

We will initially develop that part of the application that controls a 'motor' using the mechanism of virtual functions. Then we will add code to the program that allows a user to select a motor type to be driven.

The principal advantage of using virtual functions is the ability to write programs that can automatically bind a function to its associated object type at run-time. This allows us to write a very generic program. We start writing such a program by selecting a variable that can represent any of the objects in the hierarchy. The ideal variable will be associated with the `Motor` class; the base class for all the real motor classes in the hierarchy. A pointer (which is a variable) to this class can point to any of the objects of its derived classes as explained below.

C++   Base Class Pointers

A base class pointer can point to objects of its class or it can point to any objects of its derived classes. When we use a base class pointer to point to an object from a derived class and a virtual function is called through this pointer, the corresponding member function of that derived class will be selected and called.

Therefore, we can create a pointer to the `Motor` class as shown below and use it to point to any of the real motor classes derived from it:

```
Motor *MotorPtr;
```

Our particular program will carry out the following steps:

1. Drive the motor forward at a speed of 150 until a key is pressed.
2. Drive the motor forward at a speed of 255 until a key is pressed.
3. Reverse the motor at a speed of 150 until a key is pressed.
4. Reverse the motor at a speed of 255 until a key is pressed.
5. Stop the motor (braking).
6. Turn off power to the motor.

The code that implements these requirements is shown in Listing 8-10. Here we use the function `kbhit()` to detect a key press and the function `getch()` to clear the keyboard buffer after the key press.

Listing 8-10 Generic code to control 'a Motor'.

```
Motor *MotorPtr;

// Insert statements to choose a specific motor here

//..... Motor control part starts here .....
MotorPtr->SetSpeed(150);
while(!kbhit()) MotorPtr->Forward();
getch(); // clear keyboard buffer

MotorPtr->SetSpeed(255);
while(!kbhit()) MotorPtr->Forward();
getch();

MotorPtr->SetSpeed(150);
```

```
while(!kbhit()) MotorPtr->Reverse();
getch();

MotorPtr->SetSpeed(255);
while(!kbhit()) MotorPtr->Reverse();
getch();

cout << endl << "   Braking Applied!" << endl;
while(!kbhit()) MotorPtr->Brake();
getch();

MotorPtr->Off();
//..... Motor control part ends here .....
```

The user must be given a list of motor types to be able to choose a motor to operate. The code to implement this task is given in Listing 8-11.

Listing 8-11 Statements to display a menu of Motors on the screen.

```
int Selection;

clrscr();

cout << endl << "   MOTOR MENU";
cout << endl << "   ~~~~~~~~~~" << endl;
cout << "   1  DC Motor" << endl;
cout << "   2  UPFS" << endl;
cout << "   3  UPHS" << endl;
cout << "   4  BPFS" << endl;
cout << "   5  BPHS" << endl;
cout << "   6  QUIT" << endl;
cout << endl;
cout << "   Select the MOTOR Number: ";

cin >> Selection;
```

Having selected the motor, we need to use dynamic memory allocation to create the object type that corresponds to the motor selected. Listing 8-12 shows the dynamic memory allocation segment of the program.

Listing 8-12 Dynamic memory allocation for the selected Motor.

```
switch(Selection)
{
```

```
case 1: MotorPtr = new DCMotor;
    break;
case 2: MotorPtr = new StepperMotor(UPFS);
    break;
case 3: MotorPtr = new StepperMotor(UPHS);
    break;
case 4: MotorPtr = new StepperMotor(BPFS);
    break;
case 5: MotorPtr = new StepperMotor(BPHS);
    break;
case 6: return;

default: cout << endl;
    cout << "   Unspecified Motor type....";
    cout << " PRESS a key to END Program!";
    getch();
    exit(1); // Exits the program
}

if(MotorPtr == NULL)
{
    cout << "Memory allocation failed " << endl;
    getch();
    exit(1);
}
```

At the end of this program segment, the pointer `MotorPtr` should be initialised to point to a valid object in memory. If not, the program will exit because a matching motor type could not be found, or memory allocation has failed. Once the pointer is initialised, the program segment given in Listing 8-10 can be executed. The complete `main()` function is given in Listing 8-13.

Listing 8-13 The `main()` function to control 'a Motor'.

```
void main()
{
    Motor *MotorPtr;
    int Selection;

    clrscr();

    cout << endl << "    MOTOR MENU";
    cout << endl << "    ~~~~~~~~~~" << endl;
    cout << "   1  DC Motor" << endl;
```

```cpp
cout << "   2  UPFS" << endl;
cout << "   3  UPHS" << endl;
cout << "   4  BPFS" << endl;
cout << "   5  BPHS" << endl;
cout << "   6  QUIT" << endl;
cout << endl;
cout << "   Select the MOTOR Number: ";

cin >> Selection;

switch(Selection)
{
case 1: MotorPtr = new DCMotor;
    break;
case 2: MotorPtr = new StepperMotor(UPFS);
    break;
case 3: MotorPtr = new StepperMotor(UPHS);
    break;
case 4: MotorPtr = new StepperMotor(BPFS);
    break;
case 5: MotorPtr = new StepperMotor(BPHS);
    break;
case 6: return;

default: cout << endl;
    cout << "   Unspecified Motor type....";
    cout << " PRESS a key to END Program!";
    getch();
    exit(1); // Exits the program
}

if(MotorPtr == NULL)
{
    cout << "Memory allocation failed " << endl;
    getch();
    exit(1);
}
cout << "**********************************" << endl;
cout << "* CONNECT BOARD POWER SUPPLY NOW *" << endl;
cout << "**********************************" << endl;
cout << endl;
cout << "   After connecting power,";
cout << " press a key to continue " << endl;
getch();
```

```
        cout << "   Keypress changes Speed/Rotation (&
Braking)." << endl;

        //..... Motor control part starts here .....
        MotorPtr->SetSpeed(150);
        while(!kbhit()) MotorPtr->Forward();
        getch(); // clear keyboard buffer

        MotorPtr->SetSpeed(255);
        while(!kbhit()) MotorPtr->Forward();
        getch();

        MotorPtr->SetSpeed(150);
        while(!kbhit()) MotorPtr->Reverse();
        getch();

        MotorPtr->SetSpeed(255);
        while(!kbhit()) MotorPtr->Reverse();
        getch();

        cout << endl << "   Braking Applied!" << endl;
        while(!kbhit()) MotorPtr->Brake();
        getch();

        MotorPtr->Off();
        //..... Motor control part ends here .....
        // Free the memory occupied by the 'Motor' object
        delete MotorPtr;
}
```

The compiler should have seen the definition of the entire class hierarchy and the definition of all member functions when it comes time to compile the main() function in Listing 8-13. We will defer explaining the complete program until virtual destructors have been discussed.

8.6.1 Virtual Destructors

The destructor of the class is called indirectly whenever the delete operator is used on an object of the class as discussed in Section 5.3.8. Since we have not declared any destructors in our motor class hierarchy, the only destructors available to our classes are the default destructors generated by the compiler for each class in the hierarchy. The selected motor has a corresponding 'motor' object instantiated in the body of the program's switch statement (Listing 8-13). When the program is finished using the 'motor' object, it frees the memory occupied by the 'motor' object using the following statement:

```
delete MotorPtr;
```

Since `MotorPtr` is a pointer to the abstract base class `Motor`, the previous statement will call the default destructor of the `Motor` class and will only de-allocate the space occupied by a `Motor` class object. The `delete` statement will not free the memory space occupied by the actual object in use (such as `DCMotor`), thereby generating a memory leak. We can demonstrate this event using an example program that has a simple class structure and a simple `main()` function as shown in Listing 8-14. Note: to further simplify the example, the `Base` and `Derived` classes do not have data members. We have also included `cout` statements within the body of the two destructors to show when each destructor is called. If the destructors did not have these `cout` statements, they would be identical to the default destructors generated by the compiler.

Listing 8-14 Use of non-virtual destructors.

```cpp
#include <conio.h>
#include <iostream.h>

class Base
{
    public:
        Base(){}
        ~Base()
        {
            cout << "Base type object deleted" << endl;
        }
};

class Derived : public Base
{
    public:
        Derived(){};
        ~Derived()
        {
            cout << "Derived type object deleted " << endl;
        }
};

void main()
{
    Base *BasePtr;

    BasePtr = new Derived;    // BasePtr points to an object
                              // of type Derived.
```

```
    delete BasePtr;    // Deletes object
}
```

The pointer identifier `BasePtr` is declared to be of type `Base`. However, it is used to point to a dynamically allocated class object that is of type `Derived`. We use the following statement with the intention of deleting the dynamically allocated object of class `Derived`:

```
delete BasePtr;
```

We would expect this statement to call the destructor of the derived class. However, the following message is displayed when this program is executed:

```
Base type object deleted
```

This indicates that the destructor of the `Derived` class has not been called as intended to destroy the dynamically allocated `Derived` type object. We change the program to operate correctly by making the destructor `~Base()` virtual. The modified program listing is shown in Listing 8-15.

Listing 8-15 Use of virtual destructors.

```
#include <conio.h>
#include <iostream.h>

class Base
{
    public:
        Base(){}
        virtual ~Base()
        {
            cout << "Base type object deleted" << endl;
        }
};

class Derived : public Base
{
    public:
        Derived(){};
        ~Derived()
        {
            cout << "Derived type object deleted " << endl;
        }
};

void main()
```

```
{
    Base *BasePtr;

    BasePtr = new Derived;    // BasePtr points to an object
                              // of type Derived.
    delete BasePtr;    // Deletes object
}
```

Note that in Listing 8-15 the keyword `virtual` is added in front of the destructor name `~Base()`. This provides the link to all virtual destructors down to the next level of the class hierarchy so the correct destructors will be called. If you run this program you will see the following printed on the screen:

```
Derived type object deleted
Base type object deleted
```

This demonstrates that the `delete` statement has called the `Derived` class destructor and the `Base` class destructor, properly relinquishing the memory allocated for the `Derived` and `Base` class objects. Note: when an object of a derived class is instantiated, the constructor function of the base class is called first followed by a call to the constructor of the derived class. The derived class inherits the members of the base class that are instantiated in this manner. Therefore, it is important to include virtual destructors so that any memory allocation from within the derived class and its base class is properly relinquished.

Now we can return our attention to the motor control program. To allow the `delete` statements to properly de-allocate the dynamically allocated objects, we must provide a set of destructors; one destructor for each class of the motor class hierarchy, and we must make them virtual destructors. The bodies of these destructors can be empty. We simply need to establish a network of virtual destructors throughout the class hierarchy so that proper late binding will take place for the destructors. The modified class definitions are given in Listing 8-16 through to Listing 8-18.

C++   Virtual Destructor Names

In a class hierarchy all virtual functions must have the same function signature; i.e. they must have the same function name, same number of formal arguments and the same types of formal arguments in each virtual function. However, virtual destructors have different names throughout the hierarchy. Despite having different destructor function signatures, late binding will enable the correct set of destructors to be deployed in response to a `delete` statement.

Listing 8-16 AbstractMotor class with virtual destructor.

```
class AbstractMotor
{
    private:
        int Speed;

    public:
        AbstractMotor();
        void SetSpeed(int speed);
        int GetSpeed();
        virtual void Off()=0;
        virtual void Forward()=0;
        virtual void Reverse()=0;
        virtual void Brake()=0;
        virtual ~AbstractMotor(){}
};
```

Listing 8-17 ParallelPort class with virtual destructor.

```
class ParallelPort
{
    private:
        unsigned int BaseAddress;
        unsigned char InDataPort1;

    public:
        ParallelPort();
        ParallelPort(int baseaddress);
        void WritePort0(unsigned char data);
        void WritePort2(unsigned char data);
        unsigned char ReadPort1();
        virtual ~ParallelPort(){}
};
```

Listing 8-18 Motor class with virtual destructor.

```
class Motor : public AbstractMotor, public ParallelPort
{
    public:
        Motor(int baseaddress=0x378);
        void Off();
        virtual void Forward()=0;
```

```
          virtual void Reverse()=0;
          virtual void Brake()=0;
          virtual ~Motor(){}
};
```

Now that we have proper destructors in our classes, we can write a complete program using virtual functions to control a motor and free memory as intended. Such a program is shown in Listing 8-19. Note that virtual destructors are not added to the `DCMotor` class or the `StepperMotor` class since these two classes are the terminal classes of the hierarchy. However, the program would still function properly if virtual destructors had been added to these two classes.

> **NOTE**
>
> Ensure the interface board is **unpowered** before connecting any type of motor. This needs to be done for the following reason.
>
> Before first running the program, the port controlling the motor will not be under control of the program and may be in an unknown state. This unknown state can be such that the port's logic states would drive the transistors to short-circuit the motor's power supply (damaging the transistors and possibly the power supply). The program instructs the user to apply power to the board once it has set these bits to a safe state. When the program ends, it sets the used bits of the port to a safe state to prevent any damage to the transistors or the power supply.
>
> Connect a DC motor to the interface board as given in Table 8-7 and Table 8-9. Stepper motors are connected to the interface board as shown in Figure 8-25 and Figure 8-26.
>
> If the motor does not drive as expected, first check for incorrect connections.

Listing 8-19 Complete program to control 'a Motor' using Virtual Functions.

```
// **************************************************
// Program to operate a Motor using Virtual Functions.
// **************************************************
#include <dos.h>
#include <conio.h>
#include <stdio.h>
#include <stdlib.h>
#include <iostream.h>

class ParallelPort
{
    private:
```

```cpp
        unsigned int BaseAddress;
        unsigned char InDataPort1;

    public:
        ParallelPort();
        ParallelPort(int baseaddress);
        void WritePort0(unsigned char data);
        void WritePort2(unsigned char data);
        unsigned char ReadPort1();
        virtual ~ParallelPort(){}
};

ParallelPort::ParallelPort()
{
    BaseAddress = 0x378;
    InDataPort1 = 0;
}

ParallelPort::ParallelPort(int baseaddress)
{
    BaseAddress = baseaddress;
    InDataPort1 = 0;
}

void ParallelPort::WritePort0(unsigned char data)
{
    outportb(BaseAddress,data);
}

void ParallelPort::WritePort2(unsigned char data)
{
    outportb(BaseAddress+2,data ^ 0x0B);
}

unsigned char ParallelPort::ReadPort1()
{
    InDataPort1 = inportb(BaseAddress+1);
// Inverting Most significant bit to compensate
// for internal inversion by printer port hardware.
    InDataPort1 ^= 0x80;
// Filter to clear unused data bits D0, D1 and D2 to zero.
    InDataPort1 &= 0xF8;
    return InDataPort1;
}
```

```cpp
class AbstractMotor
{
   private:
        int Speed;

    public:
        AbstractMotor();
        void SetSpeed(int speed);
        int GetSpeed();
        virtual void Off()=0;
        virtual void Forward()=0;
        virtual void Reverse()=0;
        virtual void Brake()=0;
        virtual ~AbstractMotor(){}
};

AbstractMotor::AbstractMotor()
{
   Speed =0;
}

void AbstractMotor::SetSpeed(int speed)
{
   Speed = speed;
   if(Speed > 255) Speed = 255; // Limit upper value
   if(Speed < 0) Speed = 0; // Limit lower value
}

int AbstractMotor::GetSpeed()
{
   return Speed;
}

class Motor : public AbstractMotor, public ParallelPort
{
    public:
        Motor(int baseaddress=0x378);
        void Off();
        virtual void Forward()=0;
        virtual void Reverse()=0;
        virtual void Brake()=0;
        virtual ~Motor(){}
};

Motor::Motor(int baseaddress):ParallelPort(baseaddress)
```

```
{
    Off();
}

void Motor::Off()
{
    WritePort0(0x00);
}

class DCMotor : public Motor
{
    public:
        DCMotor(int baseaddress=0x378);
        virtual void Forward();
        virtual void Reverse();
        virtual void Brake();
};

DCMotor::DCMotor(int baseaddress):Motor(baseaddress)
{
}

void DCMotor::Forward()
{
    int j;
    for(j = 0; j < GetSpeed(); j++)
        WritePort0(0x09);
    for(;j < 256; j++)
        WritePort0(0x00);
}

void DCMotor::Reverse()
{
    int j;
    for(j = 0; j < GetSpeed(); j++)
        WritePort0(0x06);
    for(;j < 256; j++)
        WritePort0(0x00);
}

void DCMotor::Brake()
{
    WritePort0(0x0C);
}
```

```
enum MOTORTYPE {UPFS, UPHS, BPFS, BPHS};

class StepperMotor : public Motor
{
    private:
        MOTORTYPE MotorType;
        unsigned char Switching[8];
        int CycleIndex;
        int MaxIndex;

    public:
        StepperMotor(MOTORTYPE motortype = UPFS,
                            int baseaddress=0x378);
        virtual void Forward();
        virtual void Reverse();
        virtual void Brake();
};

StepperMotor::StepperMotor(MOTORTYPE motortype,
                int baseaddress): Motor(baseaddress)
{
    MotorType = motortype;
    CycleIndex = 0;

    switch(MotorType)
    {
        case UPFS: MaxIndex = 4;
                        Switching[0] = 0x11;
                        Switching[1] = 0x12;
                        Switching[2] = 0x22;
                        Switching[3] = 0x21;
                        break;
        case UPHS: MaxIndex = 8;
                        Switching[0] = 0x01;
                        Switching[1] = 0x11;
                        Switching[2] = 0x10;
                        Switching[3] = 0x12;
                        Switching[4] = 0x02;
                        Switching[5] = 0x22;
                        Switching[6] = 0x20;
                        Switching[7] = 0x21;
                        break;
        case BPFS: MaxIndex = 4;
                        Switching[0] = 0x99;
                        Switching[1] = 0x69;
```

```
                                    Switching[2] = 0x66;
                                    Switching[3] = 0x96;
                                    break;
                case BPHS: MaxIndex = 8;
                                    Switching[0] = 0x99;
                                    Switching[1] = 0x09;
                                    Switching[2] = 0x69;
                                    Switching[3] = 0x60;
                                    Switching[4] = 0x66;
                                    Switching[5] = 0x06;
                                    Switching[6] = 0x96;
                                    Switching[7] = 0x90;
        }
    }


    void StepperMotor::Forward()
    {
        if(++CycleIndex == MaxIndex) CycleIndex = 0;
        WritePort0(Switching[CycleIndex]);
        delay(259-GetSpeed());
    }


    void StepperMotor::Reverse()
    {
        if(--CycleIndex == -1) CycleIndex = MaxIndex -1;
        WritePort0(Switching[CycleIndex]);
        delay(259-GetSpeed());
    }


    void StepperMotor::Brake()
    {
        switch(MotorType)
        {
            case UPFS: case UPHS:
                        WritePort0(0x11);
                        break;
            case BPFS: case BPHS:
                        WritePort0(0x99);
        }
    }


    void main()
    {
        Motor *MotorPtr;
        int Selection;
```

```
clrscr();

cout << endl << "    MOTOR MENU";
cout << endl << "    ~~~~~~~~~~" << endl;
cout << "   1  DC Motor" << endl;
cout << "   2  UPFS" << endl;
cout << "   3  UPHS" << endl;
cout << "   4  BPFS" << endl;
cout << "   5  BPHS" << endl;
cout << "   6  QUIT" << endl;
cout << endl;
cout << "   Select the MOTOR Number: ";

cin >> Selection;

switch(Selection)
{
case 1: MotorPtr = new DCMotor;
    break;
case 2: MotorPtr = new StepperMotor(UPFS);
    break;
case 3: MotorPtr = new StepperMotor(UPHS);
    break;
case 4: MotorPtr = new StepperMotor(BPFS);
    break;
case 5: MotorPtr = new StepperMotor(BPHS);
    break;
case 6: return;

default: cout << endl;
    cout << "   Unspecified Motor type....";
    cout << " PRESS a key to END Program!";
    getch();
    exit(1); // Exits the program
}

if(MotorPtr == NULL)
{
    cout << "Memory allocation failed " << endl;
    getch();
    exit(1);
}
cout << "**********************************" << endl;
cout << "* CONNECT BOARD POWER SUPPLY NOW *" << endl;
```

```
cout << "*********************************" << endl;
cout << endl;
cout << "   After connecting power,";
cout << " press a key to continue " << endl;
getch();
cout << endl;
cout << "   Keypress changes Speed/Rotation";
cout << " (& Braking)." << endl;

//..... Motor control part starts here .....
MotorPtr->SetSpeed(150);
while(!kbhit()) MotorPtr->Forward();
getch(); // clear keyboard buffer

MotorPtr->SetSpeed(220);
while(!kbhit()) MotorPtr->Forward();
getch();

MotorPtr->SetSpeed(250);
while(!kbhit()) MotorPtr->Reverse();
getch();

MotorPtr->SetSpeed(255);
while(!kbhit()) MotorPtr->Reverse();
getch();

cout << endl << "   Braking Applied!" << endl;
while(!kbhit()) MotorPtr->Brake();
getch();

MotorPtr->Off();
//..... Motor control part ends here .....
// Free the memory occupied by the 'Motor' object
delete MotorPtr;
}
```

Don't be overwhelmed by the length of this program. In general, programmers create header files and library files to hide all the code that is shown ahead of the `main()` function. Had we done the same, the program to control any of the motors in our list would be the size of the main function.

Observe the motor's behaviour without the effect of dynamic braking by commenting out the call to the `Brake()` function in Listing 8-19. This is best seen if the motor shaft has some inertial load connected to it.

If we did NOT use virtual functions, then the code to control the motors (shown in

Listing 8-10) will need to be included within each `case` of the `switch` statement in the `main()` function of Listing 8-13. The pointer used within each `case` will need to be declared to point specifically to an object for that particular `case`, and all functions will be linked at programming time (early binding). A `main()` function equivalent to that of Listing 8-13, but without the use of virtual functions is given in Listing 8-20.

Listing 8-20 `main()` function WITHOUT using virtual functions.

```
void main()
{
    int Selection;
    DCMotor* DCMotorPtr;
    StepperMotor* UPFSStepperMotorPtr;
    StepperMotor* UPHSStepperMotorPtr;
    StepperMotor* BPFSStepperMotorPtr;
    StepperMotor* BPHSStepperMotorPtr;

    clrscr();

    cout << endl << "    MOTOR MENU";
    cout << endl << "    ~~~~~~~~~~" << endl;
    cout << "    1   DC Motor" << endl;
    cout << "    2   UPFS" << endl;
    cout << "    3   UPHS" << endl;
    cout << "    4   BPFS" << endl;
    cout << "    5   BPHS" << endl;
    cout << "    6   QUIT" << endl;
    cout << endl;
    cout << "    Select the MOTOR Number: ";

    cin >> Selection;
    cout << endl;

    switch(Selection)
    {
    case 1: DCMotorPtr = new DCMotor;
        if(DCMotorPtr == NULL)
        {
            cout << "Memory allocation failed " << endl;
            exit(1);
        }
        cout << "***********************************" <<
endl;
```

```cpp
        cout << "* CONNECT BOARD POWER SUPPLY NOW *" <<
endl;
        cout << "*********************************" <<
endl;   cout << endl;
        cout << "   After connecting power,";
        cout << " press a key to continue " << endl;
        getch();

        cout << "   KEYPRESS changes SPEED/ROTATION (&
Braking)." << endl;
        // DCMotor control part starts here
        DCMotorPtr->SetSpeed(150);
        while(!kbhit()) DCMotorPtr->Forward();
        getch(); // clear keyboard buffer

        DCMotorPtr->SetSpeed(255);
        while(!kbhit()) DCMotorPtr->Forward();
        getch();

        DCMotorPtr->SetSpeed(150);
        while(!kbhit()) DCMotorPtr->Reverse();
        getch();

        DCMotorPtr->SetSpeed(255);
        while(!kbhit()) DCMotorPtr->Reverse();
        getch();

        while(!kbhit()) DCMotorPtr->Brake();

        DCMotorPtr->Off();
        // DCMotor control part ends here
        // Release memory occupied by the DCMotor object
        delete DCMotorPtr;
        break;

    case 2: UPFSStepperMotorPtr = new StepperMotor(UPFS);
        if(UPFSStepperMotorPtr == NULL)
        {
            cout << "Memory allocation failed " << endl;
            exit(1);
        }
        cout << "*********************************" <<
endl;
        cout << "* CONNECT BOARD POWER SUPPLY NOW *" <<
endl;
        cout << "*********************************" <<
```

```
endl;   cout << endl;
        cout << "   After connecting power,";
        cout << " press a key to continue " << endl;
        getch();
        cout << "   KEYPRESS changes SPEED/ROTATION (&
Braking)." << endl;
        // UPFS Motor control part starts here
        UPFSStepperMotorPtr->SetSpeed(150);
        while(!kbhit()) UPFSStepperMotorPtr->Forward();
        getch(); // clear keyboard buffer

        UPFSStepperMotorPtr->SetSpeed(220);
        while(!kbhit()) UPFSStepperMotorPtr->Forward();
        getch();

        UPFSStepperMotorPtr->SetSpeed(250);
        while(!kbhit()) UPFSStepperMotorPtr->Reverse();
        getch();

        UPFSStepperMotorPtr->SetSpeed(255);
        while(!kbhit()) UPFSStepperMotorPtr->Reverse();
        getch();

        while(!kbhit()) UPFSStepperMotorPtr->Brake();

        UPFSStepperMotorPtr->Off();
        // UPFS Motor control part ends here
        // Release memory occupied by the DCMotor object
        delete UPFSStepperMotorPtr;
        break;

    case 3: UPHSStepperMotorPtr = new StepperMotor(UPHS);
        if(UPHSStepperMotorPtr == NULL)
        {
            cout << "Memory allocation failed " << endl;
            exit(1);
        }
        cout << "*********************************" <<
endl;
        cout << "* CONNECT BOARD POWER SUPPLY NOW *" <<
endl;
        cout << "*********************************" <<
endl;    cout << endl;
        cout << "   After connecting power,";
        cout << " press a key to continue " << endl;
```

```
        getch();
        cout << "   KEYPRESS changes SPEED/ROTATION (&
Braking)." << endl;
        // UPHS Motor control part starts here
        UPHSStepperMotorPtr->SetSpeed(150);
        while(!kbhit()) UPHSStepperMotorPtr->Forward();
        getch(); // clear keyboard buffer

        UPHSStepperMotorPtr->SetSpeed(220);
        while(!kbhit()) UPHSStepperMotorPtr->Forward();
        getch();

        UPHSStepperMotorPtr->SetSpeed(250);
        while(!kbhit()) UPHSStepperMotorPtr->Reverse();
        getch();

        UPHSStepperMotorPtr->SetSpeed(255);
        while(!kbhit()) UPHSStepperMotorPtr->Reverse();
        getch();

        while(!kbhit()) UPHSStepperMotorPtr->Brake();

        UPHSStepperMotorPtr->Off();
        // UPHS Motor control part ends here
        // Release memory occupied by the DCMotor object
        delete UPHSStepperMotorPtr;
        break;

    case 4: BPFSStepperMotorPtr = new StepperMotor(BPFS);
        if(BPFSStepperMotorPtr == NULL)
        {
            cout << "Memory allocation failed " << endl;
            exit(1);
        }
        cout << "**********************************" <<
endl;
        cout << "* CONNECT BOARD POWER SUPPLY NOW *" <<
endl;
        cout << "**********************************" <<
endl;    cout << endl;
        cout << "   After connecting power,";
        cout << " press a key to continue " << endl;
        getch();
        cout << "   KEYPRESS changes SPEED/ROTATION (&
Braking)." << endl;
        // BPFS Motor control part starts here
```

```cpp
        BPFSStepperMotorPtr->SetSpeed(150);
        while(!kbhit()) BPFSStepperMotorPtr->Forward();
        getch(); // clear keyboard buffer

        BPFSStepperMotorPtr->SetSpeed(220);
        while(!kbhit()) BPFSStepperMotorPtr->Forward();
        getch();

        BPFSStepperMotorPtr->SetSpeed(250);
        while(!kbhit()) BPFSStepperMotorPtr->Reverse();
        getch();

        BPFSStepperMotorPtr->SetSpeed(255);
        while(!kbhit()) BPFSStepperMotorPtr->Reverse();
        getch();

        while(!kbhit()) BPFSStepperMotorPtr->Brake();

        BPFSStepperMotorPtr->Off();
        // BPFS Motor control part ends here
        // Release memory occupied by the DCMotor object
        delete BPFSStepperMotorPtr;
        break;

    case 5: BPHSStepperMotorPtr = new StepperMotor(BPHS);
        if(BPHSStepperMotorPtr == NULL)
        {
            cout << "Memory allocation failed " << endl;
            exit(1);
        }
        cout << "**********************************" <<
endl;
        cout << "* CONNECT BOARD POWER SUPPLY NOW *" <<
endl;
        cout << "**********************************" <<
endl;   cout << endl;
        cout << "   After connecting power,";
        cout << " press a key to continue " << endl;
        getch();
        cout << "   KEYPRESS changes SPEED/ROTATION (&
Braking)." << endl;
        // BPHS Motor control part starts here
        BPHSStepperMotorPtr->SetSpeed(150);
        while(!kbhit()) BPHSStepperMotorPtr->Forward();
        getch(); // clear keyboard buffer
```

```
        BPHSStepperMotorPtr->SetSpeed(220);
        while(!kbhit()) BPHSStepperMotorPtr->Forward();
        getch();

        BPHSStepperMotorPtr->SetSpeed(250);
        while(!kbhit()) BPHSStepperMotorPtr->Reverse();
        getch();

        BPHSStepperMotorPtr->SetSpeed(255);
        while(!kbhit()) BPHSStepperMotorPtr->Reverse();
        getch();

        while(!kbhit()) BPHSStepperMotorPtr->Brake();

        BPHSStepperMotorPtr->Off();
        // BPHS Motor control part ends here
        // Release memory occupied by the DCMotor object
        delete BPHSStepperMotorPtr;
        break;

    case 6: return;

    default: cout << endl;
        cout << "   Unspecified Motor type....";
        cout << " PRESS a key to END Program!";
        getch();
        exit(1); // Exits the program
    }
}
```

As you can see, coding without the use of virtual functions can be quite inefficient. Note: the delete operator was used in Listing 8-10, Listing 8-13, and Listing 8-20 to relinquish the dynamically allocated memory that stored the 'motor' object.

8.7 Keyboard Controls

We can enhance control of the motor by making use of the PC keyboard. To do this the program must be able to detect each press of the keys used for motor control purposes. These key presses can be detected using several methods. The easiest method is to use the getch() or the getche() functions. Another option is to use the library function kbhit().

The functions getch() and getche() as their names imply, can be used to read the character corresponding to the key pressed. The function getche() has the extra capability of being able to display the character on-screen as it is read. This is known as echoing to the screen (the added 'e' stands for 'echo'). However, there is a disadvantage to using these two functions being that they wait for a key press. Execution of a getch() function will not allow following statements to be executed until a key is pressed. This delay will prevent us from generating the fast changing signals needed for motor control.

The kbhit() function operates differently. It does not wait for a key press. It checks if a key has been pressed. If a key has been pressed, the function will return true, if not it will return 0, meaning false. Because it does not wait for a key press, the program will execute continuously as intended. Note that kbhit() does not clear the keyboard buffer (ready for the next time it needs to be used).

The built-in library also provides another function named bioskey() used to detect normal keys with an ASCII code (see Appendix B) or *extended keys* such as function keys, INS, DEL keys etc. Extended keys are identified by a two-byte key code whereas normal keys are identified by a one-byte key code.

The function bioskey() can operate in different modes depending on the parameter passed to it. It uses one integer parameter whose value determines if it will detect a key press like the kbhit() function or read the pressed key. Note that this function *does* clear the keyboard buffer when used to read a key.

We can read a special location in memory using the peekb() function to detect keys including Right Shift, Left Shift, Ctrl, Alt, Caps lock, Scroll lock, etc. Note that the peekb() function cannot detect some keys such as the arrow keys.

The byte read from this memory location has flags (bits, shown in Figure 8-27) that are used to indicate which keys are pressed at any given time. The peekb() function is the fastest to execute and will generate minimum disruption to the continuous execution of the program. Therefore, using peekb() will enable smooth operation of the motors. If slower functions were used to read the keyboard, motor control could be erratic.

We will use the peekb() function to add keyboard control to our program using the following combinations of keys:

- Pressing the Ctrl key will drive the motor forward.
- Pressing the Alt key will drive the motor in reverse.
- Pressing the Right Shift key will increase speed.
- Pressing the Left Shift key will decrease speed.
- Pressing both Left Shift and Right Shift keys brakes the motor.
- Pressing No keys will switch motor power off.
- Pressing the Insert key will end the program.

The status of the Shift keys, Control key, Alt key and the Insert key can be determined by reading a special memory location using peekb(). This 8-bit memory location is at the *segment:offset* address of 0x40:0x17. When the respective key is pressed, its bit will be 1. For an explanation of segment:offset addressing, see *Technical Reference: Personal Computer AT by IBM Corporation.*

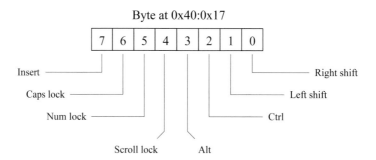

Figure 8-27 Memory byte at 0x40:0x17 used to store key status.

A `while` loop is implemented in the program segment shown in Listing 8-21 which will run continuously provided the variable `Quit` is 0. This listing can be viewed as a replacement for Listing 8-10.

Listing 8-21 Motor control using a keyboard.

```
Motor *MotorPtr;

int Quit = 0;
unsigned char key = 0;
int SpeedLock = 1;

while(!Quit)
{
  key = peekb(0x40,0x17); // Read control key byte.
  if(key & 0x80) // Test Insert key ON (MSBit '1').
    Quit = 1;  // Exit the program.

  else
  {
    // If both shift keys are released SpeedLock is
    // released

    if(!(key & 0x01) && !(key & 0x02))
        SpeedLock = 0;

    key &= 0x0F; // Filter out bits corresponding to
```

```
                    // just SHIFT, ALT & CTRL keys.

switch(key)
 {
    case 0x04 :
    MotorPtr->Forward();
    break;

    case 0x08 :
    MotorPtr->Reverse();
    break;

    case 0x01 :
    if(!SpeedLock)
    {
        MotorPtr->SetSpeed(MotorPtr->GetSpeed()+4);
        SpeedLock = 1;
    }
    break;

    case 0x02 :
    if(!SpeedLock)
    {
        MotorPtr->SetSpeed(MotorPtr->GetSpeed()-4);
        SpeedLock = 1;
    }
    break;

    case 0x03 :
    MotorPtr->Brake();
    break;

    case 0x05 :
    if(!SpeedLock)
    {
        MotorPtr->SetSpeed(MotorPtr->GetSpeed()+4);
        SpeedLock = 1;
    }
    MotorPtr->Forward();
    break;

    case 0x06 :
    if(!SpeedLock)
    {
        MotorPtr->SetSpeed(MotorPtr->GetSpeed()-4);
```

```
            SpeedLock = 1;
        }
        MotorPtr->Forward();
        break;

        case 0x09 :
        if(!SpeedLock)
        {
            MotorPtr->SetSpeed(MotorPtr->GetSpeed()+4);
            SpeedLock = 1;
        }
        MotorPtr->Reverse();
        break;

        case 0x0B :
        if(!SpeedLock)
        {
            MotorPtr->SetSpeed(MotorPtr->GetSpeed()-4);
            SpeedLock = 1;
        }
        MotorPtr->Reverse();
        break;

        case 0x00 :
        MotorPtr->Off();
    }
}
delete MotorPtr;
```

The peekb() function is used within the while loop to read the contents of the memory location 0x40:0x17. This value is stored in the variable key. We first check if the Insert key has been pressed, used as input to end program operation. We can detect if the Insert key has been pressed by carrying out an AND operation with 0x80. If the Insert key has been pressed, the result will be a non-zero value and therefore we set Quit to 1. As a result, the while loop will end followed by normal program termination.

If the Insert key has not been pressed, then program operation will continue. Hence, we now only need to read the lower four bits of the byte shown in Figure 8-27. These bits represent the remaining keys assigned for use by the program since we are not using Caps lock, Num lock, or Scroll lock keys. By filtering out the unused upper four bits of key (setting them to zero), we will have unique byte values for our respective key-press combinations. These bits are filtered out using an AND operation of the byte with 0x0F. The resulting value is stored back in key and then tested in a switch statement. All cases that can be implemented are listed in the switch statement where appropriate actions are taken. The main()

function of Listing 8-19 has been modified as shown in Listing 8-22 to implement keyboard controls.

The program has been given a 'speed-locking' mechanism that uses the identifier SpeedLock to control changes in motor speed. This mechanism will only allow speed to be changed once by a small increment for each press of the respective control key. Repeated reading of the speed control key while it is held pressed is not allowed to produce further change in speed, hence the term 'speed-locking'. Without this control in place, excessive changes in motor speed would occur during the many executions of the while loop when the key is pressed.

Motor speed can only be adjusted with the initial press of a speed control key (Shift keys). At this time the variable SpeedLock will be set. Once SpeedLock is set, no further changes in speed can be made until that speed control key has been released. Releasing the key frees or 'unlocks' the speed-locking mechanism by clearing the value of SpeedLock (i.e. !SpeedLock is now true) and allows another increment in speed to be effected.

Speed-locking is implemented in the program in all speed control statements inside the switch statement block. Each speed control statement requires SpeedLock to be false (SpeedLock = 0) before its respective if statement can be executed. SpeedLock will be false whenever both speed control keys are released. When one of the speed control keys is pressed, its associated speed control statement will be selected within its if statement. This speed control statement increments or decrements the value of Speed by 4. The next statement in this if block turns speed-locking back on by setting the value of SpeedLock. Speed-locking will remain in effect until the speed control key is again released.

IMPORTANT: Do not use the following program to drive any motor without first observing the power-up procedure and motor connections as explained in the Note Box on page 243. If the motor does not drive as expected, first check the motor is correctly wired.

Listing 8-22 The complete program with keyboard controls

```
// ****************************************************
// Program implements virtual functions and keyboard
// controls to operate a Motor.
// ****************************************************
#include <dos.h>
#include <conio.h>
#include <stdio.h>
#include <stdlib.h>
#include <iostream.h>

class ParallelPort
{
```

```cpp
        private:
            unsigned int BaseAddress;
            unsigned char InDataPort1;

        public:
            ParallelPort();
            ParallelPort(int baseaddress);
            void WritePort0(unsigned char data);
            void WritePort2(unsigned char data);
            unsigned char ReadPort1();
            virtual ~ParallelPort(){}
};

ParallelPort::ParallelPort()
{
    BaseAddress = 0x378;
    InDataPort1 = 0;
}

ParallelPort::ParallelPort(int baseaddress)
{
    BaseAddress = baseaddress;
    InDataPort1 = 0;
}

void ParallelPort::WritePort0(unsigned char data)
{
    outportb(BaseAddress,data);
}

void ParallelPort::WritePort2(unsigned char data)
{
    outportb(BaseAddress+2,data ^ 0x0B);
}

unsigned char ParallelPort::ReadPort1()
{
    InDataPort1 = inportb(BaseAddress+1);
// Inverting Most significant bit to compensate
// for internal inversion by printer port hardware.
    InDataPort1 ^= 0x80;
// Filter to clear unused data bits D0, D1 and D2 to zero.
    InDataPort1 &= 0xF8;
    return InDataPort1;
}
```

```
class AbstractMotor
{
   private:
        int Speed;

   public:
        AbstractMotor();
        void SetSpeed(int speed);
        int GetSpeed();
        virtual void Off()=0;
        virtual void Forward()=0;
        virtual void Reverse()=0;
        virtual void Brake()=0;
        virtual ~AbstractMotor(){}
};

AbstractMotor::AbstractMotor()
{
   Speed =0;
}

void AbstractMotor::SetSpeed(int speed)
{
   Speed = speed;
   if(Speed > 255) Speed = 255; // Limit upper value
   if(Speed < 0) Speed = 0; // Limit lower value
}

int AbstractMotor::GetSpeed()
{
   return Speed;
}

class Motor : public AbstractMotor, public ParallelPort
{
   public:
        Motor(int baseaddress=0x378);
        void Off();
        virtual void Forward()=0;
        virtual void Reverse()=0;
        virtual void Brake()=0;
        virtual ~Motor(){}
};
```

```cpp
Motor::Motor(int baseaddress):ParallelPort(baseaddress)
{
    Off();
}

void Motor::Off()
{
    WritePort0(0x00);
}

class DCMotor : public Motor
{
    public:
        DCMotor(int baseaddress=0x378);
        virtual void Forward();
        virtual void Reverse();
        virtual void Brake();
};

DCMotor::DCMotor(int baseaddress):Motor(baseaddress)
{
}

void DCMotor::Forward()
{
    int j;
    for(j = 0; j < GetSpeed(); j++)
        WritePort0(0x09);
    for(;j < 256; j++)
        WritePort0(0x00);
}

void DCMotor::Reverse()
{
    int j;
    for(j = 0; j < GetSpeed(); j++)
        WritePort0(0x06);
    for(;j < 256; j++)
        WritePort0(0x00);
}

void DCMotor::Brake()
{
    WritePort0(0x0C);
}
```

```
enum MOTORTYPE {UPFS, UPHS, BPFS, BPHS};

class StepperMotor : public Motor
{
    private:
        MOTORTYPE MotorType;
        unsigned char Switching[8];
        int CycleIndex;
        int MaxIndex;

    public:
        StepperMotor(MOTORTYPE motortype = UPFS,
                             int baseaddress=0x378);
        virtual void Forward();
        virtual void Reverse();
        virtual void Brake();
};

StepperMotor::StepperMotor(MOTORTYPE motortype,
                int baseaddress): Motor(baseaddress)
{
    MotorType = motortype;
    CycleIndex = 0;

    switch(MotorType)
    {
        case UPFS: MaxIndex = 4;
                        Switching[0] = 0x11;
                        Switching[1] = 0x12;
                        Switching[2] = 0x22;
                        Switching[3] = 0x21;
                        break;
        case UPHS: MaxIndex = 8;
                        Switching[0] = 0x01;
                        Switching[1] = 0x11;
                        Switching[2] = 0x10;
                        Switching[3] = 0x12;
                        Switching[4] = 0x02;
                        Switching[5] = 0x22;
                        Switching[6] = 0x20;
                        Switching[7] = 0x21;
                        break;
        case BPFS: MaxIndex = 4;
                        Switching[0] = 0x99;
```

```
                              Switching[1] = 0x69;
                              Switching[2] = 0x66;
                              Switching[3] = 0x96;
                              break;
              case BPHS: MaxIndex = 8;
                              Switching[0] = 0x99;
                              Switching[1] = 0x09;
                              Switching[2] = 0x69;
                              Switching[3] = 0x60;
                              Switching[4] = 0x66;
                              Switching[5] = 0x06;
                              Switching[6] = 0x96;
                              Switching[7] = 0x90;
       }
}

void StepperMotor::Forward()
{
    if(++CycleIndex == MaxIndex) CycleIndex = 0;
    WritePort0(Switching[CycleIndex]);
    delay(259-GetSpeed());
}

void StepperMotor::Reverse()
{
    if(--CycleIndex == -1) CycleIndex = MaxIndex -1;
    WritePort0(Switching[CycleIndex]);
    delay(259-GetSpeed());
}

void StepperMotor::Brake()
{
    switch(MotorType)
    {
        case UPFS: case UPHS:
                        WritePort0(0x11);
                        break;
        case BPFS: case BPHS:
                        WritePort0(0x99);
    }
}

void main()
{
    Motor *MotorPtr;
```

```cpp
int Quit = 0;
unsigned char key = 0;
int SpeedLock = 1;
int Selection;

clrscr();
cout << endl << "   MOTOR MENU";
cout << endl << "   ~~~~~~~~~~" << endl;
cout << "   1  DC Motor" << endl;
cout << "   2  UPFS" << endl;
cout << "   3  UPHS" << endl;
cout << "   4  BPFS" << endl;
cout << "   5  BPHS" << endl;
cout << "   6  QUIT" << endl;
cout << endl;
cout << "   Select the MOTOR Number: ";

cin >> Selection;
switch(Selection)
{
case 1: MotorPtr = new DCMotor;
    break;
case 2: MotorPtr = new StepperMotor(UPFS);
    break;
case 3: MotorPtr = new StepperMotor(UPHS);
    break;
case 4: MotorPtr = new StepperMotor(BPFS);
    break;
case 5: MotorPtr = new StepperMotor(BPHS);
    break;
case 6: return;

default: cout << endl;
    cout << "   Unspecified Motor type....";
    cout << " PRESS a key to END Program!";
    getch();
    exit(1); // Exits the program
}

if(MotorPtr == NULL)
{
    cout << "Memory allocation failed " << endl;
    getch();
    exit(1);
```

```
    }
cout << "*********************************" << endl;
cout << "* CONNECT BOARD POWER SUPPLY NOW *" << endl;
cout << "*********************************" << endl;
cout << endl;
cout << "   After connecting power,";
cout << " press a key to continue " << endl;
getch();

while(!Quit)
{
    key = peekb(0x40,0x17); // Read control key byte.
    if(key & 0x80) // Test Insert key ON (MSBit '1').
        Quit = 1;  // Exit the program.

    else
    {
    // If both shift keys are released SpeedLock is
    // released
        if(!(key & 0x01) && !(key & 0x02))
            SpeedLock = 0;

        key &= 0x0F; // Filter out bits corresponding to
                    // just SHIFT, ALT & CTRL keys.

        switch(key)
        {
            case 0x04 :
            MotorPtr->Forward();
            break;

            case 0x08 :
            MotorPtr->Reverse();
            break;

            case 0x01 :
            if(!SpeedLock)
            {
              MotorPtr->SetSpeed(MotorPtr->GetSpeed()+4);
              SpeedLock = 1;
            }
            break;

            case 0x02 :
            if(!SpeedLock)
```

```
{
  MotorPtr->SetSpeed(MotorPtr->GetSpeed()-4);
  SpeedLock = 1;
}
break;

case 0x03 :
MotorPtr->Brake();
break;

case 0x05 :
if(!SpeedLock)
{
  MotorPtr->SetSpeed(MotorPtr->GetSpeed()+4);
  SpeedLock = 1;
}
MotorPtr->Forward();
break;

case 0x06 :
if(!SpeedLock)
{
  MotorPtr->SetSpeed(MotorPtr->GetSpeed()-4);
  SpeedLock = 1;
}
MotorPtr->Forward();
break;

case 0x09 :
if(!SpeedLock)
{
  MotorPtr->SetSpeed(MotorPtr->GetSpeed()+4);
  SpeedLock = 1;
}
MotorPtr->Reverse();
break;

case 0x0B :
if(!SpeedLock)
{
  MotorPtr->SetSpeed(MotorPtr->GetSpeed()-4);
  SpeedLock = 1;
}
MotorPtr->Reverse();
break;
```

```
               case 0x00 :
               MotorPtr->Off();
            }
        }
        delete MotorPtr;
    }
}
```

In our program that uses virtual functions, we were able to write the motor control portion as a generic module. This approach allows the program to bind the correct function to the object associated with the selected motor during program execution (late binding). Only after the user has selected the type of motor, will dynamic memory allocation and late binding take place to drive the motor. If the program had not used virtual functions, the programmer would need to provide extensive dedicated code to control each motor. This was seen in the modified `main()` function given in Listing 8-20.

The program that uses virtual functions to control the motors does not need dedicated motor control code for each type of motor. Instead, the code is independent of the specific motor type and will even work for motors that may be added to the hierarchy in the future. These benefits are the main advantages of using virtual functions and can also be seen as one of the greatest strengths of object-oriented programming.

8.8 Summary

This chapter presented the construction and operation of DC Motors and Stepper Motors. Various means of controlling these motors has also been described.

A class hierarchy was developed to represent all types of motor discussed at the beginning of the chapter. This was followed by a conceptual explanation of abstract classes and pure virtual functions. Class hierarchy's and multiple inheritance were also explained. The need for a set of virtual destructors in a class hierarchy was also demonstrated. Unlike constructors, destructors can be virtual. These destructors are used to free an object's dynamically allocated memory once the program no longer needs the object.

A generic program for all real motor classes of the hierarchy was developed and integrated into a `main()` function to demonstrate the concept and advantages of late binding. Keyboard controls were then incorporated into the program to improve control of motors when using the interface board.

8.9 Bibliography

Bergsman, P. , *Controlling The World With Your PC*, HighText Publications, San Diego, 1994.

Stiffler, K., *Design with Microprocessors for Mechanical Engineers*, McGraw-Hill, 1992.

PARKER HANNIFIN CORP, *Positioning Control Systems and Drives*, 1992-1993.

Lafore, R. *Object Oriented Programming in MICROSOFT C++*, Waite Group Press, 1992.

Wang, P.S., *C++ with Object Oriented Programming*, PWS Publishing, 1994.

Borland, *Borland C++ version 5, User's Guide*, Borland International, 1996.

Borland, *Borland C++ version 5, Programmers Guide*, Borland International, 1996.

Barton, J.J. and L.R. Nackman, *Scientific and Engineering C++ - An Introduction with Advanced Techniques and Examples*, Addison Wesley, 1994.

Stevens, A. *Teach Yourself C++ Fifth Edition,* MIS Press, 1997.

Schildt, H. *Teach Yourself C++ Third Edition,* Osborne/McGraw-Hill, 1998.

9

Program Development Techniques

Inside this Chapter

- Developing programs – what is involved?

- Efficient coding techniques.

- Modular development approach.

- Header Files.

- Function files.

- Project files and make files.

9.1 Introduction

So far we have learned how to develop efficient object-oriented programs where the emphasis has been on the program statements (source code). In this chapter we will learn how to plan a program using pseudo-code, organise its structure, and write the program so typographical errors can be kept to a minimum. The development process to produce modular programs will also be explained.

A modular program can be made by separating a lengthy single file program into a number of logical modules and then placing each module into its own dedicated file. This process greatly improves our ability to maintain our programs. Furthermore, it allows us to carry out modifications with greater ease and also promotes more efficient debugging of programs. An inevitable consequence of this modular approach is the multi-file program. We will learn how to create a multi-file program from a number of source files and then generate a final executable file.

9.2 Efficient Coding Techniques

The word *coding* is used is used in this chapter and refers to the writing of programming statements. No coding should commence until a detailed plan for the program is established. This plan is written as a general worded discription known as pseudo-code. If we are writing an object-oriented program, the first step should be creating the object classes and the associated class hierarchy. Using an object-oriented approach tends to result in a program with good structure. Program development should be carried out in a number of manageable steps. At each of these steps the program (or the part of the program) coded to that point can be compiled and verified for errors.

Most editors used for programming provide cutting and pasting facilities for text editing. We can minimise typographical errors during the coding process by using text cutting and pasting operations. These typographical errors tend to be the cause of most compilation errors.

Pseudo-code

Using pseudo-code to outline the basic operation of an application can assist in its development. The following example demonstrates how pseudo-code is developed and used to generate program code.

Program description:

> A crane is used to lift a weight from point A and move it to another point B. It is assumed that the crane uses three DC motors; one to lift/lower the load, another to move the load in the x direction, and the third motor to move the load in the y direction.

The pseudo-code is:

> Enter coordinates for points A and B.
> Move the crane to point A.
> Lift the load.
> Move the crane to point B.
> Lower the load.
> End.

We can translate this pseudo-code into a C++ object-oriented program. The following example program shows one implementation of a `main()` function that implements the pseudo-code:

```cpp
void main()
{
    Crane OurCrane;     // Create a Crane object.
    Point a, b;         // Create two Point objects.

    cout << "Enter coordinates of point A  ";
    cin >> a;

    cout << endl << "Enter coordinates of point B  ";
    cin >> b;

    OurCrane.MoveToPoint(a);
    OurCrane.LiftLoad();
    OurCrane.MoveToPoint(b);
    OurCrane.LowerLoad();
}
```

This `main()` function uses a `Point` object. It also uses a `Crane` object and implies that `MoveToPoint()`, `LiftLoad()` and `LowerLoad()` be member functions of the `Crane` class. This operation requires two points and three motors; one motor lifts and lowers the load, one motor drives in the X direction and another motor drives in the Y direction.

The corresponding `Point` class would be as follows:

```cpp
class Point
{
    private:
        int X;
        int Y;

    public:
        Point();
        Point(int x, int y);
        void SetX(int x);
```

```
        void SetY(int y);
        int GetX();
        int GetY();
};
```

A suitable class definition for the `Crane` class would be:

```
class Crane
{
    private:
        Point A, B;
        DCMotor LiftMotor, Xmotor, Ymotor;

    public:
        Crane();
        void SetPointA(Point a);
        void SetPointB(Point b);
        void MoveToPoint();
        void LiftLoad();
        void LowerLoad();
};
```

All member functions of the `Crane` class that generate movement will need to use the `Forward()`, `Reverse()`, and `Brake()` functions of the `DCMotor` class.

As demonstrated here, structured programming starts with good pseudo-code written with the program steps outlined in a structured manner. Each of these steps forms a statement in the `main()` function. The member functions of the class definition are expanded to include the necessary details for the function to perform their planned tasks. The following section describes a good approach for coding the `AbstractMotor` class and its member functions.

Creating functions starting from class definitions.

A few basic principles should be kept in mind when coding object classes and their member functions. In general, the members of a class can be listed anywhere within the scope of the class definition (between the open and close braces) under different access attributes. However, keeping the member data together and separate from the member functions can facilitate coding.

The object class will often have a user-defined constructor that is used to initialise all its data members. Having placed data members together allows them to be copied as a block and placed into the body of the constructor for initialisation. They can also be copied and pasted elsewhere to code the definitions of other functions. The bodies of the member functions can be left empty and the source code then compiled to verify that it conforms to the syntax used by the C++ language.

Consider the `AbstractMotor` class definition from Listing 8-1 in Chapter 8 that has been reproduced in Listing 9-1.

Listing 9-1 AbstractMotor class definition.

```
class AbstractMotor
{
    private:
        int Speed;

     public:
        AbstractMotor();
        void SetSpeed(int speed);
        int GetSpeed();
        virtual void Off()=0;
        virtual void Forward()=0;
        virtual void Reverse()=0;
        virtual void Brake()=0;
};
```

In this class definition the data and functions are kept separate. It allows us to copy the set of non-pure virtual functions (that need to be defined) to a position just below the class definition as shown in Listing 9-2. Note that the pure virtual functions do not need to be copied since they will not have a function definition. It is good practice to keep the pure virtual functions together, preferably at the end of the member function declarations. This makes it easier to copy and paste the group of member function declarations we need to define.

Listing 9-2 Set of member functions copied to be defined.

```
class AbstractMotor
{
    private:
        int Speed;

    public:
        AbstractMotor();
        void SetSpeed(int speed);
        int GetSpeed();
        virtual void Off()=0;
        virtual void Forward()=0;
        virtual void Reverse()=0;
        virtual void Brake()=0;
};
AbstractMotor();
void SetSpeed(int speed);
int GetSpeed();
```

This file cannot be compiled in its present form since it has incomplete and therefore incorrect syntax. It must be modified to give the functions the required basic syntax. To do this, each function name just copied must be qualified with the class name followed by the double colon operator (::). In the case of the AbstractMotor() function shown in Listing 9-2, we need to have:

AbstractMotor::AbstractMotor()

The semicolon at the end of each line must be removed and an open brace and a close brace added to provide the start and the end of each member function. These changes to Listing 9-2 are shown in Listing 9-3.

Listing 9-3 Skeletal member functions for the AbstractMotor class.

```
class AbstractMotor
{
    private:
        int Speed;

    public:
        AbstractMotor();
        void SetSpeed(int speed);
        int GetSpeed();
        virtual void Off()=0;
        virtual void Forward()=0;
        virtual void Reverse()=0;
        virtual void Brake()=0;
};

AbstractMotor::AbstractMotor()
{
}

void AbstractMotor::SetSpeed(int speed)
{
}

int AbstractMotor::GetSpeed()
{
}
```

This file can now be compiled. Although it does not have any statements within the bodies of the member functions, correct program syntax has been applied. Note that the contents of the file cannot be linked to form an executable file without having a main() function. However, compiling will generate object code. There is an advantage to compiling the file at this early stage; we can verify the structure of the class has correct syntax.

Skeletal functions

Function definitions with empty bodies (as shown in Listing 9-3) can be referred to as skeletal functions. The bodies of these functions could be coded at this early stage of program development. However, they are intentionally left empty to simplify the task of establishing good program structure and correct syntax. It can even be advantageous to write an entire program using skeletal functions to verify its conceptual operation and structure.

Filling-in the constructors' bodies

The next stage is to code the bodies of the member functions, starting say with the constructor definition. As mentioned earlier, the most common purpose of the constructor is to initialise the data members of the class. This can be done by copying and pasting the entire set of data members into the body of each constructor of the class. In the example discussed above, we only have one data member. Listing 9-4 shows the constructor that has this data member copied and pasted into its body. Note that this function is incomplete and does not carry out its intended task.

Listing 9-4 Copying the member data declarations into the body of the constructor.

```
AbstractMotor::AbstractMotor()
{
    int Speed;
}
```

This single statement shown in Listing 9-4 needs to be corrected as shown in Listing 9-5. Once this is done the constructor can initialise this data member and operate as intended.

Listing 9-5 The constructor with the syntax error eliminated.

```
AbstractMotor::AbstractMotor()
{
    Speed = 0;  // copied and pasted data members are
                // initialized as required.
}
```

Good habits with parenthesis, square brackets and braces

Pairs of parentheses (), square brackets [], and braces { } are used extensively in C++ programming. We'll use the general term brackets to describe all three types. When writing text we sometimes place an open parenthesis and forget to place the close parenthesis. People use their intelligence to check and correct missing parentheses, however, a compiler lacks this ability. As a result, the programmer must correct for missing brackets. The misuse of brackets generally happens when the programmer loses track of a close bracket, and this generates errors when compiling the program. It can be difficult to determine the position in the program where the missing bracket should be placed once the program statements have been added. This situation is further complicated when nested brackets are used.

The best approach to avoid these problems is to place the matching open and close brackets simultaneously. The following techniques can be used to efficiently implement a class definition:

Step 1:

```
class AbstractMotor
{
};
```

Step 2:

```
class AbstractMotor
{
    private:
        int Speed;

    public:
        AbstractMotor();
        void SetSpeed(int speed);
        int GetSpeed();
        virtual void Off()=0;
        virtual void Forward()=0;
        virtual void Reverse()=0;
        virtual void Brake()=0;
}
```

The following steps show how the function declarations are developed:

Step 1:

```
void SetSpeed();
```

Step 2:

```
void SetSpeed(int speed)
{
}
```

Step 3:

```
void SetSpeed(int speed)
```

```
{
    int Speed;
}
```

Step 4:

```
void SetSpeed(int speed)
{
    Speed = 0;
}
```

Nested Levels

The following steps show how a nested `if` statement is developed:

Step 1:

```
if()
{
}
else
{
}
```

Step 2:

```
if()
{
if()
{
}
else
{
}
}
else
{
}
```

Indentation is then used to clearly show the levels of nesting. The skeletal `if` statement shown above would then become:

```
if()
{
    if()
    {
    }
    else
    {
```

```
        }
    }
    else
    {
    }
```

Applying the same habits will help you type error-free code when using logical operators as part of conditional expressions in `if` statements:

```
if()
{
    if(() && ())
    {
    }
    else
    {
    }
}
else
{
}
```

We can complete the conditional expression as shown in this example:

```
if(b > 0)
{
    if((a != 0) && (b/a > n))
    {
    }
    else
    {
    }
}
else
{
}
```

9.3 Modular Programs

Each program developed in previous chapters has all its program statements contained in one file. While this is satisfactory for smaller programs, it becomes less practical as programs grow in size and complexity. Larger programs have specific portions of their code separated into modules and stored as separate files. Because more than one file needs to be compiled and linked, the program becomes known as a *multiple file* program.

There are additional reasons why programs are developed in a multiple file format.

In the case of object-oriented programming, the files can be treated as modules where each module contains the code for one object class. As a result, when distributing software relating to a particular object, only the code for that object needs to be distributed. Software developers typically supply object code modules (unreadable to the users) accompanied by files that allow the object code to be used and linked with a program. This helps to prevent the developers' object code being illegally used or misused.

When a particular object is used, only the file that corresponds to that one object needs to be included in the user code. This helps to minimise the size of a program. Had multiple objects been part of the file, the additional unused functions for those objects would also be compiled, slowing the compilation process.

9.3.1 Separating Software into Modules

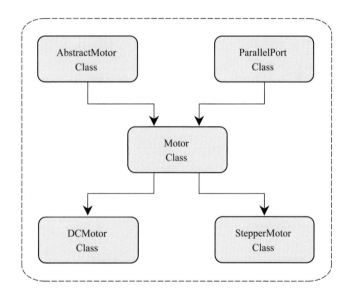

Figure 9-1 Class hierarchy for Motor Driver class (Chapter 8).

We will use the motor driver program developed in Chapter 8 (class hierarchy shown in Figure 9-1) to explain the process of generating a multiple file program. Each class is separated into two types of files. The class definition is placed into a header file, and the definitions of member functions for that class are placed into a function file (Figure 9-2). The remaining file for this example program (not shown) is generated from the `main()` function (note that this is not always the case).

The compiler needs to access the class definitions in the header file and function definitions in the function file. The function file (as source code; `*.cpp`) is compiled into an `.obj` file or `.lib` file by the developer. The programmer that uses these files cannot read them, and so the developer's software is protected from unauthorised copying and misuse.

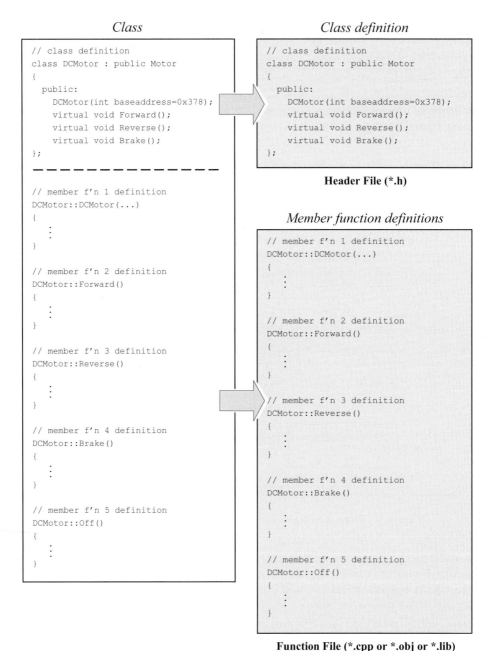

Figure 9-2 Code separation into header and function files.

Object-oriented and non-object-oriented programs differ; in one case the functions are member functions, and in the other case all the functions are non-member functions. It is also possible for an object-oriented program to use a combination of the two cases. Its header file may contain a class definition and also some non-member function declarations.

If non-member C/C++ functions are to be used in the program, it is general practice to put the declarations of such functions into a header file and then to include that header file at the top of the source file. The definitions of the non-member functions may be provided in a separate function file in much the same way as the member functions of a particular class. The function file in its `.obj`/`.lib` form will be needed at linking time.

9.3.2 Generating a Multiple File Program

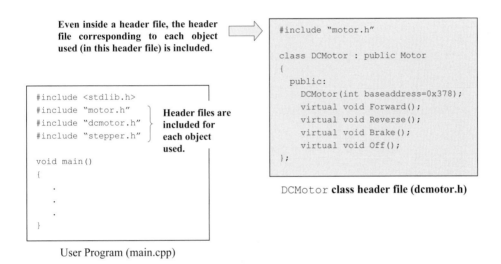

Figure 9-3 Use of include directives.

As mentioned previously, whenever an object from a class is needed within a program, its class header file needs to be included in the program before the object can be used. The compiler will first interpret the class definition (from the first header file encountered) and then check that the class has been implemented correctly throughout the remainder of the program. The program will only compile correctly if the programmer has made function calls that are compatible with the function declarations in the header files. Figure 9-3 shows the use of the `Motor`, `DCMotor`, and `StepperMotor` objects in the User Program, evident by the inclusion of their class header files.

Class definition

```
// class definition
#include "motor.h"

class DCMotor : public Motor
{
  public:
    DCMotor(int baseaddress=0x378);
    virtual void Forward();
    virtual void Reverse();
    virtual void Brake();
    virtual void Off();
```

Header File (dcmotor.h)

The structure of DCMotor class member functions must conform to the function declarations given in the dcmotor.h file.

Member function definitions

```
#include "dcmotor.h"

// member f'n 1 definition
DCMotor::DCMotor(..
{
     .
     .
     .
}

// member f'n 2 definition
DCMotor::Forward()
{
     .
     .
     .
}

// member f'n 3 definition
DCMotor::Reverse()
{
     .
     .
     .
}

     .
     .
     .
     .
     .
```

Function File (dcmotor.cpp or *.obj or *.lib)

Figure 9-4 Include header file in the function file of its class.

Figure 9-4 shows the inclusion of the header file for the DCMotor class at the start of the function file for the DCMotor class. This will ensure all member function definitions of the DCMotor class (that were derived from the Motor class and overridden as required) will conform with the function declarations stipulated in the header file dcmotor.h.

Include directives

The files that are included from the 'include' directory of the C/C++ program development software are enclosed between angle brackets angle brackets (< >):

```
#include <stdlib.h>  // For standard C library fn's
```

Header files included from the same directory or any other directory in the path are enclosed between double quotation marks (" "):

```
#include "motor.h"   // For motor class objects used.
```

The proper use of angle brackets and double quotation marks is crucial as it allows the compiler to efficiently locate the included header files during compilation and linking.

Preventing Multiple Inclusions of Header Files

A particular header file is included only once in a program. Should a header file be included more than once, the compiler will interpret this as an error and issue an error message. The compiler can interpret multiple inclusion as, for example, the redefinition of a class that is already defined by the first instance of the include file – hence the error!

The general rule "Include the header file of the object you use in the program" may lead to multiple inclusions. For example, if a program uses a DCMotor object and a StepperMotor object, we must include dcmotor.h as well as stepper.h files. They both have motor.h included in them. As a result motor.h will be included twice. There is a mechanism that enables us to practice the general rule above and at the same time avoid multiple inclusions of the same header file. The procedure uses a 'status flag', explained as follows:

Before the preprocessor begins to include header files, the flag will be inactive (file not included).

The first time a particular header file is presented for inclusion, the flag will be tested, and the result will indicate that the header file has not been included. The header file will be included this time and the flag will then be activated, indicating inclusion has now taken place.

When this header file is presented for inclusion on subsequent occasions, the flag will be tested and its status (this time; file included) will direct the preprocessor to ignore this file, preventing any multiple inclusions.

Sentries for header files

Sentries in header files are compiler directives for the preprocessor. They implement the function of the 'status flag' just described and prevent the compiler from including the same header file more than once.

The following example uses the AbstactMotor class to show how sentries are added to a header file.

Listing 9-6 `AbsMotor.h` header file.

```
#ifndef AbsmotorH
#define AbsmotorH

class AbstractMotor
{
   private:
       int Speed;

   public:
       AbstractMotor();
       void SetSpeed(int speed);
       int GetSpeed();
       virtual void Off()=0;
       virtual void Forward()=0;
       virtual void Reverse()=0;
       virtual void Brake()=0;
       virtual ~AbstractMotor(){}
};
#endif
```

The first preprocessor directive `#ifndef` represents 'if not defined'. It is similar to an `if` statement, with its body starting at `#ifndef` and ending at the line with the directive `#endif`. Therefore, the preprocessor interprets:

`#ifndef AbsmotorH`

as 'if `AbsmotorH` is not defined'. The body of this `#if` statement will be executed only if `AbsmotorH` is not defined.

The identifier `AbsmotorH` will not be defined when proceeding to process a multiple file program for the first time. As such, when the pre-processor encounters the line `#ifndef AbsmotorH`, it will enter the body of the `#if` statement. The first line within the body is:

`#define AbsmotorH`

The `#define` directive is used to state that identifier `AbsmotorH` is to be defined. The identifier must be unique and not already used to name another header file from a different class. Improper naming of identifiers can lead to programming bugs that are difficult to find. Since the file system of your computer maintains unique names for each file, the best practice is to derive the sentry name based on the name of that header file. This approach has been used to form the name of the `AbsmotorH` sentry from the associated header file `absmotor.h`. When all remaining lines in the body are processed by the pre-processor, the `AbstractMotor` class will be interpreted by the compiler and the sentry `AbsMotorH` defined. Should the pre-processor encounter another `absmotor.h` file included in another file of that program, execution of the #ifndef AbsmotorH directive will return false. In this case the body of the #if statement will be skipped, avoiding a repeated inclusion of its contents.

9.4 Case Study - Motor Driver Program

This section will demonstrate the process of generating a multiple file program as described previously using the motor driver program developed in Chapter 8. We will first create the software modules for each object class in our program. Each module will have its own header file and function file.

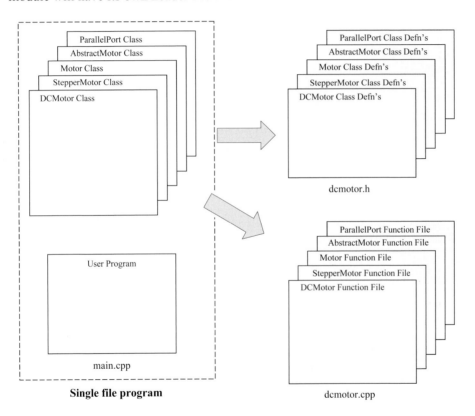

Figure 9-5 Form a multiple file program.

Figure 9-5 shows the original single file program on the left with its main function and classes. The header file and function file associated with each class is shown on the right. The program from Chapter 8 (Listing 8-19) is shown following. Each code segment has been identified from the program, copied and labelled to its appropriate file type; being a header file or function file. These files are then saved with *.h and *.cpp extensions, preferably in the same directory to minimise file search time.

Listing 9-7 Motor driver program - segmented.

absmotor.h   (header file)

```cpp
#ifndef AbsmotorH
#define AbsmotorH

class AbstractMotor
{
   private:
       int Speed;

    public:
       AbstractMotor();
       void SetSpeed(int speed);
       int GetSpeed();
       virtual void Off()=0;
       virtual void Forward()=0;
       virtual void Reverse()=0;
       virtual void Brake()=0;
       virtual ~AbstractMotor(){}
};
#endif
```

absmotor.cpp   (function file)

```cpp
#include "absmotor.h"

AbstractMotor::AbstractMotor()
{
   Speed =0;
}

void AbstractMotor::SetSpeed(int speed)
{
   Speed = speed;
   if(Speed > 255) Speed = 255; // Limit upper value
   if(Speed < 0) Speed = 0; // Limit lower value
}

int AbstractMotor::GetSpeed()
{
   return Speed;
}
```

pport.h   (header file)

```cpp
#ifndef PportH
```

```
#define PportH

class ParallelPort
{
    private:
        unsigned int BaseAddress;
        unsigned char InDataPort1;

    public:
        ParallelPort();
        ParallelPort(int baseaddress);
        void WritePort0(unsigned char data);
        void WritePort2(unsigned char data);
        unsigned char ReadPort1();
        virtual ~ParallelPort(){}
};
#endif
```

pport.cpp   (function file)

```
#include <dos.h>
#include "pport.h"

ParallelPort::ParallelPort()
{
    BaseAddress = 0x378;
    InDataPort1 = 0;
}

ParallelPort::ParallelPort(int baseaddress)
{
    BaseAddress = baseaddress;
    InDataPort1 = 0;
}

void ParallelPort::WritePort0(unsigned char data)
{
    outportb(BaseAddress,data);
}

void ParallelPort::WritePort2(unsigned char data)
{
    outportb(BaseAddress+2,data ^ 0x0B);
}

unsigned char ParallelPort::ReadPort1()
```

```
{
    InDataPort1 = inportb(BaseAddress+1);
// Inverting Most significant bit to compensate
// for internal inversion by printer port hardware.
    InDataPort1 ^= 0x80;
// Filter to clear unused data bits D0, D1 and D2 to zero.
    InDataPort1 &= 0xF8;
    return InDataPort1;
}
```

motor.h   (header file)

```
#ifndef MotorH
#define MotorH

#include "absmotor.h"
#include "pport.h"

class Motor : public AbstractMotor, public ParallelPort
{
    public:
        Motor(int baseaddress=0x378);
        void Off();
        virtual void Forward()=0;
        virtual void Reverse()=0;
        virtual void Brake()=0;
        virtual ~Motor(){}
};
#endif
```

motor.cpp   (function file)

```
#include "motor.h"

Motor::Motor(int baseaddress):ParallelPort(baseaddress)
{
    Off();
}

void Motor::Off()
{
    WritePort0(0x00);
}
```

dcmotor.h   (header file)

```
#ifndef DcmotorH
#define DcmotorH
```

```
#include "motor.h"

class DCMotor : public Motor
{
    public:
        DCMotor(int baseaddress=0x378);
        virtual void Forward();
        virtual void Reverse();
        virtual void Brake();
};
#endif
```

dcmotor.cpp   (function file)

```
#include "dcmotor.h"

DCMotor::DCMotor(int baseaddress):Motor(baseaddress)
{
}

void DCMotor::Forward()
{
    int j;
    for(j = 0; j < GetSpeed(); j++)
        WritePort0(0x09);
    for(;j < 256; j++)
        WritePort0(0x00);
}

void DCMotor::Reverse()
{
    int j;
    for(j = 0; j < GetSpeed(); j++)
        WritePort0(0x06);
    for(;j < 256; j++)
        WritePort0(0x00);
}

void DCMotor::Brake()
{
    WritePort0(0x0C);
}
```

stepper.h   (header file)

```
#ifndef StepperH
#define StepperH

#include "motor.h"

enum MOTORTYPE {UPFS, UPHS, BPFS, BPHS};

class StepperMotor : public Motor
{
    private:
        MOTORTYPE MotorType;
        unsigned char Switching[8];
        int CycleIndex;
        int MaxIndex;

    public:
        StepperMotor(MOTORTYPE motortype = UPFS,
                            int baseaddress=0x378);
        virtual void Forward();
        virtual void Reverse();
        virtual void Brake();
};
#endif
```

stepper.cpp   (function file)

```
#include <dos.h>

#include "stepper.h"

StepperMotor::StepperMotor(MOTORTYPE motortype,
                int baseaddress): Motor(baseaddress)
{
    MotorType = motortype;
    CycleIndex = 0;

    switch(MotorType)
    {
        case UPFS: MaxIndex = 4;
                        Switching[0] = 0x11;
                        Switching[1] = 0x12;
                        Switching[2] = 0x22;
                        Switching[3] = 0x21;
                        break;
        case UPHS: MaxIndex = 8;
```

```
                                Switching[0] = 0x01;
                                Switching[1] = 0x11;
                                Switching[2] = 0x10;
                                Switching[3] = 0x12;
                                Switching[4] = 0x02;
                                Switching[5] = 0x22;
                                Switching[6] = 0x20;
                                Switching[7] = 0x21;
                                break;
            case BPFS: MaxIndex = 4;
                                Switching[0] = 0x99;
                                Switching[1] = 0x69;
                                Switching[2] = 0x66;
                                Switching[3] = 0x96;
                                break;
            case BPHS: MaxIndex = 8;
                                Switching[0] = 0x99;
                                Switching[1] = 0x09;
                                Switching[2] = 0x69;
                                Switching[3] = 0x60;
                                Switching[4] = 0x66;
                                Switching[5] = 0x06;
                                Switching[6] = 0x96;
                                Switching[7] = 0x90;
        }
}


void StepperMotor::Forward()
{
    if(++CycleIndex == MaxIndex) CycleIndex = 0;
    WritePort0(Switching[CycleIndex]);
    delay(259-GetSpeed());
}

void StepperMotor::Reverse()
{
    if(--CycleIndex == -1) CycleIndex = MaxIndex -1;
    WritePort0(Switching[CycleIndex]);
    delay(259-GetSpeed());
}

void StepperMotor::Brake()
{
    switch(MotorType)
    {
```

```
            case UPFS: case UPHS:
                        WritePort0(0x11);
                        break;
            case BPFS: case BPHS:
                        WritePort0(0x99);
    }
}
```

main.cpp   (user program)

```
//**********************************************************
// Motor driver program using Virtual Functions (chapter 8).
//**********************************************************
#include <dos.h>
#include <conio.h>
#include <stdio.h>
#include <stdlib.h>
#include <iostream.h>

#include "motor.h"
#include "dcmotor.h"
#include "stepper.h"

void main()
{
    Motor *MotorPtr;
    int Selection;

    clrscr();

    cout << endl << "   MOTOR MENU";
    cout << endl << "   ~~~~~~~~~~" << endl;
    cout << "    1  DC Motor" << endl;
    cout << "    2  UPFS" << endl;
    cout << "    3  UPHS" << endl;
    cout << "    4  BPFS" << endl;
    cout << "    5  BPHS" << endl;
    cout << "    6  QUIT" << endl;
    cout << endl;
    cout << "    Select the MOTOR Number: ";

    cin >> Selection;

    switch(Selection)
    {
    case 1: MotorPtr = new DCMotor;
```

```
        break;
    case 2: MotorPtr = new StepperMotor(UPFS);
        break;
    case 3: MotorPtr = new StepperMotor(UPHS);
        break;
    case 4: MotorPtr = new StepperMotor(BPFS);
        break;
    case 5: MotorPtr = new StepperMotor(BPHS);
        break;
    case 6: return;

    default: cout << endl;
        cout << "   Unspecified Motor type....";
        cout << " PRESS a key to END Program!";
        getch();
        exit(1); // Exits the program
    }

    if(MotorPtr == NULL)
    {
        cout << "Memory allocation failed " << endl;
        getch();
        exit(1);
    }
    cout << "***********************************" << endl;
    cout << "* CONNECT BOARD POWER SUPPLY NOW *" << endl;
    cout << "***********************************" << endl;
    cout << endl;
    cout << "   After connecting power,";
    cout << " press a key to continue " << endl;
    getch();
    cout << endl;
    cout << "   Keypress changes Speed/Rotation (& Braking)."
<< endl;

    //..... Motor control part starts here .....
    MotorPtr->SetSpeed(150);
    while(!kbhit()) MotorPtr->Forward();
    getch(); // clear keyboard buffer

    MotorPtr->SetSpeed(255);
    while(!kbhit()) MotorPtr->Forward();
    getch();

    MotorPtr->SetSpeed(150);
```

```
        while(!kbhit()) MotorPtr->Reverse();
        getch();

        MotorPtr->SetSpeed(255);
        while(!kbhit()) MotorPtr->Reverse();
        getch();

        cout << endl << "   Braking Applied!" << endl;
        while(!kbhit()) MotorPtr->Brake();
        getch();

        MotorPtr->Off();
        //..... Motor control part ends here .....
        // Free the memory occupied by the 'Motor' object
        delete MotorPtr;
}
```

Main Function File

The programmer uses the objects from the class hierarchy in the development of the main() function. A programmer only needs to know how to apply the user interface of the object classes i.e. the public members of the classes. This information allows the member functions to be used in the main function file according to their specification in the class definition. The programmer using the classes does not need to know the full internal details of the member functions being used. The compiler however, needs to know the exact construct of each class used, and their base classes if any. We fulfil this requirement by including the appropriate header files.

Note that the file motor.h is the only file included in the files dcmotor.h and stepper.h. The file motor.h, in turn, has the files absmotor.h and pport.h included. Therefore, when the compiler reaches the dcmotor.h file, it has already seen the files absmotor.h, pport.h, and motor.h. These files provide all the base class definitions needed for the definition of the DCMotor and StepperMotor classes.

We have used the objects of the classes Motor, DCMotor, and StepperMotor in the main() function. Therefore, the main() function must include these three header files. These header files are quoted within double quotes which directs the compiler to search for them in the current directory or directories in the search path. Header files that are included within angle brackets (< >) have not been created by us. They reside in the 'include' directory of the C/C++ programming software.

This main() function is an ideal place to examine the effect of the sentries we included in the header files. The header files dcmotor.h and stepper.h both include motor.h. Therefore, the header file motor.h is shown included in three parts of the complete program listing; once explicitly in the main function file (main.cpp) and twice indirectly via the inclusion of dcmotor.h and stepper.h. However, this does not result in the Motor class actually being defined and included three times. The first inclusion of the file motor.h will provide the class definition and define the sentry MotorH. The pre-processor will ignore any subsequent attempts to include motor.h since MotorH has already been defined. Thererefore, the two cases of indirect inclusion of motor.h in the files dcmotor.h and stepper.h will not be processed. This does not mean that the files dcmotor.h and stepper.h cannot function. They will use the Motor class definition that the compiler has already interpreted (from the first inclusion of motor.h) and so can provide the definitions of the classes DCMotor and StepperMotor.

Creating Library Files

The linking process uses library files to produce the executable file. Library files are made by combining a number of object files together, and are normally given the file name extension .lib. C/C++ program development software provides utilities to generate library files. The library file will contain the compiled definitions of all the functions that were in each object file. The library file is generated as a binary file and so cannot be read by a programmer.

Project Files and Make Files

Compilation and linking is more complicated for multiple file programs than for our previous programs that used single source files. Nevertheless, this process can easily be automated using one of two methods. The first method creates a *project file* containing a list of the files used to form the final executable file. This project file includes all source files, library files and object files for the program. Note that project files do not contain header files. The preprocessor will include the header files when the source files are compiled.

The second method creates what is known as a *make file*. The make file is processed by a utility application known as a make utility program which operates the compiler and linker in accordance with commands contained in the make file.

Project files

We must create a project file, for example, named drive.* before being able to compile the motor drive program. The name of the file extension given to the project file is peculiar to the particular compiler being used. For example, Inprise™ Borland C++ for DOS will use .prj, and Microsoft™ Visual C++ will use .dsp as the extension for the project filename. Three different versions of managing and processing of project files are now given to work with our example program:

Version 1:

```
main.cpp
absmotor.cpp
pport.cpp
```

```
motor.cpp
dcmotor.cpp
stepper.cpp
```

When the project file has been processed, all .cpp files will be compiled individually to form .obj files. Then all .obj files will be linked with any other system related .obj or .lib files to form the final executable file.

If all files associated with the classes are available as object files (.obj), the project file would contain the following:

Version 2:

```
main.cpp
absmotor.obj
pport.obj
motor.obj
dcmotor.obj
stepper.obj
```

The main.cpp file will be compiled to form an .obj file. Then all .obj files will be linked with any other system related .obj or .lib files to form the .exe file.

It is also possible to form one .lib file combining all the files related to all classes of the hierarchy. Suppose we had a library file created named motors.lib. Then the project file would be as follows:

Version 3:

```
main.cpp
motors.lib
```

In this case, the file main.cpp will be compiled and linked with motors.lib and any other related .obj or .lib system files to form the .exe file.

Make files

The other option for automating compilation and linking is to generate a make file containing a sequence of commands used to compile and link all files needed to form an executable file. Commands can be placed in the make file to provide the required variety of options, such as compile only, link only, compile and link, etc.

A make file for the motor drive program that will generate the executable file drive.exe is given below in Listing 9-8. Replace the "CC" characters with the actual command line of the C++ compiler you are using to generate an operational make file. The file names used must be the exact file names. For example, if your compiler generates pport.o as the object file instead of pport.obj, then change all .obj file extensions to .o.

In a make file we include *dependencies* and command-line compiler commands. The dependencies must start at the left-most column of a line in a make file as shown in the following statement from the make file of Listing 9-8:

```
drive: drive1.obj pport.obj dcmotor.obj stepper.obj motor.obj
absmotor.obj
```

This statement informs the make utility that if any of the files listed after the colon have been changed, then the executable file `drive.exe` will be re-generated.

Listing 9-8 Make file example for `drive.exe`.

Makefile

```
#makefile for drive program
drive: drive1.obj pport.obj dcmotor.obj stepper.obj
          motor.obj absmotor.obj
    CC -edrive drive1.obj pport.obj dcmotor.obj
          stepper.obj motor.obj absmotor.obj
drive1.obj : drive1.cpp motor.h stepper.h dcmotor.h
motor.h : pport.h absmotor.h
stepper.h : motor.h
dcmotor.h : motor.h
    CC -c drive1.cpp
pport.obj : pport.cpp pport.h
    CC -c pport.cpp
dcmotor.obj : dcmotor.cpp dcmotor.h
    CC -c dcmotor.cpp
stepper.obj : stepper.cpp stepper.h
    CC -c stepper.cpp
motor.obj : motor.cpp motor.h
    CC -c motor.cpp
absmotor.obj : absmotor.cpp absmotor.h
    CC -c absmotor.cpp
```

Command-line compiler commands must not start at the left-most column of a line; instead they start after a tab character as shown below:

```
CC -edrive drive1.obj pport.obj dcmotor.obj
          stepper.obj motor.obj absmotor.obj
```

These commands provide information to the make utility for generating the `drive.exe` file. Note that the 'switch' `-e` informs the compiler that the executable file is to be given the name `drive.exe`. All other lines in the make file follow the rules just described. The switch `-c` represents compile only.

The make utility processes a file whose default name is `makefile`. By giving the file shown in Listing 9-8 the name `makefile`, the make utility can be invoked by entering the command `make` at the command prompt to generate `drive.exe`.

9.5 Summary

In the first planning stages of a program, the program's operation should be described using pseudo-code. This description can be refined when working towards a realisable C++ program which has the objects that are needed in the program identified and then defined. These objects can then be organised into a suitable hierarchy; this is extremely beneficial for determining if the efficient use of virtual functions can be employed.

Many typographical errors can be produced when writing a program. When this occurs the compiler will detect the errors and notify the programmer accordingly. Some of these problems can be avoided by using the copy and paste facilities available in modern editors. Copy and paste operations are readily facilitated by using good layout practices when developing a program's classes. This simplifies the process of defining functions and reduces the possibility of errors. Improper use of nested parenthesis and brackets is also a common cause of compilation and run-time errors. Several good habits have been demonstrated to help avoid these problems. Indentation also plays an important role in identifying levels of nesting, improving readability, and thereby reducing the likelihood of errors in the source file.

The modular approach to program development requires the generation of header files and functions files for each class. When distributing object classes to programmers or using object classes in a program, only the required modules need to be distributed or used. Because modular programs have multiple files, generating their executable files is more involved than for programs that use only one source file. The executable files for modular files are generated using either a project file or a make file to simplify and automate this process.

9.6 Bibliography

Meyer, B., *Object Oriented Software Construction*, Prentice Hall, 1988

Staugaard A. C. (Jr), *Structured and Object Oriented Techniques*, Prentice Hall, 1997.

Lafore, R. *Object Oriented Programming in MICROSOFT C++*, Waite Group Press, 1992

Wang, P.S., *C++ with Object Oriented Programming*, PWS Publishing, 1994.

Winston, P.H., *On to C++*, Addison Wesley, 1994.

10

Voltage and Temperature Measurement

Inside this Chapter

- Voltage-to-Frequency Conversion (VFC) using a Voltage-Controlled Oscillator (VCO).

- Temperature sensing using thermistors.

- Object class for the VCO.

- Pulse counting.

- Graphics programming.

- Programs for voltage and temperature measurement.

10.1 Introduction

This chapter describes a means of converting an analog voltage to a digital pulse-train, the frequency of which is proportional to the applied voltage. This provides an excellent means of measuring analog voltages using a single digital input. We will be using a device known as a Voltage-Controlled Oscillator (VCO) to carry out this 'analog-to-digital conversion'.

Software will be developed that measures the period of the pulse-train to quantify the applied voltage. The operation of a temperature sensitive resistor will also be described and this device will be used to measure actual temperature. This chapter also introduces graphics programming, where a graphics program is developed to display the digital pulse-train on-screen.

10.2 Converting a Voltage to a Digital Pulse-train

One of the simplest forms of analog-to-digital converter is the *voltage-to-frequency converter* (VFC). The voltage-to-frequency converter produces a digital pulse-train whose frequency is proportional to the voltage applied to the converter input. A specialised type of VFC is the *voltage-controlled oscillator* (VCO) which produces either a sinusoidal or a square waveform.

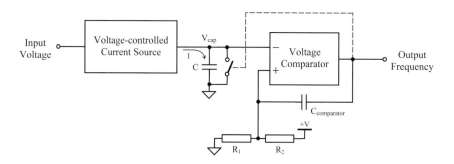

Figure 10-1 Typical voltage-to-frequency converter (VFC).

As shown in Figure 10-1, a typical VFC operates using a current source to charge a capacitor C with current that is proportional to the voltage applied to its input.

The output of the voltage comparator (initially a positive voltage) changes state when the voltage across the capacitor C, connected to its −ve input, rises to exceed the positive voltage at its +ve pin. When this happens, the output of the comparator will change polarity to a negative voltage and activate the switch closure across capacitor C. This action discharges capacitor C and also brings the comparator's resistor-capacitor circuit, connected to the comparator +ve pin, to a negative voltage.

Following these events, the voltage generated at the +ve pin by the comparator's resistor-capacitor circuit will increase, eventually exceeding the 'zero' volts across the discharged capacitor C. When this happens, the comparator output will revert to a positive voltage level. The switch across capacitor C will open, allowing current to flow into capacitor C and charge the voltage at the –ve pin until it again exceeds the positive voltage at the +ve pin. This marks the completion of one VFC cycle. This process repeats continuously, producing a digital pulse-train at the VFC output. As the input voltage increases, the level of current charging the capacitor C increases and the time to reach the voltage at the comparator +ve pin falls, leading to an increase in the frequency of the output signal.

The interface board uses a VCO housed within part of a CMOS 4046 phase-lock loop integrated circuit. The phase-lock loop device can be used for a range of purposes, however, we have configured it to use just its VCO. Note: the VCO input voltage range that generates a linear output is approximately 1.5V to 3.5V. Its output frequency range is set by two resistors and a capacitor connected to its pins.

10.3 Temperature Measurement

There are many types of electrical sensors that are sensitive to changes in temperature. These include *thermistors*, *thermocouples*, *thermally sensitive capacitors*, *semiconductor diodes*, and *quartz crystals*.

Thermistors are one of the more popular temperature sensors in use and the only sensor described in this text. They are simply resistors with very high temperature coefficients, usually having a *negative temperature coefficient* (NTC). A negative temperature coefficient is one in which the resistance decreases as the temperature of the thermistor increases. Thermistors have an exponential change in their resistance with temperature, making them a little difficult to work with. However, they are low in cost, have high sensitivity, and are small in size.

The simplest means of implementing temperature measurement with a thermistor is by connecting the thermistor in a voltage divider circuit as shown in Figure 10-2.

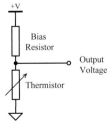

Figure 10-2 Thermistor voltage divider circuit.

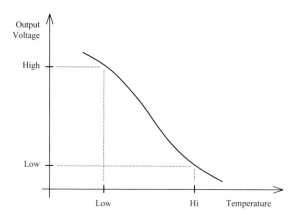

Figure 10-3 Typical curve – voltage divider output vs temperature (NTC thermistor).

The output voltage of this circuit will drop as the temperature increases for negative coefficient thermistors as shown in Figure 10-3. The shape of this curve can be brought closest to a straight line when the value of the bias resistor in the voltage divider circuit is chosen to equal the resistance of the thermistor approximately midway through its temperature range. The circuit output may need to be buffered if a significant level of current must be supplied to the electronics that senses this output voltage (in our case the VCO). Note that time is required for the body of the thermistor to reach the temperature of the object or medium it is placed into contact with. Also, the temperature of its body can be adversely affected if excessive current is passed through the device.

10.4 The Object Class vco

The output voltage from the thermistor voltage divider circuit is connected to the VCO input pin. This voltage determines the output frequency (and hence the period) of the VCO's pulse-train. The higher the voltage that is applied to the input of the VCO, the higher the frequency of the pulse-train. If we develop a method to measure the period of the pulse-train (i.e. the time to complete one cycle) we can evaluate the frequency. As expected we will be using the parallel port of the PC to interface the VCO output to the computer.

We can develop a new object to provide for future use of the VCO in other program applications. This object will need to process the pulse-train signal received at the parallel port and measure its period by detecting its signal level. By signal level, we mean the logic-high or logic-low status of the signal at any given instant. To develop this new object we can use the class ParallelPort as the base class. The class definition for the new VCO class is given in Listing 10-1.

Listing 10-1 The class definition for the VCO class, file – vco.h.

```
#ifndef VcoH
#define VcoH

#include "pport.h"

enum BITNUMBER{Bit7=0x80,Bit6=0x40,Bit5=0x20,
                          Bit4=0x10,Bit3=0x08};

class VCO : public ParallelPort
{
    private:
        long int Period;
        BITNUMBER Bit;

    public:
        VCO(int baseaddress = 0x378, BITNUMBER bit=Bit3);
        long int MeasurePeriod();
        long int GetPeriod();
        int SignalLevel();
        virtual ~VCO(){}
};
#endif
```

The VCO class has two data members and five member functions. The data member Period will store the measured period. We will use the port at address BASE+1 to read the VCO output into the computer via the parallel port. Note that the VCO output can be connected to any bit number between bit 3 and bit 7 inclusive of the port at address BASE+1. To provide the user with the flexibility to connect the VCO output to any of these bits, we pass a parameter to the constructor that specifies the bit number. The enumerated data type BITNUMBER is created for this purpose, and the enumerated identifiers are assigned integer values. The use of these values will be understood once we define all the member functions. We have used Bit3 as the default bit that will interface to the VCO output.

A strategy must be developed to measure the period of the pulse-train before defining the member functions of the VCO class. We can monitor the logic level of the VCO output by reading the port at address BASE+1 and then filtering out all unwanted bits. We can recognise the start of a pulse by detecting a transition in signal level - from low to high, or from high to low. After the transition we can begin measuring the pulse period. The signal must undergo two more transitions to complete one cycle as shown in Figure 10-4.

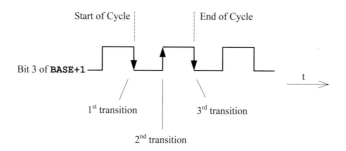

Figure 10-4 Detection of one complete cycle.

Since the signal level plays an important role, let us first establish the member function SignalLevel().

Listing 10-2 The member function SignalLevel().

```
int VCO::SignalLevel()
{
    if(ReadPort1() & Bit)
        return 1;
    else
        return 0;
}
```

The signal level will either be logic-low or logic-high at any given time. To signify these two states we can use 0 and 1 respectively. The function SignalLevel() shown in Listing 10-2, returns 0 or 1 depending on the logic level of the VCO output.

First the inherited member function ReadPort1() is used to read the port at address BASE+1. An AND operation with Bit is used to filter out all bits except the bit we have allocated for use with the VCO output. If the signal is logic-high, the AND result will be equal to Bit which is non-zero, the if condition will evaluate to a non-zero value and 1 will be returned. However, if the logic level of the signal is low, the result of the AND operation will be zero and a 0 is returned.

A signal transition can be detected if we read the port continuously, checking for a *change* in the logic level of the bit used. Measurement of the period begins as the first transition is detected and will be complete once the next two signal level transitions have been detected. The ideal means to measure the period of the pulse-train is to use real-time techniques, however, this topic is yet to be covered (in Chapter 13). Instead, we will use software loops to measure the pulse-train period by counting the number of times the port is read within a pulse-train cycle. It may be possible to evaluate the period as an actual time value if the time for one port read and filter operation was known. However, program execution times will vary for different computers, and are also affected by intervening system events.

The steps required to measure the period are as follows (assume that the signal is connected to bit 3):

1.   Initialise a counter variable to zero.
2.   Repeatedly read the signal level of bit 3 until a change is detected.
3.   Repeatedly read the signal level of bit 3 until the second change is detected while incrementing the counter after each read.
4.   Repeatedly read the signal level of bit 3 until the third change is detected while incrementing the counter after each read.
5.   Return the counter value as the period.

These steps (the *pseudo code*) can now be converted to form the function `MeasurePeriod()`. Examining the pseudo-code; steps 2, 3 and 4 are very similar. We will start by developing code for step 2. It is then possible for us to use the same code for step 3 and 4, and add the incrementing of the counter after each read.

Step 2 of the pseudo-code can be implemented as shown in Listing 10-3. `SignalLevel()` uses the inherited function `ReadPort1()` of the class `ParallelPort`.

Listing 10-3 Detecting a signal transition in the VCO output.

```
VCO Vco;
unsigned char Signal;

Signal = Vco.SignalLevel();
while(Signal == Vco.SignalLevel());
```

First, the variable `Signal` is used to store the current signal level by calling the member function `SignalLevel()`. Then the program enters a `while` loop, where the signal level is repeatedly read by calling the `SignalLevel()` function and its result compared with the previously stored value of the signal level. While they are equal, the `while` loop will continue to execute. Note that the body of the `while` loop is empty - containing only a semi-colon. When a change in signal level is detected, the `while` condition will evaluate to false and the `while` loop will terminate.

Definitions for all the member functions of the VCO class are given in Listing 10-4. The function `MeasurePeriod()` determines the period of the signal being measured and returns this value. The `GetPeriod()` function merely accesses and returns the value of the private data member `Period`. The development of the VCO class is now complete. In the next section we will learn how to use VCO objects in a program that will measure voltages.

Listing 10-4 Member function definitions of the VCO class – vco.cpp.

```cpp
#include "vco.h"

VCO::VCO(int baseaddress, BITNUMBER bit)
    :ParallelPort(baseaddress)
{
    Bit = bit;
    Period = 0;
}

long int VCO::MeasurePeriod()
{
    unsigned char Signal;
    Period = 0;

    Signal = SignalLevel();
    while(Signal== SignalLevel())
                ; // empty body

    Signal = SignalLevel();
    while(Signal== SignalLevel())
        Period++;

    Signal = SignalLevel();
    while(Signal== SignalLevel())
        Period++;

    return Period;
}

long int VCO::GetPeriod()
{
    return Period;
}

int VCO::SignalLevel()
{
    if(ReadPort1() & Bit)
        return 1;
    else
        return 0;
}
```

10.5 Measuring Voltages Using the VCO

As described in section 10.2, the voltage-controlled oscillator (VCO) is an electronic circuit that generates a square wave at a frequency proportional to the analog voltage applied to its input. Various voltage levels need to be applied to its input to verify its proper operation. The easiest way to do this is to connect the output of the interface board's potentiometer to the VCO input. We can measure the potentiometer output voltage with a voltmeter to establish a quantifiable relationship between the VCO output and the voltage applied to the VCO input. A different approach which eliminates the need for a voltmeter, is to use the output of the DAC circuit to generate a known analog voltage. We can easily control the DAC output by writing a number to it as explained in Chapter 6.

The following keyboard controls will be implemented within the `main()` function for easy use of the program:

- Pressing the up arrow will increase the output voltage of the DAC.
- Pressing the down arrow will decrease the output voltage of the DAC.
- Pressing Alt-X will exit the program.

A fragment of skeleton code is given in Listing 10-5 to implement the above steps.

Listing 10-5 Implementing keyboard control.

```
    int Quit = 0, key;

    while(!Quit)
    {
    // Insert lines to measure and display the pulse period

    if(bioskey(1)!=0) // check if a key is pressed
        {
            key = bioskey(0); // read key code

            switch(key)
            {
/* Alt-X */     case 0x2d00 : Quit = 1;
                              break;
/* Up Arrow */ case 0x4800 : //Increase DAC output
                              break;
/* Down Arrow */ case 0x5000 : //Decrease DAC output
                              ; // Empty statement
            }
        }
    }
```

The `while` loop will eventually contain code to measure and display the pulse period. This code is not shown yet - instead, a comment is included to that effect. The next statement in the `while` loop is an `if` statement that contains a `switch` statement. Program control will be transferred to the true clause of the `if` statement when the `if` condition evaluates to true - when a key is pressed. The `bioskey()` function (when passed an actual argument of 1) will return true if a key has been pressed. If no key is pressed it will return false. The `bioskey()` function will *not wait* for a key press. This means that if a key has not been pressed, the body of the `if` statement will be skipped, and the other statements of the `while` loop (such as those used to measure the pulse period and display it) will continue to execute.

If a key has been pressed, program control will be transferred to the body of the `if` statement where its true clause will be executed. Within the true clause, the first action to perform is to determine the actual key pressed. The `bioskey()` function can be used for this purpose, although to do this it must be passed an actual argument of 0. The `bioskey()` function will now retrieve the key code of the key pressed. The three key codes corresponding to the keys; up arrow, down arrow, and Alt-X are listed as cases of the `switch` statement. Within the `switch` statement; if the key code of the pressed key matches one of the cases, program control will be transferred to that case. If no matching cases are found, no action will be taken. In the case of Alt-X, the `case` statement sets the variable `Quit` to 1. This will cause the `while` loop to terminate the next time the `while` condition is evaluated. Programming statements for the two cases corresponding to the up arrow and the down arrow are not included yet. Instead, comments are used in their place.

A `main()` function that sends the DAC output to the VCO and measures the pulse period of the VCO output is given in Listing 10-6. The comment lines in Listing 10-5 have now been replaced by actual programming statements. Note that the program is written to jointly operate the VCO and the DAC. Therefore, a `DAC` object named `Dac` and a `VCO` object named `Vco` have been instantiated at the start of the `main()` function.

Listing 10-6 Measuring VCO pulse period (DAC driving VCO input) – period.cpp.

```
#include <bios.h>
#include <conio.h>

#include "dac.h"
#include "vco.h"


void main()
{
    DAC Dac;
```

```
        VCO Vco;
        int Quit = 0, key;
        unsigned char DACbyte;

        clrscr();
        Dac.SendData(0); // initialise to zero

        while(!Quit)
        {
            gotoxy(10,10);
            cprintf("The pulse period is %10lu\a",
                Vco.MeasurePeriod()/1000);

            if(bioskey(1)!=0)
            {
                DACbyte = Dac.GetLastOutput();
                key = bioskey(0);
                switch(key)
                {
/*Alt-X*/           case 0x2d00 : Quit = 1;
                                  break;
/*Up Arrow*/    case 0x4800:
                            if(DACbyte>247) // limit max value
                                DACbyte = 247;
                            Dac.SendData(DACbyte+8);
                                  break;
/*Down Arrow*/  case 0x5000 :
                            if(DACbyte<8) // limit min value
                                DACbyte = 8;
                            Dac.SendData(DACbyte-8);
                }
            }
        }
}
```

The gotoxy() function locates the cursor at screen coordinates (10,10). The first number within the above pair of parentheses is referred to as the 'X coordinate' and is measured from the left edge of the screen. The second number is referred to as the 'Y coordinate' and is measured from the top edge of the screen. Screen coordinates are shown in Figure 10-5. Therefore, the measured pulse period will be displayed *starting* at screen coordinates (10,10). The function cprintf() is similar to the printf() function we saw previously, with the exception that it does not convert the new line character combination (\n) to a new line and carriage return combination (\n\r). It is especially designed to send output to the screen. In general, `gotoxy()` is used to set the position of the cursor, and therefore there is no need for a line feed or carriage return.

The measured value of the pulse period is a 'representation' of the square wave period. This value is obtained by using the member function `MeasurePeriod()` of the `Vco` object. We use the `cprintf()` function to call the `MeasurePeriod()` function. The `cprintf()` function prints the measured value divided by 1000 on the screen.

The member function `GetLastOutput()` of the `DAC` class is called to obtain the previous value output to the DAC. This value is then used within the `switch` statement block to ensure that the byte being sent to the DAC is kept within its operating range of 0 to 255 for each press of the up or down arrow key. The two cases corresponding to the up arrow and the down arrow have been implemented using the `SendData()` function of the `DAC` class. Depending whether the up or down arrow key has been pressed, the value sent to the DAC is either incremented or decremented by 8. During execution of the `SendData()` function, the data member `LastOutput` of the `DAC` class is updated to store the value just output.

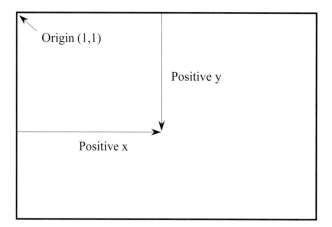

Figure 10-5 Screen coordinates in text mode.

Three code modules are required to generate an executable program for the code segment shown in Listing 10-6. These are the `ParallelPort`, `VCO`, and `DAC` modules. A project file (or make file) must be formed to compile all modules and link them together to form the executable file. The VCO module consists of the header file given in Listing 10-1 and the function file given in Listing 10-4. The `ParallelPort` class header file and its function file were formed in Section 9.4 and are repeated in Listing 10-7 and Listing 10-8, respectively.

Listing 10-7 Header file for the `ParallelPort` class - pport.h.

```
#ifndef PportH
#define PportH

class ParallelPort
{
    private:
        unsigned int BaseAddress;
        unsigned char InDataPort1;

    public:
        ParallelPort();
        ParallelPort(int baseaddress);
        void WritePort0(unsigned char data);
        void WritePort2(unsigned char data);
        unsigned char ReadPort1();
        virtual ~ParallelPort(){}
};
#endif
```

Listing 10-8 Function file for the `ParallelPort` class - pport.cpp.

```
#include <dos.h>
#include "pport.h"

ParallelPort::ParallelPort()
{
    BaseAddress = 0x378;
    InDataPort1 = 0;
}

ParallelPort::ParallelPort(int baseaddress)
{
    BaseAddress = baseaddress;
    InDataPort1 = 0;
}

void ParallelPort::WritePort0(unsigned char data)
{
    outportb(BaseAddress,data);
}

void ParallelPort::WritePort2(unsigned char data)
```

```
{
    outportb(BaseAddress+2,data ^ 0x0B);
}

unsigned char ParallelPort::ReadPort1()
{
    InDataPort1 = inportb(BaseAddress+1);
// Inverting Most significant bit to compensate
// for internal inversion by printer port hardware.
    InDataPort1 ^= 0x80;
// Filter to clear unused data bits D0, D1 and D2 to zero.
    InDataPort1 &= 0xF8;
    return InDataPort1;
}
```

So far we haven't created a header file and a function file for the DAC module. These files are given in Listing 10-9 and Listing 10-10 respectively.

Listing 10-9 The header file for the DAC class - dac.h.

```
#ifndef DacH
#define DacH

#include "pport.h"

class DAC : public ParallelPort
{
    private:
        unsigned char LastOutput;

    public:
        DAC();
        DAC(int baseaddress);
        void SendData(unsigned char data);
        unsigned char GetLastOutput();
        ~DAC(){};
};
#endif
```

Listing 10-10 The function file for the DAC class - dac.cpp.

```
#include "dac.h"
```

```
DAC::DAC()
{
    LastOutput = 0;
}

DAC::DAC(int baseaddress) : ParallelPort(baseaddress)
{
    LastOutput = 0;
}

void DAC::SendData(unsigned char data)
{
    ParallelPort::WritePort0(data);
    LastOutput = data;
}

unsigned char DAC::GetLastOutput()
{
    return LastOutput;
}
```

Executable File Generation

Required Files	Listing No.	Project File Contents
pport.cpp	Listing 10-8	pport.cpp
pport.h	Listing 10-7	
vco.cpp	Listing 10-4	vco.cpp
vco.h	Listing 10-1	
dac.cpp	Listing 10-10	dac.cpp
dac.h	Listing 10-9	
period.cpp	Listing 10-6	period.cpp

The table shown above lists all the files needed to form the executable file that should be stored in the one directory. Form a project file using the program development environment of your choice and add the files that are listed in the column titled 'Project File Contents'. Then the compiler and linker can be directed to form the executable file. Tables such as the one shown above will be provided in this text whenever modules must be combined to form an executable file.

Make the connections on the interface board as shown in Table 10-1 to Table 10-3 before executing the program. These tables list the wiring needed to control the DAC and the VCO. Set the DAC output to *unipolar* mode by fitting the jumper across the position on the board marked LINK1. Remember to connect an operational 9V battery to its terminal block (J14) to allow proper DAC operation. Note that the VCO response is linear (typically 1%) for input voltages in the range of 2.2V to 2.8V. Although linearity deteriorates outside this range, the VCO can be characterised across its entire input range and used effectively.

Table 10-1 Connections for the DAC.

BASE Address (Buffer IC, U13)	DAC0800 (U8)
D0	D0 (12)
D1	D1 (11)
D2	D2 (10)
D3	D3 (9)
D4	D4 (8)
D5	D5 (7)
D6	D6 (6)
D7	D7 (5)

Table 10-2 INPUT connections for the VCO.

LM358 (U10B)	VCO (4046, U4)
VDAC (7)	VIN (9)

Table 10-3 OUTPUT connections for the VCO.

VCO (4046, U4)	BASE+1 Address (Buffer IC, U6)
VCO OUTPUT (4)	D3

NOTE

If any malfunction occurs; first check the 9V battery is operational – its voltage should be greater than 7V when it is being used.

10.6 Graphics Programming – Square Wave Display

A program was developed in Section 10.5 that can measure the period of the square wave generated by the VCO and produce a simple numerical output. In this section we will use graphics programming to generate a graphical display so the user can visualise the signal's waveform.

The potentiometer circuit on the interface board provides a very convenient means of generating an analog voltage to apply to the input of the VCO. Varying the position of the potentiometer will change its output voltage (0V to +5V) and hence change the output frequency of the VCO. The connections that need to be made between the potentiometer and the VCO are shown in Table 10-4 and Table 10-5.

Table 10-4 INPUT connections for the VCO.

Potentiometer (POT1)	VCO (4046, U4)
OUTPUT	VIN (9)

Table 10-5 OUTPUT connections for the VCO.

VCO (4046, U4)	BASE+1 Address (Buffer IC, U6)
VCO OUTPUT (4)	D3

10.6.1 Screen Programming

This program to be developed will display the signal from the VCO as a waveform inside a fixed area of the screen. The waveform being displayed will trace across the screen similar to the trace of an oscilloscope. This must happen in real-time; the changes shown on-screen matching the instantaneous changes of the VCO signal.

The standard library provides many graphics routines for our program to use. These routines can determine which graphics driver should be used, the appropriate graphics mode, the maximum number of pixels in x and y directions, etc. The screen uses an array of pixels, where each pixel is one element of the screen that is individually illuminated to form part of the picture. Because different screens contain different numbers of pixels, it is often necessary to determine the screen's pixel count before deciding the size of the display area to be used by a program.

In any graphics program running under DOS, the system must first be configured in a graphics mode that uses a graphics *driver*. A driver is a module of executable code that is used to drive the actual graphics output. These drivers can operate in different modes that set the number of pixels used in x and y directions, and set which colour palette to use. The program must set the system in a suitable graphics mode and then determine the number of pixels in the x and y directions. This information allows the program to calculate the screen coordinates needed to centre the waveform on-screen inside the area known as the *Viewport*. Figure 10-6 shows the screen coordinates and the calculations performed by the program's functions to generate the waveform.

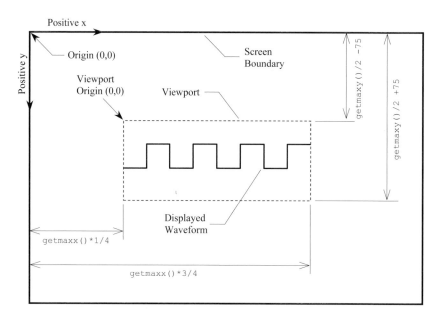

Figure 10-6 Arrangement to display the VCO output.

The program will use half of the x pixel-range and one hundred and fifty pixels in the vertical direction for its Viewport, centred on-screen in both the horizontal and vertical directions. We have separated the high and low levels of the waveform by 50 pixels in the vertical direction. The functions `getmaxx()` and `getmaxy()` are functions that can be used to determine the number of pixels in the x and y directions. The Viewport is now established (with its origin located at its upper left-most corner).

The waveform can be plotted as a line joining sequential points, explained as follows (note: the positive y direction is *down*). When the port is read; if the signal level is high, the `Y` coordinate of the current point will be `50` pixels in the y direction from the Viewport Origin. If the signal level is low, the `Y` coordinate will be `100` pixels in the y direction from the Viewport Origin. The first VCO value read will be plotted 0 pixels from the Viewport Origin in the x direction. The next point will be plotted at `x=1`, the following at `2`, and so on. When we reach the end in the x direction, we must re-start plotting from `x=0`, but not before erasing the current waveform being displayed. This plotting process will repeat continuously until the program detects a keypress and then terminates.

The required program steps can be listed as follows:

1. Initialise graphics and set the graphics mode.
2. Determine the maximum number of pixels in the x and y directions.
3. Configure the Viewport.
4. Enter a `while` loop conditioned on `!kbhit()`. If any key is pressed, terminate the program.

5. Read the port and obtain the output signal level of the VCO.

6. Plot the pixel according to the signal level (high or low) and increment the x pixel count.

7. If the x pixel count has not reached the end of its range, return to step 4. Else, reset the x pixel count to restart plotting, clear the view port, and return to step 4.

A program that carries out the above set of tasks is given in Listing 10-11. The appearance of the program's display is basic and could be improved by adding some finishing touches.

Listing 10-11 Graphically display the VCO output – trace.cpp.

```
/************************************************
The frequency of the pulse-train being output by the
voltage-controlled oscillator will change as we change
the analog input voltage to the VCO circuit. The
Potentiometer (POT1) on the interface board generates
the input voltage to the VCO and the program reads the
pulse-train being output by the VCO.  This pulse-train
is graphically displayed on-screen.
************************************************/
#include <graphics.h>
#include <stdlib.h>
#include <iostream.h>
#include <conio.h>
#include <dos.h>

#include "vco.h"

void main()
{
    VCO Vco;
    int i=0; // controls plotting in the x range
    int SignalLevel;
    int Driver = DETECT, GraphicsMode, ErrorCode;
    int X, Y;

// set to graphics mode
    initgraph(&Driver, &GraphicsMode, "");

// check for error codes
    ErrorCode = graphresult();
    if (ErrorCode != grOk)
    {
```

```
        cout << "Graphics error:   "
            << grapherrormsg(ErrorCode) << endl;
        cout << "Press any key to halt:" <<  endl;
        getch();
        exit(1);
    }

    X = getmaxx();
    Y = getmaxy();
    rectangle(X/4-1, Y/2-76,X*3/4+1,Y/2+76); // border
    setviewport(X/4, Y/2-75,X/4*3,Y/2+75,1);

    while(!kbhit())
    {
        SignalLevel = Vco.SignalLevel();
        if(SignalLevel == 0) // low level
            lineto(i,100);
        else                 // high level
            lineto(i,50);

        i++;
        delay(2);

        if(i > X/2) // half screen = Viewport width
        {
            i = 0;
            while(Vco.SignalLevel()); // wait for low level

            // wait for signal level to go high again
            while(!Vco.SignalLevel());
            clearviewport();
        }
    }
}
```

Executable File Generation		
Required Files	**Listing No.**	**Project File Contents**
pport.cpp	Listing 10-8	pport.cpp
pport.h	Listing 10-7	
vco.cpp	Listing 10-4	vco.cpp
vco.h	Listing 10-1	
trace.cpp	Listing 10-11	trace.cpp

The file `graphics.h` is needed for all graphics-related routines such as `initgraph()`, `grapherror()`, `moveto()`, `lineto()`, etc, and for use of the constants `DETECT` and `grOk`. The class `VCO` is used to create an instance of `VCO` named `Vco` as in the previous program. Several variables of type `int` are declared inside `main()`. Variables `X` and `Y` will initially be used to store the maximum number of pixels in x and y directions respectively. At later stages in the program they will be used for other purposes. Variable `i` is used to control plotting of pixels in the Viewport's x range. It will be reset to 0 for the start of a new plot when the trace reaches the end of the Viewport range.

The variables `Driver` and `GraphicsMode` are explained together with the `initgraph()` function. The first parameter to `initgraph()` must specify the type of graphics driver. The driver could be for the Colour Graphics Adapter (CGA), Enhanced Graphics Adapter (EGA), Video Graphics Array (VGA), etc. If the value of `Driver` is set to 1, then CGA is specified; if it is set to 2, EGA is specified. A description of these constants should be found in the documentation for `initgraph()`. A number of graphics modes will be available for each driver to generate the resolution (number of pixels) and the colour palette used. For example; 16 colour, 640 x 480 screen resolution is specified by assigning `GraphicsMode` the value 2. If `GraphicsMode` is assigned the value 1, the screen will use 16 colours and a resolution of 640 x 320. However, when the value of `Driver` is set to `DETECT` (predefined to be 0), the program will automatically detect the driver suitable for the computer's graphics card and set the resolution to the highest available. In this situation, `GraphicsMode` does not need to be assigned a value. The third parameter to `initgraph()` is a string specifying the path to the graphics driver, in this case the location of the file `EGAVGA.BGI`. If the graphics driver is in your current directory (the directory where you have your executable file) then this field can have an empty string. Note that, when the call to `initgraph()` is made, the first two arguments are preceded by the `&` character. This is needed because `initgraph()` takes these arguments as pointers (i.e. an address value).

To determine if `initgraph()` has successfully completed its task, we call `graphresult()` and store the value returned by `graphresult()` in `ErrorCode`. If `ErrorCode` is not equal to the predefined constant `grOk`, then an error has occurred. A message corresponding to the value in `ErrorCode` can be generated by calling `grapherrormsg()`. The true clause of the `if` statement will display the error messages and then call the `exit()` function to terminate the program. If no errors occurred, program execution will proceed to carry out the next task – to determine the maximum number of pixels in x and y directions. A rectangle will be drawn just one pixel outside the chosen Viewport followed by configuration of the Viewport. As explained previously, the Viewport is the area where the waveform will be displayed. Once the Viewport is established, the origin (0,0) becomes the upper left-corner of the Viewport.

The `while` loop, conditioned on `!kbhit()`, is used to continuously display the waveform on-screen. The `lineto()` function uses the new coordinate frame of the Viewport. The value of i will be zero when beginning to plot a new trace. A line will be drawn from the previous screen position to the new vertical position determined by the signal level. The x plot position is then incremented for the next plot. When the value of i reaches the end of its range in the x direction (Viewport width = X/2), i is reset to zero for a new plot. The remaining code synchronises the plotting so the next trace will always begin on a low level. The Viewport is then cleared to erase the current trace.

Note that since interrupts are enabled, some of the pulses displayed on-screen may have wider widths due to time consumed by interrupt service routines.

10.7 Temperature Measurement

We measure temperature indirectly by using the analog voltage generated by the thermistor resistive-divider circuit. The voltage being generated drops in a non-linear manner as the temperature increases whan a negative temperature coefficient thermistor is used in the resistive-divider circuit (as shown in Figure 10-3). To simplify our programming let us approximate the curve by a straight line. We can develop a program that will measure the actual temperature using the same wiring as described in Section 10.6 (except the POT output is replaced with the Thermistor output, VTH). A typical thermistor/VCO relationship is shown in Figure 10-7.

10.7.1 Thermistor Calibration

The program needs to measure the cycle time of the VCO output and interpret this value as temperature. The first task is to calibrate the thermistor. This is done by subjecting the thermistor to known temperatures such as that of ice, the body, and say boiling water to obtain measures of corresponding cycle times. Then we can establish a calibration equation or calibration table which can be used to extrapolate or interpolate values of temperature (within linearity limits of the thermistor/VCO circuit response). Note that the output of the thermistor circuit may extend well beyond the linearity range of the VCO (approximately 1.5V to 3.5V). If the voltage applied to the VCO input is outside its linear range, the output from the VCO will be a distorted measure of the thermistor output. However, the temperature measuring system made up of the thermistor and VCO can still be calibrated and used, but with less accuracy.

The calibration equation can be determined as follows (refer to Figure 10-7). We can add a few extra statements to the program in Listing 10-6 to include a means of entering an upper temperature and a lower temperature. The corresponding cycle times can then be read and a calibration equation can be established. If no upper and lower temperatures are entered (HiTemp and LoTemp), the program will display the cycle time as did the program in Listing 10-6. If calibration has been performed correctly, the program will display the actual temperatures. This feature requires some logic to be built into the program. We can use flags to detect whether upper and lower temperatures have been entered. If both flags are set; that is, if both temperatures have been entered, we can establish the calibration equation. Then we can display temperature instead of cycle times. We will adhere to using the same keys as for Listing 10-6; Alt-X to quit the program, up arrow to enter an upper temperature, and down arrow to enter a lower temperature.

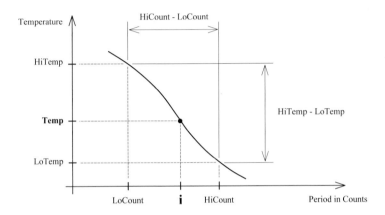

Figure 10-7 Typical curve - Thermistor circuit/VCO output (calibration).

The program steps can be listed as follows:

1. Initialise a counter to zero.
2. Repeatedly check the signal level until a change is detected.
3. Repeatedly check the signal level until the second change is detected while incrementing the counter after each read.
4. Repeatedly check the signal level until the third change is detected while incrementing the counter after each read.
5. Check if the calibration temperature for upper temperature and lower temperature has been entered (respective flags are both set).

   If they are both entered, use the calibration equation and display the temperature, else display the cycle time.
6. Check if a key has been pressed. If no key has been pressed return to step 1.
7. If the pressed key is Alt-X, exit the program.
8. If the up arrow key is pressed, read upper temperature. Return to step 1.
9. If the down arrow key is pressed, read lower temperature. Return to step 1.

Some of these steps can be expanded further as shown below:

   8.1 Ask user to enter the upper temperature and store value entered.
   8.2 Store the cycle time.
   8.3 Set the flag confirming the upper calibration temperature has been read.

9.1   Ask the user to enter the lower temperature and store value entered.

9.2   Store the cycle time.

9.3   Set the flag confirming the lower calibration temperature has been read.

The program that implements these steps is given in Listing 10-12.

Listing 10-12 Temperature measurement using thermistor & VCO – temp.cpp.

```
/*****************************************************
This program uses the thermistor on the interface board
to generate a voltage for input to the VCO, and then
repeatedly reads the cycle time of the VCO's output pulse-
train. It also allows you to calibrate the thermistor so
the program can display the actual temperature.
*****************************************************/
#include <iostream.h>
#include <bios.h>
#include <conio.h>
#include "vco.h"

void main()
{
    VCO Vco;
    int Quit=0, HiFlag = 0, LoFlag = 0;
    int key = 0;
    float HiTemp, LoTemp, Temp;
    long int HiCount, LoCount;

    clrscr();
    while(!Quit)
    {
        Vco.MeasurePeriod();
        clrscr();
        gotoxy(10,10);

        if((HiFlag == 1) && (LoFlag == 1))
        {
            Temp = LoTemp+(HiTemp-LoTemp)*
                (Vco.GetPeriod()-LoCount)/(HiCount-LoCount);

            cprintf("The temperature is:%7.1 lf (deg)\a",Temp);
        }
        else
            cprintf("The pulse period is: %10lu\a",
```

```
                    Vco.GetPeriod()/1000);

          if(bioskey(1)!=0)
          {
              key = bioskey(0);
              switch(key)
              {
/* Alt-X */        case 0x2d00 : Quit = 1;
                                 break;

/* Up Arrow */   case 0x4800 : gotoxy(10,5);
                                 cout << "Enter Upper
                                            Calibration Temp: ";
                                 cin  >> HiTemp;
                                 HiCount = Vco.GetPeriod();
                                 HiFlag = 1;
                                 break;

/* Down Arrow */ case 0x5000 : gotoxy(10,6);
                                 cout << "Enter Lower
                                            Calibration Temp: ";
                                 cin  >> LoTemp;
                                 LoCount = Vco.GetPeriod();
                                 LoFlag = 1;
              }
          }
      }
}
```

Executable File Generation		
Required Files	**Listing No.**	**Project File Contents**
pport.cpp	Listing 10-8	pport.cpp
pport.h	Listing 10-7	
vco.cpp	Listing 10-4	vco.cpp
vco.h	Listing 10-1	
temp.cpp	Listing 10-12	temp.cpp

The variable HiFlag is used to denote the upper calibration temperature has been entered, and similarly the variable LoFlag to denote the lower calibration temperature has been entered. Although HiFlag and LoFlag they are declared as integer variables, they will only be used with values of 0 or 1. The variables HighTemp and LowTemp will store the upper temperature and the lower temperature entered during calibration. The value of the pulse period (measured in counts) will be stored in the variable `HiCount` for the upper calibration temperature, and in the variable `LowCount` for the lower temperature. The actual temperature to be displayed is stored in the variable `Temp`.

The first `if` statement within `main()` tests whether both temperatures have been entered by checking the values of the flags `HiFlag` and `LoFlag`. If both flags are set, the temperature will be calculated using the calibration equation and printed on-screen.

Be aware of the importance of correctly specifying mathematical operations when calculating the value `Temp` in the program's formula:

```
Temp = LoTemp + (HiTemp-LoTemp)*
              (Vco.GetPeriod()-LoCount)/(HiCount-LoCount);
```

Note that `Vco.GetPeriod()`, `HiCount` and `LoCount` are long integer type, whereas `Temp`, `LoTemp` and `HiTemp` are float type. If we had placed a set of brackets around the expression shown on the lower line, the compiler would cast this part result to become a long integer number (incorrect – it should be a floating point number). Likewise, rearranging the order of mathematical operations can cause the compiler to implicitly cast part-expressions and change the result of an expression.

If both temperatures have not been entered yet, the calibration equation will not be used and the period is printed on the screen instead. The `switch` statement block is used to detect key presses for the up and down arrow keys, and the Alt-X key combination. If you press the up arrow key, you will be prompted to enter the upper temperature which will be stored in the variable `HiTemp`. The current value of `Vco.GetPeriod()` (returns the data member `Period`) will be stored in variable `HiCount`. The flag `HiFlag` will then be set to one. The equivalent procedure will be followed when the down arrow key is pressed to enter the lower calibration temperature.

Note: the thermistor requires time to reach the temperature of the body it is placed into contact with. Therefore, sufficient time must be allowed before pressing the up/down arrows to enter each calibration temperature. The program can be verified after it has been calibrated. Subject the thermistor to known temperatures and the program should display values close to those temperatures.

10.8 Summary

In this chapter we have described the operating principle of the Voltage-controlled Oscillator (VCO). The VCO produces a pulse-train having a frequency that is proportional to the voltage applied to its input. By measuring the frequency (or period as we did) the voltage/frequency relationship can be used to generate a measurement of voltage. In this way the VCO can be used as a simple and inexpensive alternative to an analog-to-digital converter.

A new object class named `VCO` was developed using the `ParallelPort` as the base class. Software methods have been described to continuously check the level of an incoming digital signal while incrementing a counter, and thereby measure the period of the waveform. Graphics programming was introduced to display the resulting waveform, followed by the development of a program that uses the thermistor on the interface board with the VCO to measure the actual temperature.

10.9 Bibliography

Bentley, J., *Principles of Measurement Systems*, Second edition, Longman Scientific & Technical, Essex, 1988.

Horowitz, P. and Hill, W., *The Art of Electronics*, Cambridge University Press, Cambridge, 1989.

NS CMOS, *CMOS Logic Databook*, National Semiconductor Corporation, 1988.

Webb, R.E., *Electronics for Scientists*, Ellis Horwood, New York, 1990.

Wobschall, D., *Circuit Design for Electronic Instrumentation*, McGraw-Hill, 1987.

Lafore, R. *Object Oriented Programming in MICROSOFT C++*, Waite Group Press, 1992.

Wang, P.S., *C++ with Object Oriented Programming*, PWS Publishing, 1994.

Winston, P.H., *On to C++*, Addison Wesley, 1994.

11

Analog-to-Digital Conversion

Inside this Chapter

- Analog-to-Digital Conversion (ADC) explained.

- Types of ADCs.

- Sampling Signals.

- An Object Class for the ADC.

- Voltage and Temperature Measurement using the ADC.

11.1 Introduction

This chapter explains the principles of *analog-to-digital conversion* and the operation of several commonly used types of analog-to-digital converters. This is followed by a discussion of the limitations encountered when sampling and converting signals.

Transducers measure physical quantities such as temperature, pressure, flow rate, and distance. Analog transducers typically output current, voltage, or charge, which form some mathematical relationship with the measured physical quantity. This mathematical relationship can be obtained using the calibration process we described in the previous chapter. An *analog-to-digital converter* (ADC) is typically used to interface these analog signals to a digital computer. Signal conditioning circuitry transforms the analog currents or charge into voltages that are sampled by the ADC system and converted to digital bit patterns.

Software is used to control the ADC on the interface board and read its output. This is made possible by deriving an object from the `ParallelPort` class and then encapsulating the functionality of the ADC. This new object will be used in our programs to measure analog voltages.

11.2 Analog-to-Digital Conversion

Analog-to-digital conversion is the process of sampling and then converting an analog signal, usually a voltage, to a multi-bit digital number that is proportional to the amplitude of the analog signal. Analog-to-digital conversion is used in many applications ranging from encoding of voice-generated signals in telecommunication systems, to data acquisition and control systems. Figure 11-1 shows the block diagram for a typical (8-bit) ADC. Conversion is initiated by activating the 'Start Conversion' input of the converter. At completion of the conversion process the 'Conversion Complete' output of the converter will change logic state. This signal is used to notify the controlling device that data conversion is complete, and valid data can now be read.

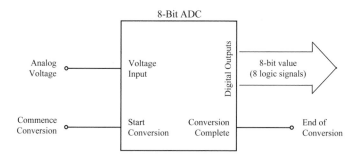

Figure 11-1 Block diagram of an 8-bit ADC.

The time that elapses from the start of conversion to the valid output of the digital code is referred to as the *conversion time*. The Conversion Complete output of the ADC can be ignored if the device requesting the converted data delays its reading of the data by a longer period than the conversion time.

An analog voltage signal has an infinite number of possible voltage levels within its range. The analog voltages are converted to digitally coded numbers by sampling and converting the analog signal into a fixed number of possible digital states or levels. This process is known as *quantisation*. For example, a 3-bit ADC can digitise an analog voltage and create digital numbers from zero to seven, which represents the analog voltage over a set range (say 0 to 3.5V) as shown in Figure 11-2 and Table 11-1. In this example the analog signal has been divided up or quantised into eight levels.

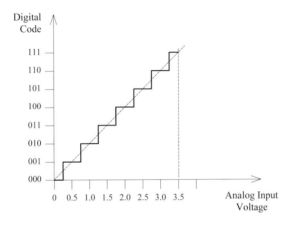

Figure 11-2 Ideal ADC Conversion.

Table 11-1 Quantisation of analog voltages to 3-bit code.

Quantised Analog Input Voltage	3-bit ADC Digital Code	Decimal
0.0 V	000	(0)
0.5 V	001	(1)
1.0 V	010	(2)
1.5 V	011	(3)
2.0 V	100	(4)
2.5 V	101	(5)
3.0 V	110	(6)
3.5 V	111	(7)

The digital code produced by the ADC will be correct and not contain any error when the analog input voltage corresponds exactly with a quantised voltage level. However, the ADC will incur an error known as *quantisation error*, when the input voltage is not exactly equal to the nearest quantised voltage level. For example, referring to Figure 11-2; zero quantisation error occurs when the analog input is equal to 0V, 0.5V, 1.0V, etc. The quantisation error is a maximum (equal to ½ a quantisation level) when the analog input voltage lies halfway between two quantisation levels – 0.25V for the previous example.

Another type of error can occur with analog-to-digital conversion, known as *monotonic error*. If the input voltage to the converter increases in discrete quantised levels, the digital output code should also increase by the same number of increments. If this does not happen, then monotonic error has occurred, reducing the useful resolution of the converter.

The digital output code from an ADC is produced in either serial or parallel format. The converter shown in Figure 11-1 uses parallel format with all 8 output bits available. Parallel output converters usually have faster operating times than the serial output types but require additional connections to the digital system. However, ADCs with serial output require more work to control than parallel output types. Some converters contain an internal analog multiplexer, allowing multiple analog input channels to be processed (at a proportionately slower speed).

At the start of the conversion process the input voltage must be sensed by the converter's input stage circuitry. The *output impedance* of the external circuit that is providing the input signal to the ADC must be sufficiently low compared to the ADC's *input impedance* for the ADC to function properly. The converter will operate over a limited range of input voltage, and this too must be considered when scaling the source of input voltage before connecting it to the ADC.

11.3 Conversion Techniques

Several popular analog-to-digital conversion techniques are implemented with electronic circuitry including voltage-to-frequency converters, *single slope* ADC, *dual slope* ADC, *successive approximation* ADC, and *flash* ADC. Some converters use a combination of methods to take advantage of the independent benefits of each approach. For example, the high-speed flash technique is combined with successive approximation to produce a 'low cost' but very fast converter. The voltage-to-frequency technique has been mentioned previously (Chapter 10) and is not normally considered for use, due to its relatively slow speed. The other converter techniques are widely used and are explained as follows.

Single Slope ADC

This converter uses a constant current source charging a capacitor, a voltage comparator, and a counter with clock source and control logic as shown in Figure 11-3.

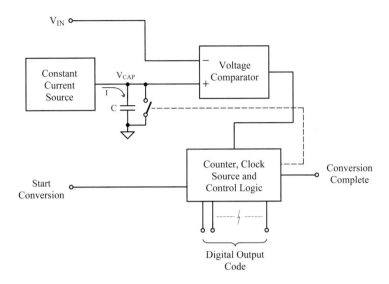

Figure 11-3 Simplified single slope ADC.

The conversion process begins immediately after the 'Start Conversion' input is driven to its active state and proceeds in two stages as follows:

Stage 1 - Initialisation

> The discharging switch across capacitor C closes (activated by the control logic), discharging the capacitor to zero volts. Next the counter is reset to a value of zero, the counter logic opens the switch, and counting commences.

Stage 2 - Integration

> The constant current source drives current into capacitor C, generating a ramping voltage at the comparator +ve input (this process is known as integration). When this ramping voltage exceeds the positive input voltage (V_{IN}) present on the comparator −ve input, the comparator will toggle state from low to high. This change in state of the comparator will signal the counter logic to cease counting, at which time conversion is complete. The 'Conversion Complete' output pin will then be switched to its active state to indicate end of conversion to external devices. The counter output code will now represent the analog input voltage. A larger magnitude of input voltage will require a longer time period for the ramping voltage to reach its level, producing a larger digital output value.

The cycle described as Stage 1 and 2 will repeat when the next 'Start Conversion' pulse arrives.

The conversion speed of the single slope ADC is relatively slow although its accuracy is reasonably good, being affected by the long-term stability of the counter's clock, the stability of the constant current source, and the quality of the capacitor, ideally having low *dielectric absorption*. When capacitors with high dielectric absorption are discharged and then removed from the discharging circuit, some charge will remain stored inside the capacitor on polarised dielectric interfaces. This charge generates an unwanted error voltage. Capacitors with very low dielectric absorption will have negligible voltage across them after being discharged. A further advantage of the single slope converter is that noise on the input voltage signal is averaged out during the process of integration.

Dual Slope ADC

This converter is similar to the single slope ADC except that two ramping stages are employed during conversion to greatly improve accuracy. Figure 11-4 shows the block diagram for such a converter.

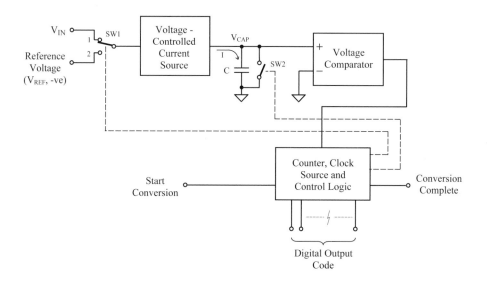

Figure 11-4 Simplified dual slope ADC.

Once the 'Start Conversion' input is asserted, the conversion process will proceed in three stages as follows:

Stage 1 – Initialisation

> The ADC is 'zeroed' by closing SW2 to fully discharge the integrating capacitor C to zero volts.

Stage 2 – Integrate Up using 'V_{IN}'

> At the start of this stage the counter is reset to a count of zero, and SW1 is set to position 1, connecting the input voltage (V_{IN}) to the voltage-controlled current source. SW2 is opened, allowing the current source, controlled by the input voltage signal V_{IN}, to charge the integration capacitor C, producing an upwards ramping voltage shown as 'A' in Figure 11-5. The ramping is

allowed to proceed for a fixed time period (usually the maximum count value of the counter – to maximise conversion accuracy). At the end of this time period, the counter is reset to a count of zero and the final stage of conversion will begin.

Stage 3 – Integrate Down using 'Reference Voltage'

SW1 is moved to position 2, allowing the precise *negative* reference voltage to control the current source. This produces a negative current of constant value, which progressively discharges capacitor C until the voltage at the comparator +ve input falls just below the ground potential (0V) connected to the comparator's –ve input. When this occurs the comparator output changes state and stops the counter. The count value reached during this stage represents the analog input voltage (V_{IN}), being proportional to its magnitude.

Figure 11-5 shows the voltage waveform generated during the two integration stages. The lower set of rising and falling voltages across the capacitor shown as 'C' and 'D' are generated when the input voltage V_{IN} is a lower value. Note that the down ramping voltages marked as 'B' and 'D' have the same slope since they are generated by a constant current of the same value (controlled by V_{REF}).

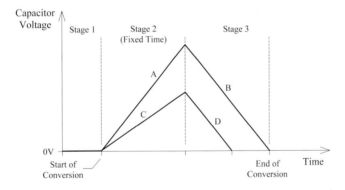

Figure 11-5 Dual slope ADC voltage waveform.

The advantage of the dual slope ADC compared with the single slope converter is its improved accuracy, largely determined by the stability of the reference voltage. Unlike the single slope ADC, the dual slope ADC is not affected by any long-term drift in clock frequency since the same clock is used for timing Stage 2 and Stage 3. The dual slope ADC shares similar noise immunity characteristics as the single slope converter and requires the use of a good quality capacitor with low dielectric absorption. This converter is relatively slow but very accurate – up to around 18-bit resolution. Other converters cannot match this converter for accuracy at low cost and this is one of the reasons it is widely used in instruments such as precision digital multimeters.

Successive Approximation ADC

This converter is very popular due to its relatively fast conversion speed, good accuracy and low cost. Figure 11-6 shows the block diagram for this converter.

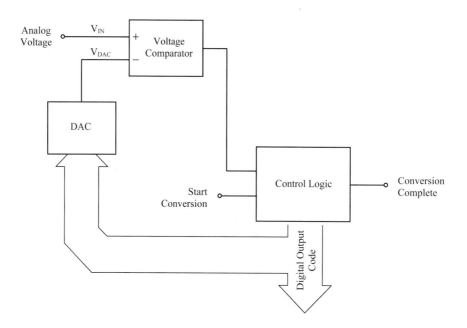

Figure 11-6 Simplified successive approximation ADC.

Conversion begins immediately after the 'Start Conversion' input is driven to its active state and proceeds as follows:

Stage 1 – Initialisation

The control logic clears all control logic output bits.

Stage 2 – Successive Approximation Process

The converter digital output code is formed sequentially during a series of tests, where the analog input voltage is compared against the analog value of a digital code, this code constructed during the conversion cycle itself.

Each bit value is tested sequentially against the analog input voltage, starting with the most significant bit (MSB). For example, using an 8-bit converter, the digital test code would be 1000 0000, representing half the quantised voltage range. This code is fed to the input of the digital-to-analog converter (DAC), producing the analog voltage V_{DAC}. Voltage V_{DAC} is tested against voltage V_{IN} using the comparator. If V_{IN} is greater in magnitude than V_{DAC}, the comparator output will be HIGH and the control logic will keep the current bit, in this case the MSB and the output code would then be

1000 0000. If V_{IN} was lower in magnitude than V_{DAC}, the comparator output would be LOW and the control logic would discard this bit, the output code would then be 0000 0000.

The next test would involve the next lower bit. This bit would be added to the code output from the control logic's previous bit value test, producing the code 1100 0000 (assuming V_{IN} was greater in the first test). This code is then passed to the DAC to generate a new value of V_{DAC} to be used for this current bit value test. This process continues as before for this bit and in-turn for the remaining bits. At the conclusion of all bit testing, the conversion complete output will change to its active state to provide external indication. This type of converter has significant benefit being that conversion time is fixed and reasonably fast.

The interface board is fitted with a successive approximation ADC, namely the ADC0804. A few additional input control pins are used with this device, one being the /RD pin (where / signifies active low) used to control reading of output data bits. The other pin is the /CS input which is used in combination with the /RD pin to allow the output data bits to be placed into *tri-state* mode. When a device's output pin is in a tri-state mode it will have high-impedance connections to other interconnected circuitry. The other two states are the low voltage state (logic-LOW) and the high voltage state (logic-HIGH). The tri-state feature of the ADC0804 is used when connecting the device into microprocessor-type systems where the data bus is shared with other devices. These systems require the ADC to be 'disconnected' at specific times to allow other devices to share the data bus.

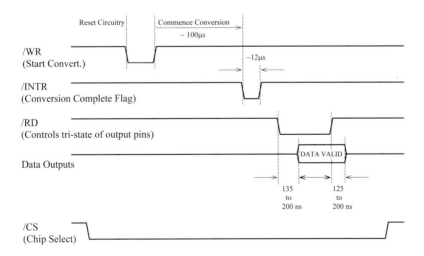

Figure 11-7 ADC0804 conversion timing.

Starting a conversion and then later reading the output data takes place as shown in Figure 11-7. Note that the /CS signal must be held low for the duration of the conversion cycle. In many applications it can be held low permanently (when tri-stating is not needed).

Caution: the analog input voltage to the converter must not exceed +5V or drop below 0V. If the input voltage is outside the specified operating range the device will likely suffer permanent damage.

Note also that the conversion complete signal (/INTR) is low for only a brief period of time. When using a PC to monitor this signal, bear in mind that the software running on the PC through the parallel port is relatively slow and might be unreliable when detecting the change in level of the /INTR signal. Generally, this signal is *latched* using hardware means (latching refers to detecting an event and then indicating its occurrence). The interface board does not have latching circuitry included for this purpose. Our programs will not detect the /INTR signal after issuing the start conversion signal. Instead, they use a software delay to wait for a longer time period than the conversion time before reading the output code.

The ADC0804 converter is constructed internally with separate digital and analog grounds to minimise effects of noise to its analog circuitry. These grounds must be connected together externally at one point as shown on the interface board schematic diagram (Appendix, Figure A-26). Here you will see two different types of ground symbol connected together, the hollow triangular symbol being analog ground and the lined ground symbol, digital ground.

Flash ADC

This converter is the fastest of all types of analog-to-digital converters and also the most expensive. Applications using flash converters include digital signal processing, video signal processing, and other types of waveform analysis as used in digital oscilloscopes. The converter does not need a 'Start Conversion' input; instead conversion is a continuous process taking place as shown in Figure 11-8.

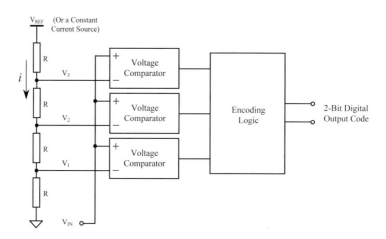

Figure 11-8 Simplified 2-bit flash ADC.

Using a 2-bit ADC for explanation purposes, the 2-bit ADC will quantise an analog input voltage to four possible levels. For example, if the ADC input voltage range was to be 0 to 3V, then the four quantisation levels would be 0V, 1V, 2V, and 3V. V_{REF} would be set to a level of 4V. The analog input voltage is tested against three of the four quantisation levels using three comparators. The fourth quantisation level in this case being 0V does not need testing.

The first comparator, shown with V_1 connected to its –ve input, tests for the analog input voltage V_{IN} exceeding 1V. If V_{IN} exceeds V_1, the comparator output will be high and this will be sensed by the encoding logic. Otherwise the comparator output will be low. The other two comparators test if the analog input exceeds 2V, and 3V, respectively. The encoding logic converts the individual comparator output signals into valid n-bit logic output, where n in this case is 2. This conversion process is continuous and is extremely fast – the conversion time being the addition of the delays generated by the comparators and the encoding logic.

The other significant characteristic of a flash ADC is related to conversion speed - its extremely short *aperture interval*. The aperture interval is the time taken for the converter to 'read' the analog voltage level during conversion. For the flash converter, this time is equal to the interval when the comparator outputs are latched (stored in the encoding logic) and does not include the remaining conversion time when encoding takes place. This characteristic of the converter makes it ideal for use in applications having 'fast' changing signals and means it doesn't need to use a *sample and hold* circuit – explained later in this chapter.

As the conversion resolution increases, so too does the number of comparators. For an n-bit device, $2^n - 1$ comparators are needed. For converters beyond the range of 8 to 10 bits, the devices become quite expensive and relatively large. One way to improve resolution is to cascade converters – for example, using four 6-bit units would create an 8-bit flash converter.

11.4 Measuring Voltages with an ADC

It is beneficial to understand some of the basic concepts of signal processing before using an ADC for measuring dynamic signals. These concepts include *slew rate*, *sample and hold*, *aliasing*, and *equivalent time sampling*. Consider a repetitive triangular waveform as shown in Figure 11-9.

Imagine we are to sample this changing input voltage using an 8-bit ADC having a conversion time of 100μs (the interface board's ADC0804 ADC). At the start of conversion, the input voltage to the converter will either be ramping up or down depending on the point in time conversion was initiated. Examine the case when the input voltage is ramping up from 0V at the rate of 5V per 0.5 second (10V/sec). Knowing the rate of change of input voltage to be 10V/sec, implies that over a 100μs period the analog input voltage will rise by 1mV.

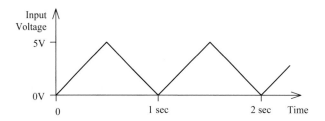

Figure 11-9 Triangular waveform.

The 8-bit converter quantises the analog voltage into 256 levels over an input range of say 5V (for maximum accuracy), where each level is equal to 5V/256 = 19mV. In order to use the converter's full 8-bit accuracy, the analog input voltage must not change by more than ½ a quantisation level during the conversion interval, being 9.5mV in 100µs (95V/sec). For the extremely slow changing input signal of 10V/sec used in this example, conversion accuracy is maintained since the analog input signal changes only 1mV during the conversion period (10V/sec) and the converter can tolerate up to 9.5mV change at its analog input (95V/sec) before losing resolution accuracy.

Should the period of the triangular waveform be changed to 0.1 seconds, the rate of change of the analog signal will be 5V/0.05 seconds (100V/sec). In this case the converter cannot quite keep up – it can work at full resolution accuracy for analog input voltages having slew rates of less than 95V/sec. As you can see, a 0.1 second period or 10 Hz signal is around the frequency limit for this converter (and the ADC0804!) when digitising a triangle shaped waveform. The triangular waveform is the least demanding of all waveforms to digitise, the sine wave being the next most demanding. Digitising a sine wave is treated as follows.

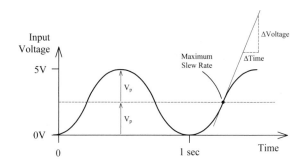

Figure 11-10 Sinusoidal waveform.

A sinusoidal changing input voltage is expressed as:

$$v(t) = V_p \sin\omega t + V_p$$

where V_p is half the *peak-to-peak* voltage,
ω is the circular frequency (radians/sec)
($\omega = 2\pi f$, where f is the frequency in Hertz)

The rate of change of v (t), or its slew rate is:

$$dv/dt = \omega V_p \cos\omega t \quad \text{this being a maximum when } \cos\omega t = 1$$
$$\rightarrow \quad \text{Max slew rate} = \omega V_p$$
$$= 2\pi f V_p$$

The ADC 'sees' the changing input voltage for nearly the whole conversion interval, T_c. To use the full-bit accuracy of the converter, this input voltage change must be less than ½ a quantisation level (or half a least significant bit, LSB) during the conversion interval T_c. Remember that one quantisation level is equal to the converter input voltage range divided by the number of quantisation states possible for the converter. For an *n*-bit converter there can be a maximum of 2^n states, so half a quantisation level can be expressed as:

½ Quantisation Level = ½ (Converter Input Range / 2^n)

The maximum slew rate of the analog input voltage to the converter is:

Converter max   = (Converter Input Range / 2^{n+1}) / T_c
input slewrate

Note: In order to optimise the accuracy and maximum digitising frequency range of the converter, scale the maximum analog input voltage up to use the full converter input range where possible.

Equating the input signal slew rate for a sinusoidal signal to the limiting converter slew rate creates the expression:

Max slew rate   < Max slew rate
input voltage   ADC
$$2\pi f V_p < (\text{Converter Input Range} / 2^{n+1}) / T_c$$
$$\rightarrow \quad f < \text{Converter Input Range} / \pi V_p T_c 2^{n+2}$$

In order to improve the accuracy and allow digitisation of much higher frequency signals, a circuit known as a *sample and hold* is often used, placed between the analog input signal and the input to the converter. Since many ADCs do not contain an internal sample and hold, external sample and hold devices are often used with converters.

Sample and Hold

This circuit stores a 'snapshot' of the changing analog voltage signal and presents an 'unchanging' buffered version of the signal to the input of the ADC. The

'Sample Command' signal from the host controller is connected to the input of the sample and hold to synchronise sampling. Sampling and holding takes place in two stages explained as follows and shown in Figure 11-11 and Figure 11-12:

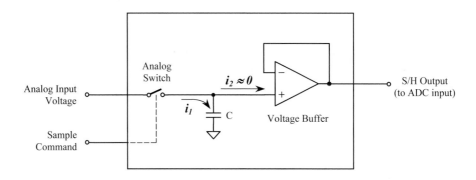

Figure 11-11 Simplified sample and hold circuit.

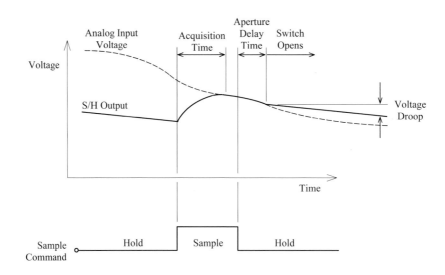

Figure 11-12 Sample and hold timing.

Stage 1 – Sampling Input Signal

Once the sample command input is activated to 'Sample', the analog switch closes and the capacitor is charged to the same voltage as the analog input voltage. The time taken for the voltage across the capacitor to 'reach' the input voltage (within limits) is referred to as the *acquisition time*.

Stage 2 – Holding Sampled Voltage

At the end of the sampling period, the 'Sample Command' is toggled to the 'Hold' state to store the sampled signal for the ADC input. Unfortunately there is a delay in opening the analog switch known as *aperture delay*. This delay causes the output of the sample and hold (S/H) to follow the input voltage for the aperture delay time period, creating an error in sampled voltage and hence possible errors in the digital code produced by the ADC. The ADC conversion commences during the hold period that follows the sampling process.

Ideally the output of the sample and hold (S/H) remains fixed in amplitude over the entire ADC conversion interval. In practice the S/H output drops over time producing what is known as voltage *droop*. The droop occurs as charge stored on the capacitor is lost during the hold period, drawn into the neighbouring S/H circuitry connected to the capacitor, and also lost through the capacitor itself.

Aliasing

When sampling a repetitive waveform, it is possible to produce various sets of data values depending on the sample rate as shown in Figure 11-13 to Figure 11-15. Considering a sinusoidal waveform; should the sample rate be less than half the signal cycle or period, then a waveform similar to that shown in Figure 11-13 will be reconstructed from the data values produced by sampling and conversion. The reconstructed signal has a different frequency from the original sampled signal and is termed an *alias* signal. Beware: in this case the alias signal has the same amplitude and sinusoidal shape as the input signal and can be mistaken to be a proper representation of the actual input signal.

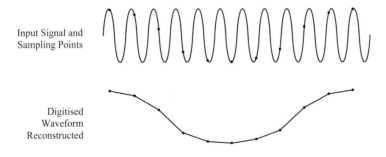

Input Signal and
Sampling Points

Digitised
Waveform
Reconstructed

Figure 11-13 Aliased reconstruction – sample rate too low.

Note: triangular-shaped waveforms will be reconstructed from digitised samples made at twice the signal frequency as shown in Figure 11-14. These reconstructed waveforms will have different amplitude depending on the position in the cycle when sampling begins.

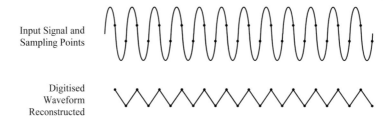

Figure 11-14 Digitising sample rate at 2 samples/signal cycle.

As the sample rate increases, the reconstructed waveform starts to resemble the original signal as shown in Figure 11-15.

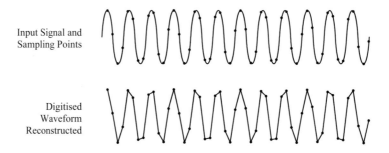

Figure 11-15 Digitising sample rate approximately 5 samples/signal cycle.

Real and Equivalent Time Sampling

All waveforms shown above have been sampled in *real-time*, meaning that data points are collected and stored sequentially as they are digitised. Repetitive signals of high frequency can be sampled and reconstructed using *equivalent time sampling*, where groups of sample sets are stored in memory and then used to generate complete waveform reconstruction. The resultant constructed waveform represents the originally sampled signal as shown in Figure 11-16. This technique is often utilised in digital oscilloscopes. When the user sets the oscilloscope time-base to sample high-speed repetitive waveforms, equivalent time sampling is used to create a pseudo sampling rate much greater than that of the oscilloscope's digitiser.

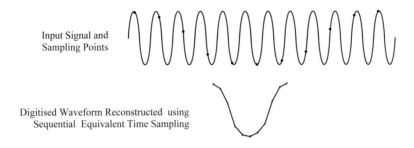

Input Signal and
Sampling Points

Digitised Waveform Reconstructed using
Sequential Equivalent Time Sampling

Figure 11-16 Equivalent time sampling.

Equivalent time sampling cannot be applied to the digitisation process when working with non-repetitive signals as shown in Figure 11-17. Instead, these signals need to be sampled and stored at a sufficiently high rate to provide enough detail in the reconstructed signal. Under these conditions many digital oscilloscopes are often challenged to provide adequate sampling rate and sufficient high-speed memory to store the digitised data. These two factors have a significant influence on the price of digital oscilloscopes.

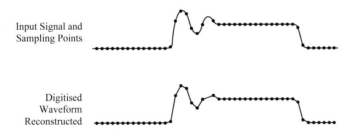

Input Signal and
Sampling Points

Digitised
Waveform
Reconstructed

Figure 11-17 Real-time sampling of a non-repetitive waveform.

11.5 An Object Class for the ADC

In the last few chapters we learnt to design object classes to suit various objects such as the parallel port, the DAC, motors and the VCO. In a similar manner we can develop a software object for the ADC. The principle purpose of an analog-to-digital converter is to convert an analog voltage applied at its input to an integer bit number that can be read by the computer.

The conversion process for most analog-to-digital converters involves the following steps:

1. Start an analog-to-digital conversion.
2. Wait for the conversion to complete.
3. Read the converted data.

Some analog-to-digital converter subsystems have a multiplexed analog input (i.e. more than one analog input where only one analog input is switched to the ADC at any given time). In such cases the above set of steps must be preceded by a "Select Input Channel" operation. The ADC used with our interface board does not have a multiplexer to use with multiple analog input channels. Therefore, we will not need to incorporate channel selection.

We must design our object class to have a member function to implement the steps listed above. The ADC on the interface board is designed to communicate through the parallel port. Therefore, the `ParallelPort` object forms an ideal base class for the new ADC class.

The ADC class needs to have only one private data member to store the digital value read from the ADC. Apart from the constructors, the ADC class must have a function to carry out the analog-to-digital conversion and store the resulting digital value into the private data member. A function is also needed to provide access to this private data member. A class definition that encapsulates this data and functions is given in Listing 11-1.

Listing 11-1 The header file for the ADC class – adc.h.

```
#ifndef AdcH
#define AdcH

#include "pport.h"

class ADC : public ParallelPort
{
    private:
        unsigned char ADCValue;

    public:
        ADC(int baseaddress=0x378);
        unsigned char ADConvert();
        unsigned char GetADCValue();
};
#endif
```

The function `ADConvert()` is the most involved of the three functions and is discussed first. We must decide which parts of the parallel port will be used and their purpose before before being able to write the C++ statements for this function.

Figure 11-18 shows a block diagram of the ADC0804 with its input pins (on the left side) and output pins. The pin labels and descriptions are given in Table 11-2. These labels are used on the schematic diagram and can also be found near the ADC on the interface board.

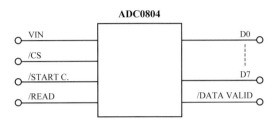

Figure 11-18 ADC0804 block diagram.

Table 11-2 Interface pins of the ADC.

Pin Label[†]	Input	Output	Function
VIN	•		Analog input voltage
/CS	•		Chip select (activates device)
/START C.	•		Start a conversion
/READ	•		Enable reading the digital data
/DATA VALID		•	Indicates conversion complete
D0-D7		•	Output digital data bits

[†]Pin labels with a prefix of '/' are active low.

The dots marking the Input and Output columns of Table 11-2 identify each signal as an input or an output with reference to the ADC. Now we need to evaluate a means of interfacing these signals to the parallel port.

Operation of signals

The voltage signal to be converted by the ADC is connected to its analog input pin labelled VIN. For testing purposes we can generate a suitable analog input voltage for the ADC using either the on-board potentiometer, the thermistor circuit, or the DAC operating in unipolar mode (jumper fitted across the header position marked LINK1).

The ADC's input pins are configured as follows. The *chip select* pin (/CS) must be at logic-low for the ADC to operate. The *read enable* pin (/READ) must be held logic-low to enable reading the digital data from the ADC. The chip select signal is typically generated by the *address decoding* circuit of a hardware system that has several devices sharing a data bus. The interface board does not share a data bus, so we can permanently activate the above two signals; i.e. connect them directly to GND. This reduces the number of signals we need to interface with the parallel port.

The computer must control the start conversion signal connected to the *start conversion* pin (/START C.). An analog-to-digital conversion is initiated by applying an active-low *pulse* to this pin. We can generate an active-low pulse by driving a high-level signal momentarily to a low-level, returning the signal to a high state.

Immediately after the ADC completes its conversion operation, the *data valid* pin (/DATA VALID) will produce a brief low-level pulse. Since this low-level pulse is short in duration, it may not be possible to detect it and therefore determine the precise moment conversion was completed. Typically, this pulse is latched using hardware to ensure that a program will reliably detect the end of conversion. The electronic circuitry on the interface board has been kept to a minimum and as such does not include a latch circuit. Therefore, our best option is to allow sufficient time for the conversion to complete before reading the digital data. In addition, using this approach will free us from the need to interface the /DATA VALID signal.

Configuration of Port data bits to interface the ADC

We now need to assign the data bits that will interface the ADC to the parallel port. When using the ADC it is possible that the DAC will be used to provide programmable input voltages to the ADC. We will assume this to be the case when allocating our data bits. The digital input and output requirements for the ADC and DAC are shown in Table 11-3.

Table 11-3 DAC & ADC digital input and output pins.

DAC	ADC	
Digital Inputs	**Digital Inputs**	**Digital Outputs**
D0	/CS	D0
D1	/RD	D1
D2	/START C.	D2
D3		D3
D4		D4
D5		D5
D6		D6
D7		D7
		/DATA VALID

Digital Inputs to the DAC: The eight digital input pins to the DAC need to be driven by parallel port output signals. Therefore, it makes sense to use output data bits (D0 to D7) of the port at address BASE to drive the DAC inputs (D0 to D7).

Digital Inputs to the ADC: To drive the ADC input pin /START C., we can use data bit D0 of the port at address BASE+2. The ADC data bus is configured to an active state by connecting /CS and /READ directly to GND using interconnecting leads. We can do this because our ADC's output data is not connected to a shared data bus.

Digital Outputs from the ADC: The software must read eight digital output signals from the ADC being sent through the parallel port (/DATA VALID not used). The parallel port has five input signals (D3 to D7) available from the port at address BASE+1. Note that we have not used port address BASE+2 in input mode as it can be unreliable at higher data transfer rates.

The interface board has been fitted with a four-channel 2-to-1 multiplexer (as shown in Figure 11-19) to provide extra capability for transfer of data to the port. If we make use of this device, we can transfer eight data bits to the port using only four signals. We do this by separating the eight data bits from the ADC into two groups of four bits. The first group is selected by the multiplexer and the port then reads these four data bits. This is followed by selection of the second group of four bits that are then read by the port. Note that we need one output data bit from the port to control the multiplexer's selection operation. Since the eight bits from the ADC represent one byte of data, driving the multiplexer's Select input low will select the low nibble (D0 to D3). Conversely, driving the Select input high will select the high nibble (D4 to D7).

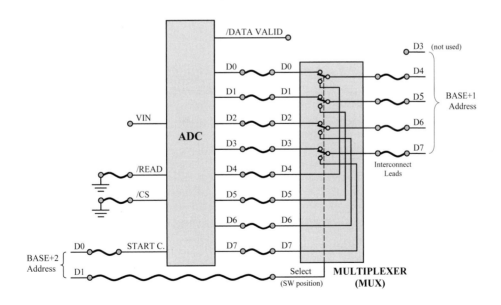

Figure 11-19 Complete ADC system using the Multiplexer.

Now that we know how to read the eight bits of data from the ADC using only four transmission signals, we can establish the configuration for the remainder of the parallel port data bits. We can use four input data bits (D4 to D7) of the port at address BASE+1 to read the four output signals from the multiplexer that transmits the ADC output byte as two nibbles.

If you should decide to modify the program to detect the narrow output pulse /DATA VALID from the ADC, then connect a lead from this pin to data bit D3 of the port at BASE+1 and write extra program statements to read its status.

Digital Input to the Multiplexer: We can drive the Select input of the multiplexer to control which nibble at its input pins is switched to its output by using an output data bit (D1) of the port at address BASE+2.

A summary of all connections for interfacing the parallel port to the ADC, to the DAC, and to the Multiplexer is given in Table 11-4. This table does not provide the internal connections needed on the interface board between the ADC and the Multiplexer - they are shown in Figure 11-19.

Table 11-4 Parallel Port interface connections for the DAC, ADC, and MUX.

BASE Address		BASE+1 Address		BASE+2 Address	
D0	DAC, D0	D3	(ADC, /DATA VALID)	D0	ADC, /START C.
D1	DAC, D1	D4	MUX, D4	D1	MUX, Select
D2	DAC, D2	D5	MUX, D5		
D3	DAC, D3	D6	MUX, D6		
D4	DAC, D4	D7	MUX, D7		
D5	DAC, D5				
D6	DAC, D6				
D7	DAC, D7				

Note: 1. ADC inputs /CS and /READ must be connected to GND using interconnect leads.

2. ADC output /DATA VALID is not used for our program.

3. Set DAC to Unipolar mode by fitting the jumper across header position marked LINK1.

We are now in a position to define the member function ADConvert(). Listing 11-2 shows one possible definition of the function.

Listing 11-2 Member function ADConvert().

```
unsigned char ADC::ADConvert()
{
// Declare variables to store nibbles.
    unsigned char LowNibble, HighNibble;

// Start conversion pulse.
    WritePort2(0x01); // set /START C to high
    WritePort2(0x00); // pull /START C to low
```

```
    WritePort2(0x01); // set /START C back to high

// Set Select signal of multiplexer (D1) to logic-high and
// maintain /START C high. This operation takes more time
// than the conversion of the ADC, so we do not need to
// check for signal /DATA VALID. */
    WritePort2(0x03); // 0000 0011

// Conversion finished by this time.
// Read high nibble and nullify low nibble.
    HighNibble = ReadPort1() & 0xF0;

// Set Select signal of multiplexer (D1) to logic-low.
    WritePort2(0x01); // 0000 0001

// Read low nibble, move data bits across into position
// and nullify high nibble.
    LowNibble = (ReadPort1() >> 4) & 0x0F;

// Form complete byte.
    ADCValue = HighNibble + LowNibble;

    return ADCValue;
}
```

The three statements from Listing 11-2 shown in bold typeface need explanation. Note that when reading the port at address BASE+1, only the bits D4-D7 carry data coming from the ADC. The data from the 8-bit ADC is read into the PC using these four bits in two stages; first the high nibble (four bits) is read and stored, followed by reading and storing the low nibble. Then the high and low nibbles are added to obtain the complete 8-bit result (ADCValue).

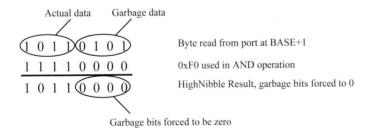

Figure 11-20 Reading the high nibble and filtering out unwanted bits.

We use the inherited member function `ReadPort1()` to read the high nibble through the port at address `BASE+1` and then clear all unused bits that contain unpredictable (garbage) data (lower four bits) by carrying out an AND operation with `0xF0`. This operation is shown in Figure 11-20.

When reading the low nibble, we first read the port and then shift these data bits by 4 locations to the right to reside in the low nibble of the final data byte. After shifting we carry out an AND operation with `0x0F` to clear all bits in the high nibble that can contain unpredictable data. This is shown in Figure 11-21.

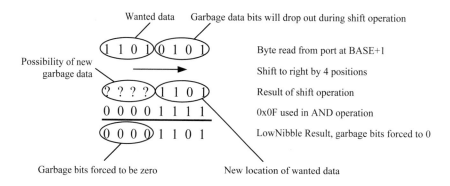

Figure 11-21 Reading the low nibble and filtering out unwanted bits.

Now we have an 8-bit number (unsigned char) named `LowNibble`, which has some data in the lower four bits and definitely zeros in the upper four bits. We also have an 8-bit number named `HighNibble`, which has some data in the upper four bits and definitely zeros in the lower four bits. Then we add these two bytes together to form one complete byte named `ADCValue` which has all 8 bits carrying the data from the analog-to-digital converter. Figure 11-22 shows the formation of `ADCValue`.

```
0 0 0 0 1 1 0 1     LowNibble
1 0 1 1 0 0 0 0     HighNibble
1 0 1 1 1 1 0 1     ADCValue
```

Figure 11-22 Add low & high nibbles to form the ADC output.

The function given in Listing 11-2 can be re-written in a slightly more efficient form as given in Listing 11-3. The data member `ADCValue` has been used to combine an operation and eliminate the need for variables `LowNibble` and `HighNibble`.

Listing 11-3 A more efficient version of `ADConvert()`.

```cpp
unsigned char ADC::ADConvert()
{
// Start conversion pulse.
    WritePort2(0x01); // set /START C to high
    WritePort2(0x00); // pull /START C to low
    WritePort2(0x01); // set /START C back to high

// Set Select signal of multiplexer (D1) to logic-high and
// maintain /START C high. This operation takes more time
// than the conversion of the ADC, so we do not need to
// check for signal /DATA VALID. */
    WritePort2(0x03); // 0000 0011

// Conversion finished by this time.
// Read high nibble and nullify low nibble.
    ADCValue = ReadPort1() & 0xF0;

// Set Select signal of multiplexer (D1) to logic-low.
    WritePort2(0x01); // 0000 0001

// Read low nibble and assemble the final 8-bit number.
    ADCValue += (ReadPort1() >> 4) & 0x0F;

    return ADCValue;
}
```

The complete definition of the ADC class must include the definitions of its member functions as given in Listing 11-4. The member function `GetADCValue()` provides access to the final 8-bit number `ADCValue` for functions outside the ADC class.

Listing 11-4 Member function definitions of the ADC class – adc.cpp.

```cpp
#include "adc.h"

ADC::ADC(int baseaddress) : ParallelPort(baseaddress)
{
    ADCValue = 0;
}

unsigned char ADC::ADConvert()
{
```

```
        WritePort2(0x01);   // Start C. pulse
        WritePort2(0x00);
        WritePort2(0x01);

        WritePort2(0x03); // Set Mux to read high nibble.
        ADCValue = ReadPort1() & 0xF0;

        WritePort2(0x01); // Set Mux to read low nibble.
        ADCValue += (ReadPort1() >> 4) & 0x0F; // Read, combine.

        return ADCValue;
}

unsigned char ADC::GetADCValue()
{
        return ADCValue;
}
```

11.6 Measuring Voltage Using the ADC

Recall that in Chapter 10 we developed a program to measure an analog voltage using the VCO. The `MeasurePeriod()` function in the VCO program returned a number representing the pulse period (and hence input voltage) of the VCO. We should be able to use the same program with the VCO object replaced by the ADC object. The function `ADConvert()` can then generate a number proportional to the analog voltage. Note that we have used a DAC object in the VCO program to provide an analog voltage. We will keep the same DAC object operating exactly the same way to provide the analog input to the ADC (at VIN).

Listing 11-5 shows the `main()` function from Listing 10-6 reproduced with the modifications needed to use it with the ADC.

Listing 11-5 Measuring voltage using the ADC – voltage.cpp.

```
#include <conio.h>
#include <bios.h>

#include "dac.h"
#include "adc.h"

void main()
{
    DAC Dac;
```

```
    ADC Adc;
    int Quit = 0, key;
    unsigned char DACbyte;

    clrscr();
    Dac.SendData(0); // initialise to zero

    while(!Quit)
    {
        gotoxy(10,10);
        cprintf("The ADC output is: %3u",
                            (int) Adc.ADConvert());

        if(bioskey(1)!=0)
        {
            DACbyte = Dac.GetLastOutput();
            key = bioskey(0);
            switch(key)
            {
/*Alt-X*/       case 0x2d00:
                            Quit = 1;
                            Dac.SendData(0); // reset to 0
                            break;

/*Up Arrow*/    case 0x4800:
                            if(DACbyte>247) // limit max value
                                DACbyte = 247;
                            Dac.SendData(DACbyte+8);
                            break;

/*Down Arrow*/  case 0x5000:
                            if(DACbyte<8) // limit min value
                                DACbyte = 8;
                            Dac.SendData(DACbyte-8);
            }
        }
    }
}
```

We have made changes to just a few statements in the main function. The keyboard controls operate almost identically to the program in Listing 10-6 that used the VCO object. The difference being the addition of statements to limit the maximum and minimum value of the number written to the DAC (0 to 255).

Executable File Generation		
Required Files	**Listing No.**	**Project File Contents**
`pport.cpp`	Listing 10-8	`pport.cpp`
`pport.h`	Listing 10-7	
`dac.cpp`	Listing 10-10	`dac.cpp`
`dac.h`	Listing 10-9	
`adc.cpp`	Listing 11-4	`adc.cpp`
`adc.h`	Listing 11-1	
`voltage.cpp`	Listing 11-5	`voltage.cpp`

The entire definition of the DAC class must be provided before the compiler can compile the `main()` function given in Listing 11-5. The VCO object is not being used. Therefore, its class definition and function definition can be eliminated, however, its inclusion will not affect the operation of our program. When compiled, linked and run, this program will display an integer value on the screen that corresponds to the voltage applied at the VIN input of the ADC.

The program can be modified slightly to display the analog voltage instead of an integer number. The ADC produces an output that is 8 bits wide. These 8 bits can represent any value in the range 0–255 both inclusive (which forms 256 numbers). The operation of the ADC requires the full-scale range to be quantised (segmented) into 256 equal quantum levels. Each quantum then represents the full-scale voltage (5V) divided by 256. The ADC's integer output of 0 corresponds to 0 volts at VIN, and its integer output value of *i* corresponds to an applied voltage of:

volts

Note that from a C++ program's point of view, the division operation 5/256 is considered as an integer division and the result will be 0. Therefore, when including the above expression in the program, it must be typed in as:

```
5.0/256.0*i
```

Now the compiler will treat the division and multiplication operations as floating point operations, and a non-zero result will be evaluated for the expression 5.0/256.0. The statement containing the `cprintf()` function in Listing 11-5 must now be modified to include the above factor as shown below:

```
cprintf("The ADC output is: %5.2f (V)",
    5.0/256.0*Adc.ADConvert());
```

The program will now display the actual voltage applied at the analog input pin of the ADC (VIN). The connections that need to be made for this program to operate are given in Figure 11-19, Table 11-4 and Table 11-5.

> **NOTE**
>
> Ensure that the DAC is placed into unipolar mode (0 to +5V output) by fitting the jumper across the header in the position marked as LINK1. Then connect the 9V battery *before* connecting the output of the DAC circuit to the ADC input.

Table 11-5 Partial connections for the ADC.

	ADC0804 (U8)
VDAC (pin 7, U10B)	VIN
GND	/READ
GND	/CS

11.7 Measuring Temperature Using the ADC

A program was developed in Chapter 10 to measure temperature using the interface board's thermistor and VCO. That program (Listing 10-12) only requires minor changes to measure temperature using the ADC. This new modified program is shown in Listing 11-6. Note that in this case we can accurately characterise the thermistor circuit response over the full 0 to +5V range since the ADC has very good linearity over its entire input range. In comparison, the VCO has similar linearity between +2.2V and +2.8V.

Listing 11-6 Program measures temperature using ADC and thermistor – temp.cpp.

```
/************************************************
This program uses the thermistor circuit on the interface
board to generate the analog input voltage to the ADC. The
byte produced by the ADC will be proportional to the applied
voltage (temperature of the thermistor).
The program also calibrates the thermistor circuit output
using upper and lower temperature points. The calibration
equation will then interpolate a straight line through these
two points. Once calibrated, the program will be able to
display actual temperatures.
************************************************/
#include <bios.h>
```

```
#include <conio.h>
#include <iostream.h>

#include "adc.h"

void main()
{
    ADC Adc;
    int Quit=0, HiFlag = 0, LoFlag = 0;
    int key = 0;
    float HiTemp, LoTemp, Temp;
    long int HiCount, LoCount;

    clrscr();

    while(!Quit)
    {
        Adc.ADConvert();

        gotoxy(10,10);
        if((HiFlag == 1) && (LoFlag == 1))
        {
            Temp = LoTemp+(HiTemp-LoTemp)*
              (Adc.GetADCValue()-LoCount)/(HiCount-LoCount);

            cprintf("The temperature is: %6.1f (deg)", Temp);
        }
        else
            cprintf("The ADC Value is: %3u",
                        (int)Adc.GetADCValue());

        if(bioskey(1)!=0)
        {
            key = bioskey(0);

            switch(key)
            {
                case 0x2d00 :   /* Alt-X */
                    Quit = 1;
                    break;

                case 0x4800 :   /* Up Arrow */
                    gotoxy(10,5);
                    cout << "Enter Upper Calibration Temp.";
                    cin  >> HiTemp;
```

```
                HiCount = Adc.GetADCValue();
                HiFlag = 1;
                break;

            case 0x5000 :   /* Down Arrow */
                gotoxy(10,6);
                cout << "Enter Lower Calibration Temp.";
                cin  >> LoTemp;
                LoCount = Adc.GetADCValue();
                LoFlag = 1;
        }
    }
  }
}
```

Executable File Generation		
Required Files	**Listing No.**	**Project File Contents**
pport.cpp	Listing 10-8	pport.cpp
pport.h	Listing 10-7	
adc.cpp	Listing 11-4	adc.cpp
adc.h	Listing 11-1	
temp.cpp	Listing 11-6	temp.cpp

The wiring connections need to be changed slightly for this program to operate. We do not use the DAC to provide the analog voltage. Instead we use the thermistor circuit to generate a voltage that represents the temperature of the thermistor. The output of the thermistor circuit is connected to the ADC analog input (VIN) as given in Table 11-6. The remaining connections for the ADC and the MUX are shown in Figure 11-19.

Table 11-6 Thermistor circuit connections to the ADC.

Thermistor Circuit	ADC 0804 (U8)
VTH	VIN
	/READ (to GND)
	/CS (to GND)

It has been rather easy for us to change our program that used the VCO to now operate in conjunction with the ADC. We have been careful to be consistent in developing our classes so that minimum changes will be needed if they are later modified for new or existing programs. These examples are typical of the ease with which object-oriented programs can be maintained and upgraded.

11.8 Summary

This chapter described the principles of operation and use of an analog-to-digital converter. The more popular types of analog-to-digital converters and their various methods of conversion have been explained. This was followed by a discussion of the importance of a sample and hold circuit and the effects of aliasing that occurs when signals are sampled too slowly.

We used our now familiar object-oriented approach to develop software for interfacing the parallel port of the PC with the ADC. We developed a new object class named ADC using an approach consistent with that of Chapter 10 when the VCO class was developed. This object-oriented approach has allowed us to develop the voltage and temperature measuring programs that used the ADC by making minor changes to the programs written for the VCO.

11.9 Bibliography

Fluke, *The ABC's of Oscilloscopes*, Fluke Corporation, 1997.

Horowitz, P. and Hill, W., *The Art of Electronics*, Cambridge University Press, Cambridge, 1989.

Loveday, G., *Microprocessor Sourcebook*, Pitman Publishing Limited, London, 1986.

NS *DATA CONVERSION/ACQUISITION Databook*, National Semiconductor Corporation, 1984.

Stiffler, K., *Design with Microprocessors for Mechanical Engineers*, McGraw-Hill, 1992.

Webb, R.E., *Electronics for Scientists*, Ellis Horwood, New York, 1990.

Wobschall, D., *Circuit Design for Electronic Instrumentation*, McGraw-Hill, 1987.

Van Gilluwe, F., *The Undocumented PC*, Addison Wesley, 1994.

Winston, P.H., *On to C++*, Addison Wesley, 1994.

12

Data Acquisition with Operator Overloading

Inside this Chapter

- Pass parameters by value and by reference.

- Returning values by reference.

- Operator overloading.

- The Copy constructor and the assignment operator.

- File input/output.

- Friend functions.

- Pass-through objects.

- Data acquisition using the ADC.

12.1 Introduction

Some of the programs written in this book can be improved to require less memory when executing, and also operate faster by using *pass by reference* and *return by reference* mechanisms. They can also be changed to take advantage of simpler statements by using *operator overloading* and gain access to private member data through *friend functions.*

In this chapter, we will develop a data acquisition program to demonstrate how operator overloading can be used to write elegant programs that have the advantages outlined above. During data acquisition, signals are converted to data using a device such as an analog-to-digital converter. The data is then directly processed or written to a data file on a mass storage device such as a hard disk, or in some cases, sent to a standard output device such as a screen or printer.

12.2 Operator Overloading

When an operator is overloaded, the action carried out by the operator depends on the arguments the operator is associated with. For example, the results will be different if the division operator (/) is used in the following two contexts. One operation produces integer division and the other floating-point division:

```
5/2;      // the result is 2
5.0/2.0; // the result is 2.5
```

Similarly, the double left arrow operator behaves in two different ways in the following two cases:

```
int y = 200;
cout << y; // 200 is sent to the standard output.
y << 1;    // Shifts bits of y to left by 1 bit-position.
```

The action of an operator depends on the type of object it is used with. In the expression cout << y, cout is a class object of type ostream and y is an object of type int. The << operator takes appropriate action to print the value of y on the screen. However, both operands are of type integer in y << 1, and the action taken by the operator is to shift bits of y by 1 bit-position to the left.

The operators shown in Table 12-1 cannot be overloaded.

Table 12-1 Operators that cannot be overloaded.

.	?:	::	.*	sizeof

We will demonstrate operator overloading by developing program segments that overload the double right arrow (>>) and the double left arrow (<<) operators to perform the following tasks:

1. Carry out an analog-to-digital conversion using an `Adc` object of type `ADC`, and store the result in the variable named `value` of type `unsigned char`. We want to be able to use a statement of the following form to accomplish this.

    ```
    Adc >> value;
    ```

2. Carry out an analog-to-digital conversion and send the data directly to the standard output device (computer screen) using a statement of the following form. The object `cout` is of type `ostream` and `Adc` is again an object of type `ADC`.

    ```
    cout << Adc;
    ```

    Using a statement like this would simplify programming of a data acquisition system with an analog-to-digital converter where the results are to be viewed onscreen or stored in a file.

Operators can be overloaded in two different ways by writing a function using syntax that is specific to operator overloading:

1. As a member function of a class.
2. As a non-member function.

These two ways of overloading an operator will be discussed in the sections ahead. Operators can also operate as unary operators (such as ++ in the case of ++i) or as binary operators (such as + in the expression x+y). The unary operators shown in Table 12-2 operate on an object of type `ObjectX`. The binary operators operate on two objects; one of type `ObjectX`, and the other of type `ObjectY`. In this example the operator being overloaded is the @ symbol.

Table 12-2 Function headings for operator overloading.

Unary operator as a member function	`ObjectX::operator@()`
Unary operator as a non-member function	`operator@(ObjectX x)`
Binary operator as a member function	`ObjectX::operator@(ObjectY y)`
Binary operator as a non-member function	`operator@(ObjectX x, ObjectY y)`

The operators overloaded as shown above are used as follows. In the case of a unary operator, the operand must be to the right of the operator. For example, if x is an object of type `ObjectX`, the usage is:

```
@x;
```

In the case of a binary operator, the first operand must be to the left of the operator and the second operand must be to the right of the operator. If x is an object of type ObjectX and y is an object of type ObjectY, then the usage is:

```
x @ y;
```

The syntax used with the operator is the same if the operator is overloaded as a member function or a non-member function. C++ concepts such as pass by value, pass by reference and copying objects with the copy constructor need to be understood before being able to understand how operators can be overloaded. These concepts are explained in the sections ahead.

12.2.1 Passing Parameters to a Function by Value

Our previous programs have often employed functions that used parameters. At the time of calling the function, these parameters are replaced by copies of the actual arguments (the real values used in the calling function). These copies of the arguments passed to the function are created as temporary values, used by the function, and destroyed when the function exits. As a result, the actual argument used when calling the function (in the calling environment) will not be affected by any changes the function makes to its copy.

The passing of parameters to functions can be better understood by considering the following example that attempts to add up n integers that start from the number 0:

```
#include <iostream.h>

// NOTE: the result from this function cannot be used!
void FindSum(int sum, int n)
{
    for(int j = 0; j < n; j++)
        sum = sum + j;
}

void main()
{
    int Sum = 0;
    int n   = 10;

    FindSum(Sum,n); // FindSum() is called here
    cout << "The sum of " << n << " integers is " << Sum
        << endl;
}
```

This program will print the following text on the screen:

```
The sum of 10 integers is 0
```

When `FindSum()` is called it receives a copy of `Sum` and a copy of `n`. The copy of `Sum` is changed as expected inside this function. When the function exits, this copy is discarded and as a result the sum of ten integers evaluated within the function is also discarded. The outcome is that the variable `Sum` declared within the `main()` function remains unchanged (i.e. it still has the value 0).

This manner of passing parameters is known as *pass by value*, which is actually 'pass by a copy'. The two disadvantages when passing parameters by value are; i) the time taken, and ii) memory space needed to make a copy. If the passed parameter is an object that occupies a large portion of memory, an equal amount of extra memory space will be needed to make the copy, and this will take time.

12.2.2 Passing Parameters to a Function by Reference

A different way of passing parameters to a function is *by reference*. Passing parameters by reference allows a function to effect changes to a variable being used in the calling environment. The program segment given in Section 12.2.1 has been reproduced below with an apparently minor change. In this modified example, when the function `FindSum()` changes the value of `sum`, it actually changes the variable `Sum` that was declared within the `main()` function – not a copy of it. The function directly uses the variable in the calling environment (to generate a correct result) rather than working with a copy of it.

```
#include <iostream.h>

void FindSum(int& sum, int n)   // Function heading changed
{
    for(int j = 0; j < n; j++)
        sum = sum + j;
}

void main()
{
    int Sum = 0;
    int n   = 10;

    FindSum(Sum,n);
    cout << "The sum of " << n << " integers is " << Sum
        << endl;
}
```

This program will print the following line on the screen:

```
The sum of 10 integers is 45
```

The change in the program is shown in bold typeface. Instead of declaring the first parameter `sum` as an `int`, it is now declared as *reference to* `int` by changing `int` to `int&`. Therefore, when the function is called, no copy is made, and the function carries out changes to the variable in the calling environment, i.e. the variable `Sum` declared within the `main()` function.

Passing parameters by reference is memory efficient and time efficient (no need to make a copy). It also allows the function to deliver a result through reference parameters and also through return values. The disadvantage is that the passed parameters are vulnerable to inadvertent changes carried out by the function.

Use of `const` with reference parameters

The keyword `const` can be added in the parameter declaration to prevent the function from making changes to the reference variable. The keyword `const` can also be added to parameters passed by value. In either case, statements within the body of the function are not allowed to change the value of the parameter.

Note that in the previous example we cannot use the function heading:

```
void FindSum(const int& sum, int n)
```

for the simple reason that we want to change the value of `sum` to be able to obtain the correct result.

12.2.3 Preferred Ways of Passing Parameters

Passing parameters by value has the advantage of safeguarding the original values of the actual arguments in the calling environment. However, making a copy consumes time and memory. A more serious subtlety associated with pass by value is related to objects in a class hierarchy. This subtlety is demonstrated using the following example.

Consider the simple class hierarchy and the program shown in Listing 12-1.

Listing 12-1 Adverse effects of passing parameters by value.

```
//This program produces WRONG results!
#include <iostream.h>

class Base
{
    private:
        int BaseClassData;

    public:
        Base(int baseclassdata)
        {
            BaseClassData = baseclassdata;
        }
```

```cpp
    virtual int GetClassData() const // Constant function.
        {
            return BaseClassData;
        }
};

class Derived : public Base
{
    private:
        int DerivedClassData;

    public:
        Derived(int derivedclassdata,
            int baseclassdata) : Base(baseclassdata)
        {
            DerivedClassData = derivedclassdata;
        }

        int GetClassData() const // Constant function.
        {
            return DerivedClassData;
        }
};

int GetData(const Base baseObject)//Pass by value
{
    return baseObject.GetClassData();
}

void main()
{
    Base* BasePtr;
    int ClassData;

    BasePtr = new Base(100);
    ClassData = GetData(*BasePtr);
    cout << "Base class data " << ClassData << endl;
    delete BasePtr;

    BasePtr = new Derived(200, 100);
    ClassData = GetData(*BasePtr);
    cout << "Derived class data " << ClassData << endl;
    delete BasePtr;
}
```

Note that for both `GetClassData()` functions in program Listing 12-1, the keyword `const` is added at the end of the function heading as shown below:

```
int GetClassData() const
{
    return DerivedClassData;
}
```

Such functions are named *constant functions*. These functions are not allowed to modify any of the data members of their class.

Now consider the function `GetData()`:

```
int GetData(const Base baseObject)   //Pass by value
{
    return baseObject.GetClassData();
}
```

The parameter `baseObject` passed to the *non-member* function `GetData()`, is passed by value as a `const` object. As such, the object passed must not be changed by any statements within the body of the `GetData()` function. Therefore, the statement `baseObject.GetClassData()` must not make any changes to `baseObject`. This is ensured since the `GetClassData()` function has been specified as a constant function. In the next program, when we pass parameters by reference, we will pass them as `const` objects to prevent the function from changing them. This allows us to keep both programs as similar as possible and to focus on the behaviour of the two programs in terms of pass by reference and pass by value.

The `GetData()` function is intended to extract the value of the data member that belongs to a particular class. The value of data member `BaseClassData` will be returned if `baseObject` is of type `Base`, and the value of data member `DerivedClassData` is returned if `baseObject` is of type `Derived`.

In this program we call the `GetData()` function under two different circumstances. Consider the first case:

```
Base* BasePtr = new Base(100);
int ClassData = GetData(*BasePtr);
```

We would expect the function `GetData()` to call, from within its body, the member function `GetClassData()` belonging to the `Base` class. This should and does retrieve the value of member data `BaseClassData` and assign its value of 100 to variable `ClassData`. Now consider the second case:

```
Base* BasePtr = new Derived(200, 100);
int ClassData = GetData(*BasePtr);
```

Once again, we would expect the `GetData()` function to call, from within its body, the member function `GetClassData()` belonging to the `Derived` class.

This should set the value of `ClassData` to 200. However, this will not happen in this case. Instead, the program produces an unexpected (error) result by setting `ClassData` to 100. Note: the parameter is passed by de-referencing a pointer. Recall that base class pointers can point to derived class objects. Had a pointer not been used, we could not pass an object of type `Derived` as an actual argument for a parameter of type `Base`. If we simply attempted to pass a derived class object to take the place of a base class parameter, the compiler would report a type mismatch error.

In the second case described above, since the function `GetData()` is programmed to receive its parameter by value, the function will be compiled to get a *copy* of a `Base` class object rather than a copy of a `Derived` class object. Thus, the entire derived class object is *not visible* to the `GetData()` function – only the base class portion (inherited by derivation) is visible. This is a typical situation where the object type of the parameter is different from the object type of the actual argument. Note that the compiler cannot detect this situation since it occurs at run-time. The program in Listing 12-1 demonstrates this faulty behaviour producing the following result when it executes:

```
Base class data 100
Derived class data 100
```

The data from the `Derived` class is certainly not 100. The program should have stored and then retrieved the data as 200. This problem can be rectified by passing the `baseObject` parameter by reference to the `GetData()` function. If passed by reference, no copy of a base class object will be made. The entire `Derived` class object will be accessible to the function `GetData()`, and the correct result of 200 will be produced. The corrected program is shown in Listing 12-2.

Listing 12-2 Corrected version of Listing 12-1.

```cpp
//This program produces correct results.
#include <iostream.h>

class Base
{
    private:
        int BaseClassData;
    public:
        Base(int baseclassdata)
        {
            BaseClassData = baseclassdata;
        }

        virtual int GetClassData() const
        {
```

```
            return BaseClassData;
        }
};

class Derived : public Base
{
    private:
        int DerivedClassData;

    public:
        Derived(int derivedclassdata,
                int baseclassdata): Base(baseclassdata)
        {
            DerivedClassData = derivedclassdata;
        }

        int GetClassData() const
        {
            return DerivedClassData;
        }
};

int GetData(const Base& baseObject)  //Pass by reference
{
    return baseObject.GetClassData();
}

void main()
{
    Base* BasePtr;
    int ClassData;

    BasePtr = new Base(100);
    ClassData = GetData(*BasePtr);
    cout << "Base class data " << ClassData << endl;
    BasePtr = new Derived(200, 100);
    ClassData = GetData(*BasePtr);
    cout << "Derived class data " << ClassData << endl;
}
```

You will see the following result when this program executes:

```
Base class data 100
Derived class data 200
```

There is a lesson to be learned from this exercise - whenever class objects are passed to a function it is prudent to pass them by reference. As shown in Listing 12-2, the keyword `const` can be prefixed to the parameter passed by reference to protect the parameter from any inadvertent changes within the function.

12.2.4 The Copy Constructor

In previous chapters we used default constructors and standard constructors to instantiate objects. The copy constructor is a special constructor that is called when a copy of an object is created. If the developer does not provide a copy constructor, the compiler will generate one by default. This constructor makes a copy of an object by copying member-by-member from one object to the other for three situations:

(i)   When passing parameters to a function by value, a copy of the object must be created.

(ii) When parameters are returned by value, a copy of the object to be returned must be made. Again, the same copy constructor will be called to make the copy.

(iii) If an object is declared and initialised using another object passed as a parameter, the copy constructor must be called.

The assignment operator can also be used to initialise an object that was created previously using the default constructor. In principle, the assignment operator (=) must carry out the same actions as the copy constructor. If the developer does not overload the assignment operator the compiler will do so to suit the class. We will defer discussing overloading the assignment operator until operator overloading concepts have been described.

A few examples of object instantiation using various constructors are:

```
DCMotor Motor1;           // default constructor used
DCMotor Motor2(Motor1);   // copy constructor used
DCMotor Motor3;           // default constructor used
Motor3 = Motor1;          // assignment operator used
```

We will develop an example program that operates with arrays using the `IntArray` object to improve your understanding of the copy constructor. Consider the definition of the `IntArray` class given in Listing 12-3.

Listing 12-3 Header file intarray.h shows a class definition for an array of integers.

```
#ifndef IntarrayH
#define IntarrayH

class IntArray
{
    private:
```

```
        int NumInts;
        int* ArrayPointer;

    public:
        IntArray();                // Default constructor
        IntArray(int numints);  // Constructor
        ~IntArray();               // Destructor
        void EnterArray();         // Other member functions
        void PrintArray();
};
#endif
```

The `IntArray` class will instantiate an array of integers having a specified number of elements. The data member `NumInts` will store the total number of elements in the array and the pointer `ArrayPointer` will point to the dynamically allocated portion of memory containing the array of integers. The destructor `~IntArray()` will release the dynamically allocated memory. The function `EnterArray()` will prompt the user for array values, receive user input via the keyboard, and fill the array. The final function `PrintArray()` is used to print the contents of the array on the screen.

The `IntArray` class's constructor and its default constructor initialise the member data `NumInts` and `ArrayPointer`. If a parameter is passed to the constructor, memory for the array will be dynamically allocated as shown in the following constructor definitions:

```
IntArray::IntArray()   // Default constructor
{
    NumInts = 0;
    ArrayPointer = NULL;
}


IntArray::IntArray(int numints)    // Constructor
{
    if(numints <=0)
    {
        NumInts = 0;
        ArrayPointer = NULL;
    }
    else
    {
        NumInts = numints;
        ArrayPointer = new int[NumInts];
    }
}
```

Note that if the value of `numints` is 0 or negative, `NumInts` is initialised to 0 and `ArrayPointer` is initialised to the predefined constant `NULL` to indicate that the pointer is not pointing anywhere.

Suppose we use the object class `IntArray` in a `main()` function to instantiate two `IntArray` objects named A and B. We will then pass the objects A and B *by value* to a function named `AddArrays()` to add the `IntArray` object A to the `IntArray` object B as shown in the fragment of code below:

```
void AddArrays(IntArray a, IntArray b) // Pass by value
{
    // print the result of summation on-screen
}

void main()
{
    IntArray A(5);
    IntArray B(5);

    AddArrays(A,B);
    .
    .
    .
}
```

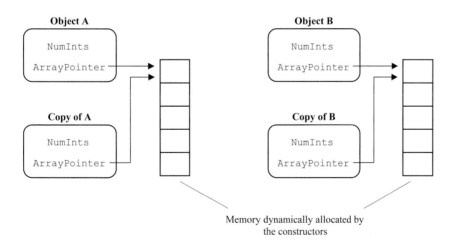

Figure 12-1 Objects A and B copied by the compiler-generated copy constructor.

When the actual arguments A and B are passed by value to the function `AddArrays()`, copies of A and B must be made. The compiler-generated copy constructor will be called to make these copies; this copy-making process is shown graphically in Figure 12-1.

The compiler-generated copy constructor has created a copy of each data member of the original objects. However, it did not make copies of the dynamically allocated memory. As a result, the pointer to the original object and the pointer to the copied object point to the same portion of memory.

The destructor `~IntArray()` is called after the function `AddArrays()` executes. The destructor will free the memory used for the temporary copies of A and B together with the original data objects that were dynamically allocated. This outcome is shown in Figure 12-2.

Figure 12-2 Result of discarding the copies of A and B.

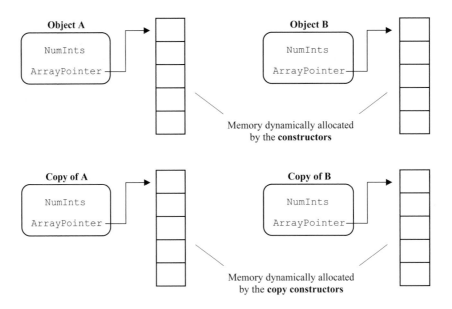

Figure 12-3 Copies of objects A and B made by a user written copy constructor.

Now the `main()` function has lost the original data it had created at the time of instantiating the objects A and B. We can overcome the problem created by the compiler-generated copy constructor if we write our own copy constructor that not only carries out member-by-member copying, but also makes copies of any portions of memory pointed to by pointer data members. Having such a copy constructor will change the situation shown in Figure 12-1 to that shown in Figure 12-3.

Now when the destructor is called to discard the copies, the original data will be unaffected and the outcome will be as shown in Figure 12-4.

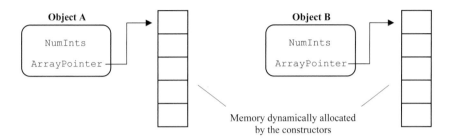

Figure 12-4 The original data areas are not destroyed by the destructor.

The header file `intarray.h` contains the definition of the copy constructor as given in Listing 12-4.

Listing 12-4 The header file intarray.h with user supplied copy constructor.

```
#ifndef IntarrayH
#define IntarrayH

class IntArray
{
private:
    int NumInts;
    int* ArrayPointer;

public:
    IntArray();
    IntArray(int numints);
    IntArray(const IntArray& intArray);
    ~IntArray();
    void EnterArray();
    void PrintArray();
};
#endif
```

The heading of the copy constructor is:

```
IntArray(const IntArray& intArray);
```

Remember that constructors do not have return values. The parameter passed to the copy constructor is the same type as the class the copy constructor belongs to and must be passed by reference. If it is passed by value, then a copy will be created by a call to the copy constructor. To pass a parameter to that copy constructor by value, another copy has to be created and so on – an endless sequence of function calls. Therefore, the parameter passed to the copy constructor *must* be passed by reference. The keyword const is used to protect the parameter passed by reference. Therefore, no changes will be made to the parameter being passed.

The complete function file is given in Listing 12-5.

Listing 12-5 The function file intarray.cpp for the IntArray class.

```
#include <iostream.h>
#include "intarray.h"

IntArray::IntArray()
{
    NumInts = 0;
    ArrayPointer = NULL;
}

IntArray::IntArray(int numints)
{
    if(numints <=0)
    {
        NumInts = 0;
        ArrayPointer = NULL;
    }
    else
    {
        NumInts = numints;
        ArrayPointer = new int[NumInts];
    }
}

IntArray::IntArray(const IntArray& intArray)
{
    NumInts = intArray.NumInts;
    ArrayPointer = new int[NumInts];
    for(int i=0; i < NumInts; i++)
        *(ArrayPointer+i) = *(intArray.ArrayPointer +i);
}
```

```
IntArray::~IntArray()
{
    if(ArrayPointer !=NULL)
    {
        delete ArrayPointer;
        ArrayPointer = NULL;
    }
}

void IntArray::EnterArray()
{
    cout << "Enter " << NumInts << " integer values."
        << endl;
    for(int i = 0; i < NumInts; i++)
        cin >> *(ArrayPointer + i);
}

void IntArray::PrintArray()
{
    for(int i =0; i < NumInts; i++)
        cout << *(ArrayPointer + i) << '\t';
    cout << endl;
}
```

A main() function can be written with an explicit call to the copy constructor and an explicit call to the destructor to destroy the object generated by the copy constructor. The program copycnst.cpp given in Listing 12-6 demonstrates that the call to the destructor *does not* destroy the original data.

Listing 12-6 This program tests the user supplied copy constructor - copycnst.cpp.

```
#include <iostream.h>
#include <conio.h>

#include "intarray.h"

void main()
{
    IntArray A(5);
    A.EnterArray();

    cout << "A " ; A.PrintArray(); // Print array A.
    getch();
    IntArray B(A);                 // Call copy constructor.
```

```
cout << "B "; B.PrintArray();   // Print array B.
B.~IntArray();                  // Destroy array B.
// Print array A again, check data is intact.
cout << "A "; A.PrintArray();
}
```

Executable File Generation		
Required Files	**Listing No.**	**Project File Contents**
copycnst.cpp	Listing 12-6	copycnst.cpp
intarray.cpp	Listing 12-5	intarray.cpp
intarray.h	Listing 12-4	

12.2.5 Overloading Operators as Member Functions

In this section we will be overloading the >> operator for the ADC class so that it can be used to carry out an analog-to-digital conversion and store the result in a variable. To do this the class definition in the header file adc.h must be changed to include the member function that overloads the operator. We will follow the syntax for function headings shown in Table 12-2 to produce the new header file for the ADC class given in Listing 12-7.

The parameter passed to the operator overloading function has been passed by reference. This allows the passed parameter to be changed within the function and the parameter in the calling environment will now hold the new value read from the ADC when the function exits.

Listing 12-7 Header file adc.h shows operator overloading as a member function.

```
#ifndef AdcH
#define AdcH
#include "pport.h"

class ADC : public ParallelPort
{
    private:
        unsigned char ADCValue;

    public:
        ADC(int baseaddress=0x378);
        unsigned char ADConvert();
        unsigned char GetADCValue();
        void operator>>(unsigned char& value);
};
#endif
```

The original `adc.cpp` file from Listing 11-4 is then modified to that given in Listing 12-8.

Listing 12-8 Function file adc.cpp associated with header file in Listing 12-7.

```
#include "adc.h"

ADC::ADC(int baseaddress) : ParallelPort(baseaddress)
{
    ADCValue = 0;
}

unsigned char ADC::ADConvert()
{
    WritePort2(0x01);
    WritePort2(0x00);
    WritePort2(0x01);

    WritePort2(0x03);
    ADCValue = ReadPort1() & 0xF0;

    WritePort2(0x01);
    ADCValue += (ReadPort1() >> 4) & 0x0F;

    return ADCValue;
}

unsigned char ADC::GetADCValue()
{
    return ADCValue;
}

void ADC::operator>>(unsigned char& value)
{
    ADConvert();
    value = ADCValue;
}
```

Note: although the >> operator needs two arguments, we have passed only one parameter. This parameter becomes the operand to the right of the >> operator. The operand to the left of the >> operator is the object of type ADC (see the `main()` function in Listing 12-9). Furthermore, within the function's body we have used the statements:

```
ADConvert();
value = ADCValue;
```

rather than the statement:

```
value = ADConvert();
```

We deliberately do this to keep the program (which overloads an operator as a member function) as similar as possible to the next program (which overloads an operator as a non-member function). The `main()` function shown in Listing 12-9 demonstrates use of the overloaded operator as a member function.

Listing 12-9 `main()` function file ovld.cpp uses the overloaded operator >>.

```cpp
#include <bios.h>
#include <conio.h>

#include "adc.h"

void main()
{
    ADC Adc;
    int Quit = 0, key;
    unsigned char Value;

    clrscr();

    while(!Quit)
    {
        gotoxy(10,10);


        Adc >> Value;
        cprintf("The ADC output is %10d\a",(int)Value);

        if(bioskey(1)!=0)
        {
            key = bioskey(0);


            if(key == 0x2d00)  Quit = 1;  // Alt-X key
        }
    }
}
```

Executable File Generation		
Required Files	**Listing No.**	**Project File Contents**
pport.cpp	Listing 10-8	pport.cpp
pport.h	Listing 10-7	
adc.cpp	Listing 12-8	adc.cpp
adc.h	Listing 12-7	
ovld.cpp	Listing 12-9	ovld.cpp

The operation of this program can be verified with the interface board by connecting the potentiometer output to the input of the ADC.

12.2.6 Overloading Operators as non-member f'ns

The operator overloading function will now be moved outside the class definition as an ordinary function to create the header file adc.h given in Listing 12-10.

Listing 12-10 File adc.h shows operator overloading as a non-member function.

```
#ifndef AdcH
#define AdcH

#include "pport.h"

class ADC : public ParallelPort
{
    private:
        unsigned char ADCValue;

    public:
        ADC(int baseaddress=0x378);
        unsigned char ADConvert();
        unsigned char GetADCValue();
};

// declaration of the non-member function
void operator>>(ADC adc, unsigned char& value);
#endif
```

Listing 12-11 shows the function file associated with the above header file.

Listing 12-11 Function file adc.cpp associated with the header file in Listing 12-10.

```
#include "adc.h"

ADC::ADC(int baseaddress) : ParallelPort(baseaddress)
{
    ADCValue = 0;
}

unsigned char ADC::ADConvert()
{
    WritePort2(0x01);
    WritePort2(0x00);
    WritePort2(0x01);

    WritePort2(0x03);
    ADCValue = ReadPort1() & 0xF0;
```

```
    WritePort2(0x01);
    ADCValue += (ReadPort1() >> 4) & 0x0F;

    return ADCValue;
}

unsigned char ADC::GetADCValue()
{
    return ADCValue;
}

void operator>>(ADC adc, unsigned char& value)
{
    adc.ADConvert();
    value = adc.GetADCValue();
}
```

Executable File Generation		
Required Files	**Listing No.**	**Project File Contents**
pport.cpp	Listing 10-8	pport.cpp
pport.h	Listing 10-7	adc.cpp
adc.cpp	Listing 12-11	ovld.cpp
adc.h	Listing 12-10	
ovld.cpp	Listing 12-9	

The operator overloading function has now been passed two parameters unlike the case described in Section 12.2.5. The first parameter becomes the operand to the left of the operator >> and the second parameter becomes the operand to the right. Since the function does not belong to any class, the first parameter must be used explicitly. We cannot use the following statement in this function because we do not have access to the private data member ADCValue:

```
value = adc.ADCValue;
```

Instead, the public member function GetADCValue() must be called. The same main() function (Listing 12-9) can now be used provided that its #include "adc.h" statement includes the header file given in Listing 12-10, and is compiled and linked with the function file given in Listing 12-11.

12.2.7 Friend Functions

Ordinary non-member functions can be declared as friends of a particular class to eliminate the difficulties associated with accessing private data members. In this case the non-member friend function will have unrestricted access to all data members of the class that it is declared in. The friend functions can be viewed as public functions of the class that do not need to be attached (tagged) to the object using the membership access operators (i.e . or ->). Careful use of `friend` functions can facilitate some tasks such as input/output streaming we will be discussing shortly. To demonstrate the use of `friend` functions, we can modify the header file given in Listing 12-10 to become that given in Listing 12-12.

Listing 12-12 Use of `friend` functions - adc.h.

```
#ifndef AdcH
#define AdcH

#include "pport.h"

class ADC : public ParallelPort
{
    private:
        unsigned char ADCValue;

    public:
        ADC(int baseaddress=0x378);
        unsigned char ADConvert();
        unsigned char GetADCValue();
        friend void operator>>(ADC adc, unsigned char& value);
};
#endif
```

Since the operator overloading function is a friend of the ADC class, it will now have access to the private data member `ADCValue`. Therefore, instead of the line:

```
value = adc.GetADCValue();
```

we can now use:

```
value = adc.ADCValue;
```

This avoids a function call and so speeds up program execution. The associated function file is shown in Listing 12-13.

Listing 12-13 Function file adc.cpp for the header file in Listing 12-12.

```
#include "adc.h"

ADC::ADC(int baseaddress) : ParallelPort(baseaddress)
{
    ADCValue = 0;
}

unsigned char ADC::ADConvert()
{
    WritePort2(0x01);
```

```
        WritePort2(0x00);
        WritePort2(0x01);

        WritePort2(0x03);
        ADCValue = ReadPort1() & 0xF0;

        WritePort2(0x01);
        ADCValue += (ReadPort1() >> 4) & 0x0F;

        return ADCValue;
}

unsigned char ADC::GetADCValue()
{
        return ADCValue;
}

void operator>>(ADC adc, unsigned char& value)
{
        adc.ADConvert();
        value = adc.ADCValue;
}
```

Executable File Generation		
Required Files	**Listing No.**	**Project File Contents**
pport.cpp	Listing 10-8	pport.cpp
pport.h	Listing 10-7	
adc.cpp	Listing 12-13	adc.cpp
adc.h	Listing 12-12	
ovld.cpp	Listing 12-9	ovld.cpp

In the previous sections we implemented operator overloading to carry out analog-to-digital conversion and stored the resulting value in a variable. We overloaded the >> operator to accomplish this 'input' operation. Next we will overload the << operator to output the result of an anlog-to-digital conversion to the screen or to a file. Before we do this, we need to have a fundamental understanding of *I/O streams* and *pass-through objects*.

12.2.8 I/O Streams

I/O streams can be viewed as a sequential transfer of one or more objects between two locations. Examples of such transfers could be data generated by the keyboard to a program, program data to screen, program data to a file, data from a file to a program, etc. C++ software provides a number of object classes and a large variety of functions to facilitate input/output streaming of objects. Output streams can be generated by calling the constructor of the `ostream` class. Similarly, an input stream can be generated by calling the constructor of the `istream` class.

The object classes, `ofstream` and `ifstream` are part of the standard C++ library and facilitate writing to and reading from various types of objects including files. To enable us to use these objects we must include the header file `fstream.h` just as we included the header file `iostream.h` file for us to use the `cin` and `cout` objects.

A simple program that demonstrates file I/O is given in Listing 12-14.

Listing 12-14 File Input/Output.

```
#include <fstream.h>

main()
{
    int Data;

    ifstream is("infile.dat");
    ofstream os("outfile.dat");

    while(is)
    {
        is >> Data;              // Number from file -> Data
        if(!is.fail())
            os << '\t' << Data; // Data -> file os
    }
    os.close();
    is.close();
    return 0;
}
```

The statements:

```
ifstream is("infile.dat");
ofstream os("outfile.dat");
```

call the `ifstream()` and `ofstream()` constructors and instantiate the two objects named `is` and `os`, respectively. Each constructor opens the file whose name has been passed as a parameter in the form of a character string. The constructors also create their own memory buffer needed to transfer data between memory and the respective file.

Assume that the file `infile.dat` contains the number data:

```
10
20
```

```
30
40
50
```

Each number needs to be separated by white space (one or more spaces, tab, line feed or a carriage return). The variable `Data` is used to temporarily store the number read from `infile.dat`.

The contents of `outfile.dat` will be:

```
10  20  30  40  50
```

Only data that has been read successfully from `infile.dat` will then be written to the file `outfile.dat`. The `ifstream` class has a member function `fail()` that is called to determine if an error has occurred when reading data from the file. The `ifstream` object `is` will evaluate to be 0 when the end of the file has been reached and cause the `while` loop to terminate.

12.2.9 Pass-through Objects

Pass-through objects are objects that enter a function as a parameter by reference and appear as a return value of the function, also by reference. We will examine returning values by reference in a moment. First, let us understand the motivation behind the use of pass-through objects. Consider the following three output streaming operations where a, b, and c are integer objects:

```
cout << a;
cout << b;
cout << c:
```

These three operations can be combined in one statement as follows:

```
cout << a << b << c ;
```

Parentheses can be used to show the precedence of output streaming as follows:

```
(((cout << a) << b) << c );
```

After streaming the object a out, the remainder of the operation will be:

```
(((cout) << b) << c);
```

The last statement to be executed is then:

```
(cout << c);
```

Consider the output streaming of any one of the objects where the overloaded operator `<<` receives two arguments, namely `cout` and an integer object. The overloaded operator will return `cout` as the return value. This is why `(cout << a)` is replaced by `cout`. Here, `cout` enters the function as a parameter and then appears as the return value. Therefore, `cout` becomes a *pass-through object* in the operator overloading function for the operator `<<`. Pass-through objects must maintain their 'life' throughout this process and so there is *no object copying* involved. Therefore, the parameter passed must be passed by reference and the value returned must also be returned by reference.

Pass-through objects are especially useful in 'chained' use of operators or functions. Another example is shown below where a, b, and c are integer objects:

```
a = b = c = 3;
```

Once again, the precedence of operations can be shown using parentheses:

```
(a = (b = (c = 3)));
```

After the first assignment operation the expression reduces to:

```
(a = (b = (c)));
```

Therefore, in the operation c = 3, c enters as a parameter and then appears as the return value, explaining why c = 3 can be replaced by c.

An analogy from day-to-day life can be used to further explain the use of returning values by reference. Consider a situation where you have a string and a set of beads. We want to thread all the beads, one at a time, onto the string. We will write a function to receive the string and a bead. The purpose of the function is to attach one bead to the string and to return the *same string* with the beads(s) it received.

We specify that a function must return a value by reference by adding & after the return value type in the function heading. This is shown in bold typeface in the Add_a_Bead function shown below:

```
String& Add_A_Bead(String& string, Bead bead)
{
    string = string + bead;
    return string;
}
```

For the two parameters of the function Add_A_Bead(), parameter string is of data type String and bead is of data type Bead. The expression string = string + bead represents attaching a bead to the string. You can attach n beads to the *same* string by executing this function n times.

If we did not return the value by reference, each time you execute the function to attach a bead, you will get a *new copy* of the string, with an increasing number of beads on each subsequent copy. If you execute such a function n times, you will create and discard n copies of the string, the first copy having one bead, the second copy having 2 beads etc. until you end up with the last copy having n beads - what a waste of time and memory! In addition, you do not have the original string - you have a copy.

12.2.10 Assignment Operator

If the developer does not provide an operator overloading function to overload the assignment operator, the compiler will generate a default overloaded assignment operator. This applies in the same way as for the copy constructor. The assignment operator is used to carry out member-by-member assignment from one *existing* object to another *existing* object. The compiler-generated assignment operator has the same weakness as the compiler-generated copy constructor; it will not copy any portions of memory pointed to by pointer type data members. The developer must provide a function to safely overload the assignment operator to overcome this deficiency.

The function heading to overload the assignment operator is:

```
IntArray& operator=(const IntArray& intarray);
```

Note that the parameter is passed in exactly the same way it was passed to the copy constructor, however, the return value is a reference to an `IntArray` object. This is necessary to be consistent with the usage of the assignment operator so a chained assignment can be carried out like that shown in this simple example:

```
int a,b,c;
a = b = c = 3;
```

The object class `IntArray` and the associated source files described in Section 12.2.4 are used below to enhance the capabilities of the class by the addition of the overloaded assignment operator. The overloaded assignment operator that has been added is shown in the header file of Listing 12-15 and the function file of Listing 12-16.

Listing 12-15 Header file intarray.h with overloaded assignment operator.

```
#ifndef IntarrayH
#define IntarrayH

class IntArray
{
private:
    int     NumInts;
    int*    ArrayPointer;

public:
    IntArray();
    IntArray(int numints);
    IntArray(const IntArray& IntArray);
    IntArray& operator=(const IntArray& intArray);
    ~IntArray();
    void EnterArray();
```

```
        void PrintArray();
};
#endif
```

Listing 12-16 Function file intarray.cpp with overloaded assignment operator.

```
#include <iostream.h>
#include "intarray.h"

IntArray::IntArray()
{
    NumInts = 0;
    ArrayPointer = NULL;
}

IntArray::IntArray(int numints)
{
    if(numints <=0)
    {
        NumInts = 0;
        ArrayPointer = NULL;
    }
    else
    {
        NumInts = numints;
        ArrayPointer = new int[NumInts];
    }
}

IntArray::IntArray(const IntArray& intArray)
{
    NumInts = intArray.NumInts;
    ArrayPointer = new int[NumInts];
    for(int i=0; i < NumInts; i++)
        *(ArrayPointer+i) = *(intArray.ArrayPointer +i);
}

IntArray& IntArray::operator=(const IntArray& intArray)
{
    if(this != &intArray)
    {
        if(ArrayPointer != NULL)
            delete ArrayPointer;// Release any allocated memory
        NumInts = intArray.NumInts;
```

```
        ArrayPointer = new int[NumInts];
        for(int i = 0; i < NumInts; i++)
            *(ArrayPointer + i) = *(intArray.ArrayPointer + i);
    }
    return *this;
}


IntArray::~IntArray()
{
    if(ArrayPointer !=NULL)
    {
        delete ArrayPointer;
        ArrayPointer = NULL;
    }
}


void IntArray::EnterArray()
{
    cout << "Enter " << NumInts << " integer values." << endl;
    for(int i = 0; i < NumInts; i++)
        cin >> *(ArrayPointer + i);
}


void IntArray::PrintArray()
{
    for(int i =0; i < NumInts; i++)
        cout << *(ArrayPointer + i) << '\t';
    cout << endl;
}
```

We have used the this pointer that points to the object itself to overload the assignment operator. A test is performed inside the if condition to check that the object to be copied is the same object. If so, there is no point copying the object. Within the true clause of the if statement a test is made to check whether ArrayPointer is pointing to any previously allocated memory area. If so the delete operator is used to release that memory. If this is not done there will be a memory leak; i.e. there will be a portion of memory that is not used and cannot be used again because it has not been released. Identical steps are then carried out as performed for the copy constructor. Finally, the object itself is returned to facilitate the chained use of the assignment operator. This is done by returning the object using the result *this (where this is the pointer to the object, and *this is the object itself).

Listing 12-17 shows a program that demonstrates the use of the copy constructor and the assignment operator with the IntArray class.

Listing 12-17 File asgnopr.cpp shows use of the assignment operator.

```
#include <iostream.h>
#include <conio.h>

#include "intarray.h"

void main()
{
// Call constructor
    IntArray A(5);
    A.EnterArray();

// Print array A
    cout << "A " ; A.PrintArray();
    getch();

// Call copy constructor
    IntArray B(A);

// Print array B
    cout << "B "; B.PrintArray();
    getch();

// Call default constructor
    IntArray C;

// Use assignment operator
    C = A;

// Print array C
    cout << "C "; C.PrintArray();
}
```

Executable File Generation		
Required Files	**Listing No.**	**Project File Contents**
asgnopr.cpp	Listing 12-17	asgnopr.cpp
intarray.cpp	Listing 12-16	intarray.cpp
intarray.h	Listing 12-15	

12.3 Data Acquisition

In this section, we will combine the concepts we have learned in this chapter to create a data acquisition program that uses operator overloading. An analog-to-digital conversion will be performed by writing a function that overloads the operator << in association with an ADC object and an output stream object such as

cout. This overloading process will enable us to output the converted value to
cout. The header file adc.h must be modified to include the new operator
overloading function. It is possible to write this function without it needing to
access any private data of the ADC class. We have chosen to write a non-member
function, and so it does not need to be declared as a friend function. The new
header file is given in Listing 12-18.

Listing 12-18 File adc.h overloads the << operator.

```
#ifndef AdcH
#define AdcH

#include <iostream.h>
#include "pport.h"

class ADC : public ParallelPort
{
    private:
        unsigned char ADCValue;

    public:
        ADC(int baseaddress=0x378);
        unsigned char ADConvert();
        unsigned char GetADCValue();
        friend void operator>>(ADC adc, unsigned char& value);
};

// declaration of the non-member functions
ostream& operator<<(ostream& os, ADC adc);
#endif
```

Observe the similarity of the following function with the Add_A_Bead()
function described earlier in Section 12.2.9:

```
ostream& operator<<(ostream& os, ADC adc);
```

The operator overloading function operates in a very similar manner, except the
operator << is now used to call the function. The function file is given in Listing
12-19.

Listing 12-19 Function file adc.cpp for the header file in Listing 12-18.

```
#include <iostream.h>
#include "adc.h"

ADC::ADC(int baseaddress) : ParallelPort(baseaddress)
{
    ADCValue = 0;
```

```
    }

unsigned char ADC::ADConvert()
{
    WritePort2(0x01);
    WritePort2(0x00);
    WritePort2(0x01);

    WritePort2(0x03);
    ADCValue = ReadPort1() & 0xF0;

    WritePort2(0x01);
    ADCValue += (ReadPort1() >> 4) & 0x0F;

    return ADCValue;
}

unsigned char ADC::GetADCValue()
{
    return ADCValue;
}

void operator>>(ADC adc, unsigned char& value)
{
    adc.ADConvert();
    value = adc.ADCValue;
}

ostream& operator<<(ostream& os, ADC adc)
{
    os << " " << (int)adc.ADConvert();

    return os;
}
```

The following programming statement demonstrates the elegance of operator overloading.

```
cout << Adc << Adc << Adc;
```

Each use of the overloaded operator << with an Adc object will perform an analog-to-digital conversion and then send the resulting output value to the standard output device. As such we will be able to use the above statement to produce three output values to the screen.

The above statement has the precedence of evaluation as shown by the parentheses in the statement below:

```
(((cout << Adc) << Adc) << Adc;)
```

The program `dataacq.cpp` shown in Listing 12-20 contains a `main()` function you can experiment with. This program will carry out three analog-to-digital conversions and send the values to the screen every second. Several analog-to-digital conversion samples can be acquired as a group and averaged to overcome the effects of noise on a signal. Note that a significant period of time is consumed to print each set of results to the screen, slowing the effective speed of acquisition.

Listing 12-20 `main()` function datacq.cpp checks operation of operator `<<`.

```
#include <conio.h>
#include <bios.h>
#include <dos.h>

#include "adc.h"

void main()
{
    ADC Adc;
    int Quit = 0;

    clrscr();

    while(!Quit)
    {
        cout << endl << Adc << Adc << Adc;

        if(bioskey(1)!=0)
            if(bioskey(0) == 0x2d00) Quit = 1; /*Alt-X*/
        delay(1000);
    }
}
```

Executable File Generation		
Required Files	**Listing No.**	**Project File Contents**
pport.cpp	Listing 10-8	pport.cpp
pport.h	Listing 10-7	
adc.cpp	Listing 12-19	adc.cpp
adc.h	Listing 12-18	
dataacq.cpp	Listing 12-20	dataacq.cpp

Operator overloading can be used for a variety of other tasks. For example, we can overload the ++ operator in the `DCMotor` class described in Chapter 8 to enable us to increment the `Speed` by 1 unit of resolution. Similarly, the `--` operator can be overloaded in the `DCMotor` class to decrement the `Speed` by 1 unit of resolution. A segment of sample code that uses such overloaded operators could be:

```
DCMotor Motor1;
Motor1++; // increase speed by 1
```

Another example would be to overload the `<<` operator in the `DAC` class so it can output the integer value `N` to the Digital-to-Analog Converter:

```
DAC Dac;
Dac << N;
```

12.4 Summary

There are several methods that can be used to pass parameters into functions and return the result from the function. Most functions receive their parameters as a copy of the argument given in the calling environment. An alternate means of passing parameters is pass by reference. When passing parameters by reference the function has access to the actual argument used in the calling environment. This saves memory space and also provides an improvement in speed. For reasons such as these, pass by reference is generally preferred when passing class objects to functions.

Similarly, most functions are written to return values as a copy of the value generated within the function. If values are returned by reference, the real object within the function is returned rather than a copy of it. This facility can be used efficiently in operator overloading, in particular with the chained use of an operator.

Friend functions are a special category of functions that have unrestricted access to all members of the class they are declared in. Although they are not member functions, they have all the privileges of a member function. Friend functions have a further advantage over member functions in that they do not need to be tagged to an object when being called (using '`.`' or '`->`' membership operators).

12.5 Bibliography

Winston, P.H., *On to C++*, Addison Wesley, 1994.

Johnsonbaugh, R and Martin Kalin, *Object-Oriented Programming in C++*, Prentice Hall, 1995.

Staugaard A. C. (Jr), *Structured and Object Oriented Techniques*, Prentice Hall, 1997.

Lafore, R. *Object Oriented Programming in MICROSOFT C++*, Waite Group Press, 1992.

Wang, P.S., *C++ with Object Oriented Programming*, PWS Publishing, 1994.

13

The PC Timer

Inside this Chapter

- What is a timer?

- The PC timer architecture.

- Programming the PC timer.

- Time measurement.

- Reflex measurement.

- Plotting with a time-base.

- Digitising a signal with a time-stamp.

13.1 Introduction

So far we have not used *real-time* for tasks that have involved timing. Recall the generation of PWM signals from Chapter 8. In these programs we generated time delays by executing software loops whose duration was unknown and dependent on the computer's speed. A hardware timer is typically used when time needs to be measured accurately. Your PC is equipped with such a timer that can be programmed to carry out various timing-related tasks. It operates independently of the PC's processor to ensure uninterrupted and accurate operation, and has spare resources for us to use in our own programs.

In general, the timer subsystem of your PC has three independent timers. More modern systems will have five independent timers. We will keep our discussion to the most general case, i.e. three timers. There are two principal functions associated with timers; timing of an event and counting events. The basic requirement for any timer is a *clock* signal; being a continuous train of pulses with a known and highly stable frequency. Having access to a steady clock allows us to write programs that can take advantage of real-time operations.

13.2 PC Timer System

Central to the timing system of all but the most recent PC's is the *8254 Programmable Interval Timer* containing three timers, named Timer 0, Timer 1 and Timer 2 as shown in Figure 13-1. The timers can be operated in several different modes controlled by gate signal level and the use of a control register (explained in section 13.2.2). These modes include single timeout, square wave generator and rate generator, discussed in section 13.2.3. They share a common clock signal driving their clock inputs, but only Timer 2 has a gate input that is free to be controlled through software.

Each of the timers contains a 16-bit *counter*. A counter can be considered as a special memory location in hardware, the value of which is incremented or decremented by each incoming clock pulse. In your PC, each clock pulse drives the counters *down* a count value. Typically, the counter's output signal will change state when it reaches zero.

Because all three timers share the same fixed clock signal, they cannot be used for *event counting*. Event counting takes place when a counter/timer is used to count external pulses applied to its clock input - often arriving at irregular intervals. Regardless of the speed of a PC, its clock frequency will be 1.1932 MHz. This enables every PC to maintain a fixed standard for timing.

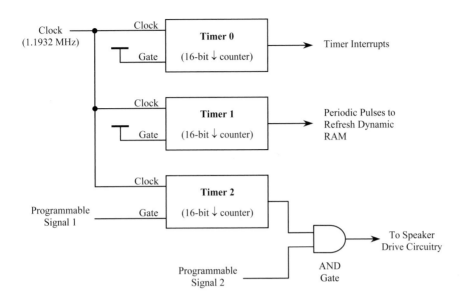

Figure 13-1 PC Timing System - 3 timers (clock input, gate, and output).

The three timers in your PC have special tasks assigned to them, explained as follows.

TIMER 0

Timer 0 is used to generate the so-called *timer interrupt*. The timer interrupt will regularly trigger the CPU to execute a special routine that updates the system time. This action takes place every 54.9 milliseconds. The gate signal of Timer 0 is held permanently at logic-HIGH, and therefore is not programmable. Since the clock and the gate are not programmable, the only timer variables that can be altered are the count value written to the counter and the mode of operation. Timer 0 has an output latch register that allows software to read the count value. Furthermore, the state of its output signal can be determined (i.e. high or low) by reading Timer 0's status register.

NOTE

An *interrupt* is a signal generated by hardware or software that is sent to the CPU to request its attention. Depending upon the priority of an interrupt, it will be attended to immediately or flagged for later attention. The interrupt generated by timer 0 has the highest priority and will be attended to immediately by the CPU.

TIMER 1

Timer 1 is used to generate a periodic signal for driving the hardware that refreshes the dynamic RAM (memory). Timer 1 also has its clock and gate signals permanently wired. Like Timer 0, it has an output latch register that allows software to read the count value. The state of its output signal can be determined by reading Timer 1's status register.

TIMER 2

Timer 2 is used to drive the speaker via an AND gate whose other input is programmable (see Figure 13-1). Its clock signal is permanently connected, although the gate is free to be controlled through software. Timer 2 also has an output latch register that allows software to read the count value. Similarly, the state of its output signal can be determined by reading Timer 2's status register.

13.2.1 Configuring the Counters

As mentioned earlier, all three timers countdown when operating. The number loaded into each timer's counter determines the timing duration. Therefore, writing a smaller number to a counter will result in a shorter time interval before it reaches zero. A particular counter can be read or written to at any time.

The counter's value is constantly changing as it operates and can be read by sampling the count value and storing the result. The hardware device that performs such a function is known as a *latch register*. Each timer has an input latch register to allow writes to its counter (load it) and also an output latch register to allow its counter's value to be read. This event is also dependent upon the status of the gate and the mode of operation.

A special *Control Register* is used to provide the facility to select which counter we wish to write to or read from, set the mode of operation, set the number format, etc. There is one more register in addition to the three Timer registers and the Control Register. This register allows access to the output of Timer 2 and the speaker gate control signal.

Timer Ports

Table 13-1 Port addresses - PC timers and speakers.

Address	Function
0x40	Timer 0 data latch register (input and output).
0x41	Timer 1 data latch register.
0x42	Timer 2 data latch register .
0x43	Control Register for Timers 0, 1 and 2.
0x61	Timer 2 output and speaker control.

The timers are part of your hardware system. To be able to program the timers, software must have access to timer subsystem hardware. This is possible by accessing the timer's ports similar to the way we accessed the parallel port hardware using its port addresses. The port addresses associated with the timer and speaker system are given in Table 13-1.

Programming the timer starts with programming the Control Register as described in the following sections.

13.2.2 The Control Register

The Control Register is used to configure a timer. It can also be used to request the counter status or to latch the current count value. The register contains eight bits that must be appropriately set to enable the following:

- Select a particular counter.
- Specify the byte(s) used to Read/Load the counter.
- Specify the mode of operation.
- Specify the counting format in binary or BCD.

The configuration of the Control Register is shown in Table 13-2.

Table 13-2 Configuration of the Control Register (address 0x43).

Bit 7	Bit 6	Bit 5	Bit 4	Bit 3	Bit 2	Bit 1	Bit 0
SC1	SC0	RL1	RL0	M2	M1	M0	BCD
Select Counter		Read/Load MSB/LSB		Mode Select			Binary or BCD

Select Counter

The counter can be selected by programming Bits 6 (SC0) and 7 (SC1) of the Control Register as shown:

SC1	SC0	Counter
0	0	Counter 0
0	1	Counter 1
1	0	Counter 2

Byte(s) used to Read/Load Counters

Bits 4 (RL0) and 5 (RL1) can be programmed to choose either Least Significant Byte (LSB) only, or the Most Significant Byte (MSB) only, or to use both bytes of the counter when counting:

Operation	RL1	RL0
Counter Latch	0	0
Read/Load LSB	0	1
Read/Load MSB	1	0
Read/Load LSB then MSB	1	1

For example; if the LSB is to be written to a counter, RL1 and RL0 will need to be set at 0 and 1 respectively. The LSB is then loaded into the counter by writing to its data register (e.g. 0x40 for Timer 0).

If the MSB is to be written to a counter, RL1 and RL0 will need to be set to 1 and 0 respectively. The MSB is then loaded into the counter by writing to its data register (e.g. 0x40 for Timer 0).

If a 16-bit number (two bytes) is to be written to a counter, both RL1 and RL0 will need to be set to 1. The 16-bit number will be loaded into the counter by carrying out two consecutive write operations to the appropriate data register (e.g. 0x40 for Timer 0) by first writing the LSB followed by writing the MSB.

Counting format (Binary/BCD)

The mode of counting can be set to binary (typically used) or BCD (Binary Coded Decimal) - we don't explain counting in BCD mode, so set the bit to 0.

BCD	Counting Operation
0	Binary
1	Binary Coded Decimal (BCD)

Timer mode of operation

Mode Select			
M2	M1	M0	Mode Name
0	0	0	Mode 0
0	0	1	Mode 1
0	1	0	Mode 2
0	1	1	Mode 3
1	0	0	Mode 4
1	0	1	Mode 5
1	1	0	Mode 2
1	1	1	Mode 3

The Timer's modes of operation are explained in the next section. They are set using the Control Register bits 1, 2 and 3 as shown in the previous table.

13.2.3 Modes of Operation of the Timers

There are up to six different modes of timer operation. Timer 0 and Timer 1 have their gates hard-wired to a logic-HIGH level. This excludes them from operating in some of the modes described below. Timer 2 is the only timer that has a controllable gate. As such it can operate in all six modes.

Mode 0: Single Timeout

In this mode, the counter generates a low-level output signal for the fixed number of clock pulses loaded into its data register. Each incoming clock pulse will decrement the count value by one count (provided the gate input is high). When the count value reaches 0, the output line will change from low to remain high. Note that the maximum period is obtained when the decimal number 65,535 is loaded into the counter. This period will be approximately 54.9 ms.

All three counters can be programmed in this mode. To initiate single timeout operation, first configure the timer to operate in this mode and then write the count value to the data register. The countdown will begin immediately after writing the count data. If the gate signal is held low, counting stops until the gate signal returns high. Timers 0 and 1 have their gate signals hard-wired to a logic-HIGH level. This is not the case for Timer 2, whose gate input can be controlled by writing to bit 0 of the port 0x61. The bits of the port at 0x61 for operating Timer 2 are shown in Table 13-3.

Table 13-3 Bits at port address 0x61 (control of Timer 2 output & speaker).

Bit7	Bit6	Bit5	Bit4	Bit3	Bit2	Bit1	Bit0
X	X	Timer 2 OUT	X	X	X	Speaker Gate	Timer 2 GATE

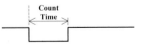

Mode 1: Re-triggerable one-shot

This mode is used to generate a low-level pulse following a trigger command from the gate (hence only Timer 2 can use this mode). The duration of the pulse is set by the count value loaded into the counter. The gate signal is briefly sent from low to high and back to low to initiate counting down. As this happens, the counter output will drop from a high state to a low state. When the count value has decremented to zero, the output will return to a high state.

The previously used count value is automatically reloaded into the counter at the end of the countdown when the counter reaches zero. Another one-shot period is generated when another high pulse is applied to the gate (hence the term 're-triggerable'). The timer can also be re-triggered during the current countdown. In this case, a new countdown will start immediately after the re-trigger. A new count value can also be written to the input latch register during the current countdown; this will not affect the current countdown. The new count value in the input latch register will be loaded for the next countdown immediately after the next re-trigger.

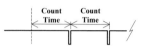

Mode 2: Rate generator

This mode is used to periodically generate a narrow low-level output pulse. When the value that was loaded into the counter reaches 1, the output changes to a low state for one clock period. The count value is then automatically reloaded to repeat the same process. All three timers can be used in this mode. Counting is stopped whenever the gate signal is low (only applies to Timer 2). The rate generator is predominantly used to generate hardware interrupts at regular intervals since the narrow pulse can be missed when detecting using software means.

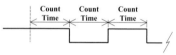

Mode 3: Square wave generator

This mode produces a continuous square wave output. Everytime the timer's counter reaches zero, the output toggles and the count value is automatically re-loaded into the counter from its input latch. Note that the gate must be high to enable down-counting and the count is decremented by *two* for each clock pulse. All three timers can operate in this mode. Timer 0 generates the timer interrupt when configured in this mode.

Mode 4: Software triggered strobe

Counting down is initiated when a count value is written to the data register. This mode produces a single narrow output pulse for one clock period when the count reaches zero (a non-periodic narrow pulse is known as a *strobe*). The counter will remain inactive until at a later time software again writes to the counter. The gate must be high at all times to enable operation.

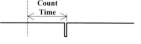

Mode 5: Hardware triggered strobe

This mode is identical to Mode 4 except triggering is carried out by hardware means. The start of a countdown is triggered by applying a brief high-level pulse to the gate. The output will pulse low for one clock period when the count reaches zero. The counter will remain inactive until the next hardware trigger. As for mode 1, this mode is only possible with Timer 2.

13.2.4 Read-back Commands

A timer 'read-back' command allows the following data to be read from a counter; the count value, status of its output signal, read/load status, configuration mode, and count mode. Two very useful 'read-back' tasks can be invoked by writing the bit patterns described below to the Control Register.

Task 1 - Multi Counter Latch:

A counter is latched by taking a 'snap shot' of the selected counter's value and transferring that count value to the counter's output latch register. A command to latch one or more counters can be issued by writing the bit pattern shown below to the Control Register at address 0x43:

Control Register at address 0x43.

Bit 7	Bit 6	Bit 5	Bit 4	Bit 3	Bit 2	Bit 1	Bit 0
1	1	0	1	CT2	CT1	CT0	0

Setting of bits CT0, CT1, and CT2 determines which counter(s) will be latched. Once latched, a program can read the count value by reading the output latch register. Addresses of latch registers are shown in Table 13-1. Two 8-bit reads of the output latch register must be carried out to read the 16-bit count value. The low byte will be obtained in the first read followed by the high byte in the second read.

Task 2 - Status of Timers:

The status information for each timer can be read via the corresponding timer data register. The command to report the status is issued by assembling the byte as shown following, and writing it to the Control Register (at address 0x43). We select the timer(s) by setting their respective control bits CT0, CT1 and CT2 of the Control Register to 1. The 8-bit status value is obtained by reading the data register of the selected timer (addresses are shown in Table 13-1). The status information is interpreted according to Figure 13-2.

Control Register at address 0x43.

Bit 7	Bit 6	Bit 5	Bit 4	Bit 3	Bit 2	Bit 1	Bit 0
1	1	1	0	CT2	CT1	CT0	0

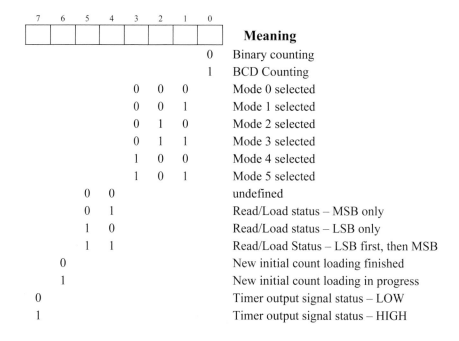

Figure 13-2 Description of the returned counter status byte.

13.3 Programming the Timer

Programs will be developed in the coming sections that take advantage of the PC timer's accurate timing. We will start by developing an object class for the PC timer. Note that although the timer hardware is independent of the parallel port hardware, the object classes we created in the preceding chapters can be used with the new timer object class developed in this chapter.

Programs have been developed to accurately measure time periods, measure a persons reflexes, and to generate a time-base for an earlier program that displayed the VCO output on-screen (see Chapter 10). The final program digitises an electrical waveform generated on the interface board, and saves the digitised values for later analysis.

As explained previously, the PC's three timers are connected to a clock signal having a frequency of 1.1932 MHz. Their counters can be loaded with a maximum value of 65,535 to produce a maximum time of 54.9 ms per countdown cycle. If we are to measure times greater than 54.9 ms, we must monitor and account for the number of down-counts. Each countdown of Timer 0 is referred to as a 'tick'. A special region of memory known as BIOS is used to store the number of ticks since mid-night. BIOS is the abbreviation for the PC's Basic Input Output System. It is a segment of software that predominantly interacts with hardware devices to carry out low-level tasks. Like many operating systems, the PC's operating system suffers from poor determinism (determinism is the term used to describe the ability of software to carry out tasks on time), and other latencies (time delays when actioning requests – some generated by hardware). As a result of these latencies, the tick count does not always get updated immediately, and so the BIOS tick count cannot be used reliably. We need to devise a reliable method to monitor and account for the number of tick counts so we can use this value in conjunction with the timer count value to measure time periods greater than one full timer countdown cycle. We will use Timer 0 for our time reference since it operates in a suitable mode.

13.3.1 Reading Timer 0 Count Value and Ticks

To read the number of counts corresponding to a given instant in time, we must read the timer ticks and also the contents of the Timer 0 counter. Section 13.2.4 describes how to latch the count value of Timer 0 (a snap shot taken) by issuing a read-back command. The latched data can be read later by reading the output register of Timer 0.

Before starting to program the timers, it is important to be aware that we will not adhere to the strict practice of structured procedure abstraction (use of functions). This will allow us to produce programs that execute fast and provide satisfactory accuracy when reading the timer. Additional time would be consumed if we called a function within one of the member functions of our PCTimer class.

13.4 The Object Class PCTimer

We will develop a class to provide a means of measuring time and generating time delays. To do this the class needs functions that perform the following tasks:

- Set a 'zero' time reference (this does not mean clearing the hardware timer in the PC).
- Generate a time delay of a specified value.
- Read the current time.

The definition of the PCTimer class that has the above capabilities is shown in Listing 13-1.

Listing 13-1 The PCTimer class - pctimer.h.

```
#ifndef PctimerH
#define PctimerH

class PCTimer
{
```

```
private:
    unsigned int  InitCount;
    unsigned long TickCount;
    unsigned int  LastCount;

public:
    PCTimer();
    void ResetTimer();
    void Delay(const double& milliseconds);
    double ReadTimer();
    void UpdateTicks();
};
#endif
```

Setting 'Zero' Time

The basic requirement when making use of time is to have a reference of 'zero' time. The data member `InitCount` is used to store the count value (remaining before the next tick) that corresponds to this 'zero time' as shown in Figure 13-4. This value of `InitCount` is established at the time of instantiating the `PCTimer` object, and can also be re-established by calling the member function `ResetTimer()`.

The `ResetTimer()` function can be used at any time to create a 'zero time' reference. It latches and reads Timer 0, then stores the count value into the data members `InitCount` and `LastCount`. `LastCount` is to temporarily store the timer's last count value so the other member functions can detect the next tick as described in the following text. The number of ticks that have occurred, stored in data member `TickCount`, is then reset to zero.

Accounting for Timer Ticks

As explained previously, to determine the exact time that corresponds to any given instant, we need to know the number of timer ticks elapsed since the 'zero' time reference was created using Timer 0. We must monitor each countdown of Timer 0 to prevent any ticks being missed. The three member functions; `Delay()`, `ReadTimer()`, and `UpdateTicks()`, all need to track and record the number of ticks that take place during their operation. Figure 13-3 shows three different scenarios when monitoring timer ticks, explained as follows.

The counter is first read and its count value stored in the member variable `LastCount`. A short time later during execution of the program, the counter is read again to test if a tick has passed, and this count value is stored in the variable `Count`. Case (a) shows the two reads of the counter occurring *within* a countdown cycle – no tick has occurred. In this case the variable `Count` is not greater than the variable `LastCount`. Case (b) shows a new countdown cycle underway when the counter is read for the second time. In this case a tick has occurred; `Count` is greater than `LastCount`. Case (c) shares the same result as case (a) in that

`Count` is not greater than the variable `LastCount`. When using the same criteria to test for a tick as used for case (a), we will make an incorrect evaluation that a tick has not occurred because the two reads of the counter in case (c) are made more than a full countdown period apart (and `Count` < `LastCount`).

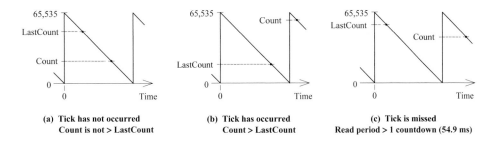

(a) Tick has not occurred   (b) Tick has occurred   (c) Tick is missed
Count is not > LastCount   Count > LastCount   Read period > 1 countdown (54.9 ms)

Figure 13-3 Accounting for timer ticks.

Evaluating Elapsed Time

At the start of the timing process, the counter is latched, read, and its value is stored in the data member `InitCount`. This value represents 'zero' time, and is the number of counts remaining before the next tick as shown in Figure 13-4.

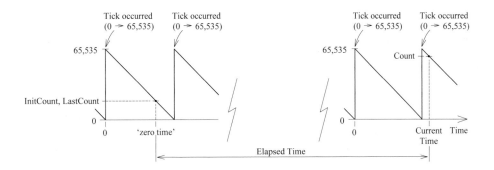

Figure 13-4 Evaluation of the elapsed time.

At the end of the 'Elapsed Time' the timer is read again and its value stored in the variable `Count`. The number of timer counts since the last tick can be evaluated by subtracting `Count` from 65,535 (i.e. 65,535 – Count). The number of tick counts within the elapsed time is stored in `TickCount`. Therefore, the total counts corresponding to the elapsed time is:

```
TotalCounts = InitCount + (TickCount-1)*65535 + 65535-Count;
```

Which can be written as:

```
TotalCounts = InitCount + TickCount*65535 - Count;
```

This is then easily converted into a value of time by knowing the frequency of the PC's clock that drives the counter.

The purpose and operation of the member functions of this class are explained as follows.

The `PCTimer()` function shown in Listing 13-2 is the constructor for the `PCTimer` object class. The constructor allocates space for all data members of the class and then takes an initial reading of the timer for the 'zero time' reference. Obtaining the 'zero time' reference is performed within the body of the constructor by using the function `ResetTimer()`.

Listing 13-2 The constructor of the PCTimer class.

```
PCTimer::PCTimer()
{
    ResetTimer();
}
```

The `ReadTimer()` function determines the total number of counts since zero time until the instant `ReadTimer()` is called. The total number of counts is calculated as described above. Finally, the total count value is converted into a value of time in milliseconds and a value of type `double` is returned. The `ReadTimer()` function does not call another member function to avoid added delays. Instead, all the code it needs is placed within its body.

The `Delay()` function is used to generate a specified delay that is passed to it as a parameter in milliseconds. When called, it reads the timer once to determine the start of the delay period, being the current number of counts since zero time. Then it adds the number of counts since zero time to the number of counts corresponding to the specified delay. The timer is then read repeatedly until the number of counts since zero time is equal to the total number of counts evaluated above. At this point in time the delay will expire.

The `UpdateTicks()` function simply performs the task of quickly monitoring the tick status during down-counting (faster than the `ReadTimer()` function). It does this using the same method as employed in the other member functions of the `PCTimer` class. For the proper functioning of the `PCTimer` class, the `Update()` function must be called at least once within a countdown, preferably every 40 ms. This completes the development of our `PCTimer` object class. The function file `pctimer.cpp` that contains these functions is given in Listing 13-3.

Listing 13-3 Member functions of the PCTimer class - pctimer.cpp.

```
#include <dos.h>
```

```cpp
#include "pctimer.h"

PCTimer::PCTimer()
{
    ResetTimer();
}

void PCTimer::ResetTimer()
{
//  Latch Timer 0
    outportb(0x43, 0xD2);

//  Read latched Count
    LastCount = InitCount = inportb(0x40)+inportb(0x40)*256;
    TickCount = 0;  // Initialise to zero.
}

double PCTimer::ReadTimer()
{
    double Time;
    unsigned long TotalCounts;
    unsigned int  Count;

//  Latch Timer 0.
    outportb(0x43, 0xD2);

//  Read latched Count
    Count = inportb(0x40) + inportb(0x40)*256;

    if(Count > LastCount)
        TickCount++;

    LastCount = Count;

    TotalCounts = ((long)InitCount + TickCount*65535L
                    -(long)Count);
    Time = (TotalCounts/1.1932)/1000.0;
    return Time;  // In milliseconds.
//  return TotalCounts;
}

void PCTimer::Delay(const double& milliseconds)
{
    unsigned int Count;
    long  StartCount, DelayCount, EndCount, TotalCount;
```

```
// Latch Timer 0.
    outportb(0x43, 0xD2);

// Read latched CountOne.
    Count = inportb(0x40) + inportb(0x40)*256;

    if(Count > LastCount)
        TickCount++;

    LastCount = Count;
    StartCount = ((long)InitCount + TickCount*65535L
                    -(long) Count);
    DelayCount = (long) (milliseconds*1.1932*1000);
    EndCount = StartCount + DelayCount;

// Repeat a loop for the duration of the period.
    do
    {
        // Latch Timer 0.
        outportb(0x43, 0xD2);

        // Read latched CountOne.
        Count = inportb(0x40) + inportb(0x40)*256;

        if(Count > LastCount)
            TickCount++;

        LastCount = Count;
        TotalCount = ((long)InitCount + TickCount*65535L
                        -(long)Count);
    }
    while (TotalCount < EndCount);
}

void PCTimer::UpdateTicks()
{
    unsigned int Count;

// Latch Timer 0.
    outportb(0x43, 0xD2);

// Read latched Count.
    Count = inportb(0x40) + inportb(0x40)*256;
```

```
    if(Count > LastCount)
        TickCount++;

    LastCount = Count;
}
```

As mentioned previously, it is essential that when using any of the member functions of the class that they are called *within* a countdown cycle. The class will function best if the PC's interrupts are disabled, however, we have chosen to leave all interrupts active to avoid unnecessary complexity in our programs. The effect of these interrupts may cause small and unforeseen time delays.

For example; An attempt is made to read the timer at a particular instant using the `ReadTimer()` function. If the timer interrupt also occurs at this time, it will force the function `ReadTimer()` to wait until the timer interrupt has been serviced. This delay will cause `ReadTimer()` to return a count value that is greater than the instantaneous value when the call to the `ReadTimer()` function was initiated. The effect of interrupts is demonstrated in one of the programs that uses a member function of the `PCTimer` object.

> **NOTE**
>
> Although updating the tick count is handled automatically by the `Delay()` function while it is active, the `ReadTimer()` and `UpdateTicks()` functions will only perform this task at the instant they are called. Therefore, if time periods need to be measured, ensure that functions `ReadTimer()` and `UpdateTicks()` are called repetitively within a full timer countdown period (54.9 ms) to correctly monitor and update the tick count value.

13.5 Measurement of Time

The `PCTimer` object class will be used to demonstrate the measurement of real-time. As explained earlier, measurement of time periods can be affected by the execution of interrupt routines. The following program will allow us to observe the delays that can be generated by the various interrupt service routines executing in the PC. The `disable()` function can be called to stop all interrupts (disabling most of the PC's peripherals, including the keyboard). Therefore, make sure that the interrupts are disabled for a minimum length of time! Calling the `enable()` function re-enables the interrupts and allows the PC's peripherals to resume operation. The program that measures time is shown in Listing 13-4.

Listing 13-4 Measurement of time – time.cpp.

```cpp
#include <iomanip.h>
#include <math.h>
#include <iostream.h>
#include <conio.h>
#include <dos.h>

#include "pctimer.h"

main()
{
    PCTimer T;
    double TimeValue[1000];
    int i;

//  disable();
    for(i = 0; i < 1000; i++)
    {
        for(int j = 0; j < 50; j++)
            sin(j);
        TimeValue[i] = T.ReadTimer();
    }
    enable();

    for(i = 1; i < 1000; i++)
    {
        cout << i << '\t';
        cout << setprecision(3) << TimeValue[i] << '\t';
        cout << setprecision(3) << (TimeValue[i] -
                                TimeValue[i-1])<< '\t';
        cout << endl;

        if( i % 20 == 0)
        {
            cout << "Press a key for more ... ";
            getch();
            cout << endl;
        }
    }

    return 0;
}
```

Executable File Generation		
Required Files	**Listing No.**	**Project File Contents**
`pctimer.cpp`	Listing 13-3	`pctimer.cpp`
`pctimer.h`	Listing 13-1	`time.cpp`
`time.cpp`	Listing 13-4	

The program measures the time consumed by each iteration of a `for` loop that executes 1000 times. Within this `for` loop is another `for` loop that repeatedly executes the `sin()` function 50 times for no real purpose except to waste time. You can alter the number of times the `sin()` function executes to change the time spent by each iteration of the external `for` loop. The `ReadTimer()` function is called within each iteration to read the time and store the time values in an array. Note that the `UpdateTicks()` function does not need to be called since the `ReadTimer()` function monitors and accounts for timer ticks, and importantly, the `for` loop will execute in less than a full countdown period.

The time value and also the time difference between two consecutive readings of the timer are displayed on-screen 20 lines at a time. Interrupts will have the effect of adding time delays of varying value to the time taken to perform the calculations. Therefore, if interrupts are active (*not* disabled), calculations will take different periods of time, and it is these differences that the program is displaying. If the interrupts *are* disabled, calculation times will be uniform. Therefore, any irregularities in the times displayed in the third column will be caused by interrupts. You can run the program twice, once with the interrupts disabled and then with the interrupts enabled to observe these effects. These results should provide some insight into the effect of interrupts when measuring time.

13.6 Reflex Measurement

In this section we will use the interface board to measure a person's hand reflexes. A program has been provided that will light up a set of LEDs on the interface board after a random time delay. The person under test will react and press the button switch on the board in response to the LEDs lighting up. The delay in time from the LEDs lighting up and the press of the button switch is a measure of a person's reflexes. Listing 13-5 shows the program that performs the reflex measurement.

Listing 13-5 Reflex measurement – reflex.cpp.

```
#include <iomanip.h>
#include <conio.h>
#include <iostream.h>
#include <stdlib.h>
```

```cpp
#include "pport.h"
#include "pctimer.h"

main()
{
    double ReflexTime;
    ParallelPort PPort;
    PCTimer T;

//  Turn off LEDs at start.
    PPort.WritePort0(0);
//  A long beep.
    cout << "\a\a\a\a\a" ;

//  Time delay of 1.5-5.0 sec.
    T.Delay(1500+rand()%3500);

//  Light up all 8 LEDs.
    PPort.WritePort0(255);

//  Reset Timer.
    T.ResetTimer();

//  Wait for button press.
    while((PPort.ReadPort1() & 0x80) == 0)
        T.UpdateTicks();

//  Read PC Timer.
    ReflexTime = T.ReadTimer();

//  Turn off LEDs.
    PPort.WritePort0(0);

    cout << "Your reflex time is ";
    cout << setprecision(3) << ReflexTime;
    cout << " ms." << endl;
    getch();

    return 0;
}
```

Executable File Generation		
Required Files	**Listing No.**	**Project File Contents**
pport.cpp	Listing 10-8	pport.cpp
pport.h	Listing 10-7	
pctimer.cpp	Listing 13-3	pctimer.cpp
pctimer.h	Listing 13-1	
reflex.cpp	Listing 13-5	reflex.cpp

When this program first executes, it turns off all LEDs and issues a long beep. It then generates a random time delay between 1.5 and 5.0 seconds, followed by lighting all eight LEDs and initialising the timer to 'zero time'. The user then reacts to the LEDs lighting up by pressing the button switch on the interface board. The program detects the button press and calls the ReadTimer() function to read the timer and return the time that has elapsed since 'zero time'. The LEDs are turned off and the reflex time is displayed on-screen.

Connect the interface board's BASE address outputs to the inputs of the LED Driver IC according to Table 13-4. The button switch connects to the BASE+1 input as shown in Table 13-5.

Table 13-4 LED connections.

BASE Address (Buffer IC, U13)	ULN2803A Pin No. (Driver IC, U3)
D0	1
D1	2
D2	3
D3	4
D4	5
D5	6
D6	7
D7	8

Table 13-5 Switch connection.

Button Switch	BASE+1 Address (Buffer IC, U6)
OUT	D3

13.7 Generating a Time-Base

In Chapter 10 we developed a program (Listing 10-11) to monitor and display the pulse-train from the the interface board's VCO (voltage-controlled oscillator). The horizontal axis of the plot (time) was generated by using software loops and not from real-time techniques. In the original program, at every instant the VCO's pulse-train was read, the trace was plotted and the value of i incremented.

Therefore, each change in `i` could be considered as a 'new reading'. Our real-time capabilities can be now used to incorporate an accurate time-base with the graphics plot. Listing 13-6 shows a modified version of the original program (changes shown in bold typeface) that uses a proper time-base for its horizontal axis.

In this program the variable i is still used for positioning the trace along the x axis. However, its meaning is different, now becoming a time unit of 10 ms. For this new arrangement, the main loop in the program plots continuously but only increments the value of `i` every 10 ms. Therefore, plotting along the horizontal axis moves by one pixel every 10 ms. This also means that the resolution of the plot is 10 ms. That is, if the signal changes in less than a 10 ms period, its change cannot be properly represented and may show as a series of vertical lines. This can be rectified by re-coding the program to use a smaller value for the delay between timer reads to suit the frequency of the incoming signal.

Listing 13-6 Graphical display of pulse-train with a real time-base - timebase.cpp.

```
/**********************************************************
The frequency of the pulse-train being output by the
voltage-controlled oscillator will change as we change
the analog input voltage to the VCO circuit. The
Potentiometer (POT1) on the interface board generates
the input voltage to the VCO and the program reads the
pulse-train being output by the VCO.  This pulse-train
is graphically displayed on-screen.
**********************************************************/
#include <graphics.h>
#include <stdlib.h>
#include <iostream.h>
#include <conio.h>
#include <dos.h>

#include "vco.h"
#include "pctimer.h"

void main()
{
    VCO Vco;
    PCTimer T;
    int i=0; // controls plotting in the x range
    int SignalLevel;
    int Driver = DETECT, GraphicsMode, ErrorCode;
    int X, Y;

// set to graphics mode
```

```
        initgraph(&Driver, &GraphicsMode, "");

// check for error codes
    ErrorCode = graphresult();
    if (ErrorCode != grOk)
    {
        cout << "Graphics error:   "
            << grapherrormsg(ErrorCode) << endl;
        cout << "Press any key to halt:" <<  endl;
        getch();
        exit(1);
    }

    X = getmaxx();
    Y = getmaxy();
    rectangle(X/4-1, Y/2-76,X*3/4+1,Y/2+76); // border
    setviewport(X/4, Y/2-75,X/4*3,Y/2+75,1);

    T.ResetTimer();
    while(!kbhit())
    {
        SignalLevel = Vco.SignalLevel();

        if(SignalLevel == 0) // low level
            lineto(i,100);
        else                 // high level
            lineto(i,50);

        if(T.ReadTimer() > 10)
        {
            T.ResetTimer();
            i++;
        }

        if(i > X/2) // half screen = Viewport width
        {
            i = 0;
            while(Vco.SignalLevel()); // wait for low level
            while(!Vco.SignalLevel());// wait for high level
            clearviewport();
            moveto(0,50);
            T.ResetTimer();
        }
    }
}
```

Executable File Generation		
Required Files	**Listing No.**	**Project File Contents**
pport.cpp	Listing 10-8	pport.cpp
pport.h	Listing 10-7	
vco.cpp	Listing 10-4	vco.cpp
vco.h	Listing 10-1	
pctimer.cpp	Listing 13-3	pctimer.cpp
pctimer.h	Listing 13-1	
timebase.cpp	Listing 13-6	timebase.cpp

The 'zero time' reference is set before starting to plot by calling the function ResetTimer(). This function is also used in the following if statement to periodically reset the timer every 10 ms:

```
if(T.ReadTimer() > 10)
{
    T.ResetTimer();
    i++;
}
```

If a 10 ms period has elapsed, the timer is reset to allow the next 10 ms period to be measured, and the index i is incremented to allow the next pixel to be plotted. Otherwise, i will remain as is and plotting will repeat at the same time position.

When the trace has reached the edge of the Viewport's plot region (X/2; half the screen width) the program enters an if statement used to setup the screen ready for a new trace. Inside this if statement the program resets the value of i to zero, and then waits for the incoming VCO signal to switch to a high level by waiting for the VCO output to change state from logic-low to logic-high using the following combination of statements:

```
while(Vco.SignalLevel()); // wait for low level
while(!Vco.SignalLevel());// wait for high level
```

This ensures the plot always starts with the same edge transition on-screen. Once the VCO signal has made the required low-to-high level transition, the screen is cleared, the cursor repositioned to the left edge, and the timer is reset to a fresh 'zero time' reference. Note: if interrupts are enabled, some of the pulses displayed may have wider widths due to time consumed by interrupt service routines.

Just as we used a real-time program to plot a waveform on the screen, we can timestamp data in real-time as it is acquired. We will generate a waveform using the VCO and Charge/Discharge circuitry on the interface board, digitise its analogue output using the ADC, timestamp these values, and store this data in a file.

13.8 Data Acquisition with Timestamp

The Charge/Discharge circuit on the interface board can be driven by a digital logic signal to generate an analog waveform that can be sampled to demonstrate the data acquisition process. Each data sample can be accurately time-stamped as it is acquired by using the `PCTimer` object in the data acquisition program.

In this section, one program will perform data acquisition, time-stamp the data as it is acquired, and store the data into a disk file for later analysis. A second program will retrieve the stored data from the disk file and process the data to determine the period of the waveform generated by the Charge/Discharge circuit.

13.8.1 The Charge/Discharge Circuit

The Charge/Discharge circuit can be driven from any digital logic signal, including one that may be generated by software using an output bit of one of the ports. However, this application uses the simple arrangement whereby the VCO drives the input of the Charge/Discharge circuit with a periodic signal as shown in Figure 13-5. The Charge/Discharge circuit has a capacitor that is charged when the VCO output becomes low and discharged when the VCO output becomes high. The analog signal output from the Charge/Discharge circuit (shown in Figure 13-6) is digitised by connecting it to the analog-to-digital converter. The program will acquire the digitised signal for more than one period and time-stamp each digitised sample. This data is then stored by writing it to a file.

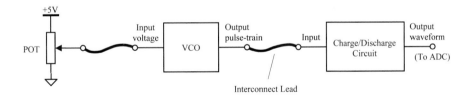

Figure 13-5 Connections for the VCO to drive the Charge/Discharge circuit.

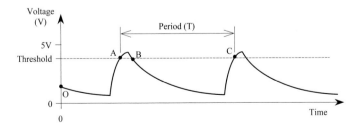

Figure 13-6 Voltage waveform generated by the Charge/Discharge circuit.

13.8.2 Programming Data Acquisition & Timestamp

In this section we will develop two programs. The first program (named TimeStmp.cpp) will sample the signal using the analog-to-digital converter, and read the time of sampling. These paired results will then be written to a disk file. This program will need to use the ADC class and the PCTimer class. Note: the VCO object is not used since the VCO circuit is only used as a signal generator to drive the input of the Charge/Discharge circuit.

The second program (named Period.cpp) will retrieve the stored data from the disk file and scan through the data to determine the period (T) of the waveform. It does not need to use any of the objects developed previously.

Program 1 – TimeStmp.cpp

The steps involved in this program that digitises the signal are:

1. Reset the timer.
2. While looping until sufficient time has elapsed (5 seconds suggested):
   - read time and store in an array.
   - read ADC and store in an array.
   - wait for sampling period (10 ms. is suggested)
3. Write the data to a file.

The program for timed acquisition of data is given in Listing 13-7.

Listing 13-7 Program to acquire data with time-stamps – timestmp.cpp.

```cpp
#include <iomanip.h>
#include <iostream.h>
#include <fstream.h>
#include <stdlib.h>
#include <conio.h>

#include "adc.h"
#include "pctimer.h"

void main()
{
    ADC Adc;
    PCTimer T;
    double Time[500];
    unsigned char Data[500];
    double TempTime;
    double Duration = 5000; // Acquisition period - 5000 ms.
    int i = 0;
    const int SamplingInterval = 10; // Milliseconds
```

```
    T.ResetTimer();
    do
    {
        TempTime = T.ReadTimer();
        if(TempTime  > i*SamplingInterval)
        {
            Data[i] = Adc.ADConvert();
            Time[i++] = TempTime;
        }
    }
    while(T.ReadTimer() < Duration);

// Create, open then write data to disk file.
    ofstream os("timestmp.dat");

    for(int j = 0; j < i; j++)
    {
        os << setprecision(3) << Time[j] << '\t';
        os << setprecision(3) << (double) Data[j] << endl;
    }
    os.close(); // Close file.
}
```

Executable File Generation		
Required Files	**Listing No.**	**Project File Contents**
pport.cpp	Listing 10-8	pport.cpp
pport.h	Listing 10-7	
adc.cpp	Listing 10-4	adc.cpp
adc.h	Listing 10-1	
pctimer.cpp	Listing 13-3	pctimer.cpp
pctimer.h	Listing 13-1	
timestmp.cpp	Listing 13-7	timestmp.cpp

This program uses the ADC class and the PCTimer class. The ADC class is used to acquire the data by controlling and reading the analog-to-digital converter. The PCTimer class is used to accurately measure the time when the signal was sampled by the ADC. The two objects Adc of type ADC and T of type PCTimer have been instantiated from these two classes.

We have decided to perform data acquisition for a 5 second duration using a sampling interval of 10 ms, generating a total of 500 samples. Two arrays named Time and Data have been created to store the respective data, each having 500 elements, with the variable i used as the array subscript. The variable named

`Duration` is used to store the overall sampling duration, and is set to 5000 milliseconds (5 seconds). The variable named `SamplingInterval` is used to define the sampling interval and is set to 10 ms. Since the sampling interval is constant during this data acquisition process, `SamplingInterval` is declared as `const`. The identifier `TempTime` is used to temporarily store the time read by reading the PC timer.

Just prior to entering the `do-while` loop, the program resets the `PCTimer` object T, thereby establishing 'zero time'. The loop will execute continuously with data read at 10 ms intervals. The `if` statement compares the current time with the time when the next sample must be taken. When the current time becomes greater than the time for the next sample, the ADC will be read, its value stored in the array `Data[]`, and the time stored in array `Time[]`. The `while` loop terminates when the time exceeds the value of `Duration`.

All the data stored in the two arrays is then written to a data file named `timestmp.dat`. The `ofstream` constructor is called to create the file by instantiating the object `os` of type `ostream`. The file name `timestmp.dat` is passed as a parameter to the `ofstream` constructor. The `for` loop initialises the integer identifier `j` to zero and continues to write the values until the value of the subscript `j` reaches the number of data elements recorded in the `do-while` loop. The file `os` is then closed by calling the member function `close()` of the `ofstream` class. When the program has completed its execution, all data will be stored as two columns separated by a tab character in the text file `timestmp.dat`. The first column contains time values and the second column contains integer values representing the analog voltage output from the Charge/Discharge circuit.

Program 2 – Period.cpp

The second program retrieves the data from the disk file and stores this data in memory. It then processes the data to determine the period of the waveform that was sampled.

The steps involved in the second program are now given:

1. Read the data file and store the data values in memory.
2. Loop to find two data points one period apart:
   - search until a data value in the second column is less than the threshold (Point O).
   - continue searching until a data value is greater than the threshold; store its corresponding time value (Point A).
   - continue searching until a number less than the threshold is found (Point B).
   - continue searching until a number greater than threshold is found; store its corresponding time value (Point C). Then quit the loop.
3. Calculate the period of the waveform, being the difference between the two times (Point C – Point A).

The program that performs the above data retrieval and data processing steps is shown in Listing 13-8 and does not need to use any of the object classes that have developed previously.

Listing 13-8 Program to determine the period of the output waveform - period.cpp

```cpp
#include <iostream.h>
#include <fstream.h>

void main()
{
    double *TimePtr;        // Pointer to Time data
    double *DataPtr;        // Pointer to ADC data
    double MaxData = 0;     // Maximum value of ADC data
    int NumData=0;          // Number of data pairs in file
    int i =0;
    int Case = 1;           // Case in initialised to 1
    int Quit = 0;           // Quit = 0 means do not quit
    unsigned char Threshold;
    double TimeA, TimeC;

//Instantiate ifstream object (is).
    ifstream is("tmddata.dat");

// Read through the file to find the number of
// data pairs and the max value of ADC data.
    while(is)
    {
        is >> *TimePtr >> *DataPtr;
        if(!is.fail())
        {
            if(*DataPtr > MaxData)
                MaxData = *DataPtr;
            NumData++;
        }
    }

// close input stream
    is.close();

// Set Threshold based on MaxData
    Threshold = MaxData - 5;

// Alocate memory for Time and ADC data
```

```
        TimePtr = new double[NumData];
        DataPtr = new double[NumData];

// Re-open the file so that reading starts from beginning.
        is.open("tmddata.dat");

// Read file and fill allocated memory
        while(is)
        {
            is >> *(TimePtr+i) >> *(DataPtr+i);
            if(!is.fail())
                i++;
        }

// scan through all array elements pointed by DataPtr.
        for(int j = 0; j < i; j++)
        {
            switch(Case)
            {
                // Search for a Data element less than
                // the threshold
                case 1: if(*(DataPtr+j) < Threshold)
                            Case = 2;
                        else
                            break;

                // Search for a Data element greater than the
                // threshold. Note the time.
                case 2: if(*(DataPtr+j) > Threshold)
                        {
                            TimeA = *(TimePtr+j);
                            Case = 3;
                        }
                        else
                            break;

                // Search for a Data element less than
                // threshold
                case 3: if(*(DataPtr+j) < Threshold)
                            Case = 4;
                        else
                            break;

                // Search for a Data element greater than
                // threshold. Note the time. Set Quit flag
```

```
        case 4: if(*(DataPtr+j) > Threshold)
                {
                        TimeC = *(TimePtr+j);
                        Quit = 1;
                }
        }
        if(Quit)
            break;
    }

// Clean up - deallocate dynamic memory
    delete TimePtr;
    delete DataPtr;

// Display the time difference on the screen.
    cout << "The VCO signal period is ";
    cout << TimeC - TimeA;
    cout << " ms." << endl;
}
```

The initial part of this program scans through the data file to determine the number of data pairs (time and ADC data) it contains, and stores this value in the variable NumData. During this process, the program also determines the maximum value of ADC data that was sampled, and stores this value in the variable named MaxData. A threshold value (calculated as five ADC units below the value of MaxData) is evaluated for use to determine the period of the measured waveform. Then the program dynamically allocates memory for time data and ADC data using the two pointer variables TimePtr and DataPtr. The data file is then closed and opened again so it can be re-read from its beginning to fill the dynamically allocated memory with its data.

The second part of the program scans through the data stored in the allocated memory, and evaluates the period of the signal as follows. It first searches for an element of ADC data (second column) that is below the threshold value. This would, for example, represent a point such as 'O' shown in Figure 13-6. Starting from this element, the program begins scanning the second column of data until an element with a value greater than the threshold is first encountered. For example, this point would represent a point such as 'A' shown in Figure 13-6. At this point the corresponding time value from the first column is stored. Scanning then continues down the second column of data while searching for an element less than the threshold value. Such a point would correspond to point 'B' in Figure 13-6. Recording of time is not needed for this point. Scanning continues down the second data column for the next number that is greater than the threshold value. This will correspond to point 'C' in Figure 13-6. The time corresponding to this instant is stored. The Quit flag will then be set since we no longer need to continue scanning the data. The period of the digitised waveform is then the difference between the time at point 'A' and 'C'.

Note that the identifier Case (initially set to 1) is used to control the different phases of scanning the data. When the value of Case is 1, the ADC data array is first searched for an element pointed to by DataPtr that is less than the threshold. Once such an element has been found, the variable Case is set to 2 to commence the next phase of scanning. The identifier Quit is used in a logical sense. It is initially set to 0, meaning 'do not quit'. The value of Quit is tested at the end of each iteration of the loop and if set, the loop will be terminated by executing the break statement. Once the last point (C) has been found, there is no need to proceed with scanning, so Quit is set to 1. The time difference representing the period of the signal is then calculated and printed on-screen.

13.9 Summary

In this chapter we learned how the built-in timer of the PC operates and how it can be used. The object class PCTimer has been developed with the capability to measure very long time periods. It has member functions to mark a time reference, accurately read the elapsed time, and also generate specific delays.

The PCTimer class operates without disabling the PC's interrupts. As such, the interrupt service routines will generate short interruptions that can contribute to minor inaccuracies when measuring time. This was demonstrated when one of our example programs made repeated measurements of a 'fixed-time' event with interrupts enabled, and later with interrupts disabled. Other programs were presented in this chapter that measured a person's reflex reaction time, generated a waveform plot using an accurate time-base, and used regular and accurate timing to digitise the electrical waveform produced by the interface board's Charge/Discharge circuit.

13.10 Bibliography

Van Gilluwe, F., *The Undocumented PC*, Addison Wesley, 1994.

IBM, *Technical Reference – Personal Computer AT*, IBM Corporation, 1985.

Auslander D.M. and Tham, C. H., *Real-Time Software for Control*, Prentice Hall, 1990.

Intel, *M8254 Programmable Interval Timer – Data Sheet*, Intel Corporation, 1986.

Appendix A - Hardware

Circuit Construction

Interface Board Bill of Materials

Circuit Construction

The interface board contains many different and independent circuit blocks to give the reader the option of working with any number of projects in any order. This flexibility also allows the reader to combine the board's various circuit elements/blocks to form a wide range of custom projects. All circuit blocks need to be powered and most must be able to interface with the PC's parallel port. To satisfy these needs, the interface board has its own power supply circuit block and a parallel port interface circuit block.

The *power supply* is the first circuit block that should be assembled, tested, and be operating properly - should any faults in the power supply be present, then only this part of the board will need to be investigated and debugged (assuming the interface pcb was properly checked to be functional). Next, the *parallel port interface* circuit block should be assembled, tested, and be operating properly before the other circuit blocks on the interface board are constructed (in any order).

IMPORTANT: The interface printed circuit board (pcb) should be checked for faults *before* proceeding to assemble and solder components onto the board. If in the very unlikely case the board does have faults, these faults can be detected quickly and simply on the unpopulated pcb and easily rectified as explained in the next section. The following sections provide instructions for the assembly of components, soldering, testing, and debugging of circuits.

Bare Printed Circuit Board

Visually check the bare printed circuit board (pcb) for any obvious short-circuit tracks or open-circuit tracks caused by faulty manufacture or handling. If any faults are detected, repair with a sharp blade and/or soldering iron as follows. Breaks in tracks can be repaired by scraping the coating from both sides of the break to expose the copper surface. Solder a piece of solid wire across the break. Short-circuits can be removed by cutting between shorted tracks with a sharp blade.

Test the power supply tracks for short-circuits by measuring the resistance between the following power and ground paths using a multimeter:

- +5V and GND.
- +9V and GND.
- 12Vunreg and GND.
- - 8V and GND.

The easiest place to probe each of these paths is at one of the pcb pads connected to that power path as shown in Figure A-1. In all cases the resistance should be at least several mega-ohms. If not, there is a short-circuit somewhere which needs to be detected visually, or by cutting/removing links to break the path up into more easily managed segments, and re-measuring with the multi-meter.

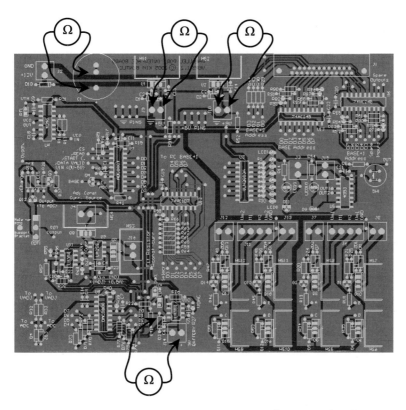

Figure A-1 Test Power Supply track paths for short-circuits.

The Assembly Process

Equipment: 1 pair of Cutters/Nippers and 1 pair of small sized long-nose pliers.

The easiest approach to take when assembling the printed circuit board (pcb) is to manage the process in several stages, assembling on a project-by-project basis if convenient. This will simplify the testing and any debugging that might be necessary. However, if the whole board is to be completely assembled in just one stage, then the following guidelines should simplify and speed the process.

Components should be assembled and soldered flush with the pcb, starting with those that are lowest in height before proceeding with taller components. This strategy will result in the following order of assembly:

1. Flat mounting diodes and resistors.
2. IC sockets/ICs.
3. LEDs.
4. Pcb pins.
5. Capacitors that are small in size.
6. Vertically mounted resistors.
7. Terminal blocks and remaining tall components.

As each component is fitted to the board, bend two of its leads over slightly to retain the part. When all components of this category are fitted, flip the board over and lay the partially complete assembly flat against the work surface (these components should now be in contact with the work surface). This will keep the parts positioned flush against the pcb, ready for soldering.

Note: It is easiest to fit a limited number of components at a time, solder them and then trim their leads close to the board surface using cutters, and then repeat these steps. This will avoid the situation occuring where a large number of component leads restrict access to a joint that needs to be soldered.

Many components must be correctly oriented when fitted to the board. These components include ICs, electrolytic capacitors, LEDs, diodes, and transistors. Figure A-2 shows the convention used to mark ICs to denote orientation with respect to pin number one. Usually a notch or dot is placed at the end of the IC where pin number one is found.

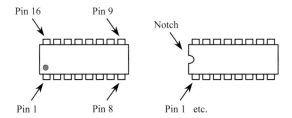

Figure A-2 Marking of IC orientation and pin numbering (top view).

When using IC sockets (recommended) make sure the socket is soldered into the board with it's 'notched' end corresponding to the IC outline marked on the board. The notched end of the socket will provide the marker for correct orientation when fitting the IC into its socket.

Other components have characteristic marking schemes to denote polarity. Electrolytic capacitors are marked with either a + or - sign on their body. To determine correct LED orientation, look through the transparent coloured body at the two terminals as shown in Figure A-3. Typically, the smaller shaped contact is the anode and the other contact the cathode.

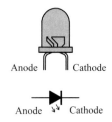

Figure A-3 LED polarity.

Diodes are marked with a line across the length of their body to represent the cathode. Transistors are correctly fitted to the pcb by complying with the pcb overlay markings that show appropriate orientation of the transistor bodies. Importantly, CMOS devices are sensitive to damage from static electricity and are preferably assembled with antistatic precautions in place.

Note: apply a thin film of heatsink paste to the flat metalised surfaces of the transistors and voltage regulators before fitting heatsinks. This will improve the transfer of heat to the heatsink.

The Soldering Process

Equipment & Materials: Soldering iron (> 15 Watts) and electronics grade solder (0.7 mm to 1.0 mm diameter, 60% tin, 40% lead with flux inside).

Soldering by hand can be described basically as a process where heat is transferred to the joint to be soldered, followed by the application of solder that then melts and flows into and around the joining materials. The heat source is removed and the molten solder solidifies, forming a connection between the component lead and the pcb pad.

Preparation: To ensure successful soldering, the soldering iron tip and the joint itself must be clean. Normally the printed circuit board and the component lead passing through the board will be sufficiently clean. Unfortunately, the same cannot be said for the soldering iron tip. To clean the iron tip, wet a piece of 'kitchen' sponge to be damp but not soaked, and then repeatedly wipe the iron tip across the sponge until the tip is in a shiny metallic state. If the iron tip cannot be brought to a shiny metallic state, try *tinning* the soldering iron tip by applying solder to it, waiting a short period of time and then wiping it against the sponge. If this fails to work, a new soldering iron tip is probably needed.

Soldering: This process takes place in three steps:

1. Lightly wet the iron tip with solder (the tip may be sufficiently wet with solder from the previous soldering operation). This improves the rate at which heat can be transferred from the iron tip to the joint.
2. Heat the joint with the iron tip for several seconds.
3. Apply solder sparingly to the heated joint - it will flow to fill the joint if the joint was heated sufficiently during step 2.

Note: When applying solder to the joint in step 3, do not be tempted to apply solder to the iron tip in order to melt the solder (this can result in a poorly soldered joint). If the joint is heated sufficiently in step 2, solder flows into the joint as required. Figure A-4 shows the shape of a correctly soldered joint before and after trimming using cutters.

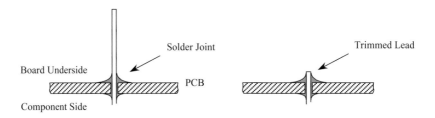

Figure A-4 Correctly soldered and trimmed joint (plated through hole).

Schematic Diagram Conventions

The following information will help you interpret the schematic circuit diagrams when studying the circuit blocks and during times of test and debugging:

- Inputs to circuits are generally drawn on the left side of the component and outputs are generally drawn on the right side of the component.
- Logic circuits or circuits having logic inputs and/or outputs sometimes have small circles placed on their input or output pins at the border of the circuit block. This denotes an *active low* pin. In the case of an input, a low logic level will activate the pin. For the case of an output, the output pin will be at a low logic level when in an active state.
- The small dots at the ends of input and output lines are pcb pins. Solid dots show connection between signal lines.
- VCC denotes logic circuit +5V.
- There are two types of grounds shown on the schematic; horizontally-lined triangular shape (digital ground) and hollow triangular shape (analog ground). They are connected together on the printed circuit board and will both be at 0V.

Testing and Debugging

Testing and debugging should take place in a systematic manner to minimise the time and effort required. The easiest means of achieving this is to reduce the size or scope of the system under test by breaking it into smaller circuit blocks and testing them separately in-turn.

When testing circuit blocks; the input voltages/currents must be set to appropriate levels for that block before the outputs are checked. If a fault in circuit function is detected, it will usually be caused by factors such as:

- incorrect wiring lead connections.
- incorrect value or type of a component(s).
- incorrect orientation of a component(s).

- poor/inadequate solder joint or wiring connection.
- short-circuit due to solder bridging.
- defective pcb track(s) caused by faulty manufacture, improper handling of the pcb (excessive flexing, scoring, etc), and damage from poor repair by the user.
- incorrect power supply voltage to circuit(s).
- damaged component(s).

The typical steps in circuit test and debugging are:

1. Check for correct wiring connections.
2. Check that the correct components are installed in the right position, with proper orientation.
3. Check correct power supply to a circuit. If the voltage to the circuit is not correct:
   - Feel the IC to check if the package is unusually warm or hot; indicating an overloaded or damaged IC that may need to be replaced. Also, check if other components connected to the power supply are also excessively hot.
   - Visually check the tracks and solder joints for any unintended short-circuits and also for unintended open-circuits or inadequate soldering. If necessary, follow this with an electrical continuity test using a multimeter as discussed below.
4. Check that voltages or currents to circuit inputs are set at the correct levels.
5. Test that circuit outputs are correct for the given state of the inputs – if this is not so:
   - The outputs are excessively loaded by; connection to other circuits or components on the board, or by short-circuited track(s).
   - The component is faulty and needs to be replaced.

CONTINUITY TESTS

Continuity tests are performed to detect short-circuits and open-circuits using a multi-meter switched to 'resistance' mode. The power to all circuitry MUST be TURNED OFF before commencing with continuity tests.

As mentioned earlier, to simplify testing and debugging, one circuit block should be assembled and then tested at a time, before proceeding to build the next circuit block. The testing and debugging of these circuit blocks is explained in the following text, presented in the order the related projects appear through the book.

Removing Components

The following instructions are recommended for removing components from the pcb (and not damaging it) with the aid of the most basic hand tools. More elaborate tools are often used if they are available, however, most readers will not have access to such equipment. To minimise any likelihood of damage to the pcb, components with more than two leads (or pins) are sacrificed during their removal. NOTE: the pcb can also be damaged if any pads (or tracks) are subject to the application of heat from a soldering iron for excessive periods of time.

Discrete components (resistors, capacitors, etc.)

If the component has pliable leads, then the simplest method to remove the component is to grip one lead using pliers and then heat its solder joint until the joint becomes molten. Once this happens, lift the lead completely from the board. Perform the same process for the other component lead(s).

If the component has stiff leads that will not allow an individual lead to be lifted from the board, the component will need to be sacrificed by snipping each of its leads using cutters.

Integrated Circuits (ICs)

IC sockets should be used for all positions on the pcb where ICs are to be placed. This allows quick and easy fitting and removal of ICs without damage to the board or the IC. Should an IC be fitted without using an IC socket, it is recommded that it be removed from the board by cutting each of its legs. Then remove each leg in-turn in a similar method as for individual leads of discrete components.

Cables and Connection Leads

Table A-1 shows the components needed for all cabling. These cables are shown in Figure A-5 and Figure A-6. We recommended you fabricate only those interconnecting leads actually needed for a particular project and purchase the D25M to D25M cable as an already manufactured unit.

Table A-1 Cable Components for all Projects.

Quantity	Component Description
1	One-to-one D25 Male to D25 Male cable
50	Pcb pin socket, suit pin 0.9 – 1.0 mm (for interconnect cables)
7 m	Hookup wire (for interconnect cables)
1 m	Heatshrink tubing; 2.5 - 3 mm diameter

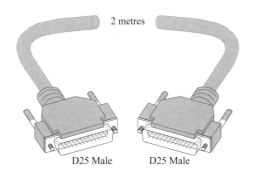

2 metres

D25 Male   D25 Male

Figure A-5 One-to-one D25 Male to D25 Male cable.

Interconnect Lead Assembly

The interconnect leads are used to connect *outputs* of circuit blocks/elements to *inputs* of circuit blocks/elements. DO NOT at any time connect an output to another output - doing so will most likely damage the components involved. Each interconnect lead (shown in Figure A-6) you need can be fabricated as follows.

Solder a pcb pin socket to one end of a 25 cm length of hookup wire. Slide a 15 mm length (approximate) of heatshrink tube along the wire and onto the socket. The tube should be positioned about halfway along the socket (far enough to prevent contact between sockets when the sockets are in use connected to adjacent pcb pins on the board). Apply heat to the heatshrink tube to shrink it in place.

Slide a second 15 mm length (approximate) of heatshrink tube onto the other end of the wire. Solder this end of the wire to the second pcb pin socket. Position and shrink the second heatshrink tube tube as described above.

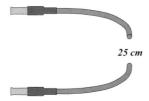

25 cm

Figure A-6 PCB interconnecting lead – socket to socket.

Interconnect Lead Testing

Test the mechanical strength of each lead's solder joints by holding a socket in each hand and pulling the lead using moderate force. Should a socket come loose, repeat the assembly operation. Electrical continuity can be tested using a multimeter set to resistance mode. Resistance between sockets should be thousandths of an Ohm at most.

Power Supply Circuitry

Assembly

Fit and solder all the components listed in Table A-2 into their position as marked on the pcb overlay and as shown in Figure A-8 and Figure A-9.

Table A-2 Power Supply - Bill of Materials.

Quantity	Component Description	Lead Spacing or Footprint	Designator
2	0.1 µF ceramic monolithic capacitor	0.2 inch	C2, 3
2	1µF ≥ 16V tantalum electrolytic capacitor	0.2 inch	C13, 14
1	4700 µF, ≥16V electrolytic capacitor	RB 0.5 or 0.35 inch	C1
1	1K resistor	¼ W	R94
1	1K8 resistor	¼ W	R52
3	1N4004 diode		D1, 2, 10
1	LM7805CT voltage regulator	TO-220	U1
1	LM7809CT voltage regulator	TO-220	U2
1	Power pack; +12V DC, 1A		
3	2 way terminal block	5mm pitch	J2, 9, 10
16	Pcb pin, 0.9 - 1.0 mm diameter		
2	Heatsink 12 °C/W		HS1, 2
2	M3 screw 6-10 mm long (or equiv.)		
2	M3 nut		
2	M3 locking washer		

Testing

Figure A-7 shows the schematic diagram for the power supply, excluding the circuitry for generating –8V DC, that is used solely by the DAC and grouped with it. The interface board can be powered by any DC power supply capable of providing voltages in the range from 13V to 18V at currents of 1A or greater. The cheapest means of providing this power is by using a 12V DC powerpack with 1A capacity. The powerpack will usually provide voltages above 12V until current draw becomes excessive (greater than approximately 1A).

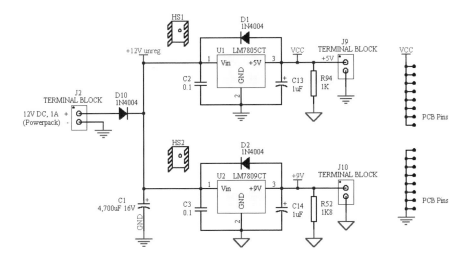

Figure A-7 Power Supply Circuit – schematic diagram.

The input voltage to both the +5V and +9V voltage regulators should be in the range +12V to +18V (this upper value depends on the powerpack or DC supply used). If not:

– Ensure the wiring between the powerpack or DC power supply, and terminal block J2 has the correct polarity.

– Check that diode D10 is fitted with correct polarity.

The output voltage of the +5V and +9V regulators should be within +4.75V to +5.25V and +8.55V to +9.45V respectively, due to their voltage tolerances. If this is not the case, check:

– Tantalum electrolytic capacitors C13, C14, and diodes D1, D2 are fitted with correct polarity.

– Regulators are fitted in the correct orientation as marked on the pcb.

– Short- and open-circuits on connecting tracks or around solder joints.

– That load resistors R52 and R94 are fitted.

NOTE

Eight pcb pins connected to GND and eight pcb pins connected to +5V are located adjacent to the power supply's two-way terminal blocks. These pins have been provided to set **input** logic levels for circuits, and to facilitate testing of circuits.

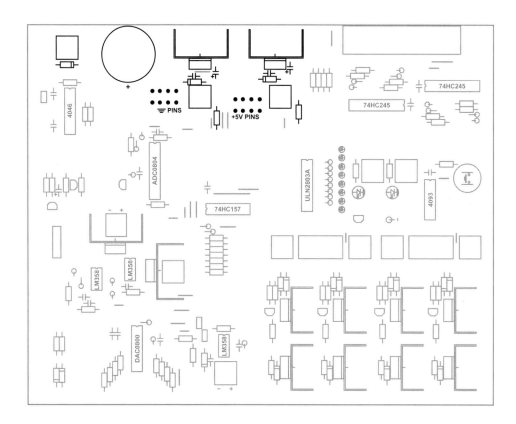

Figure A-8 Power Supply Circuit – component positions.

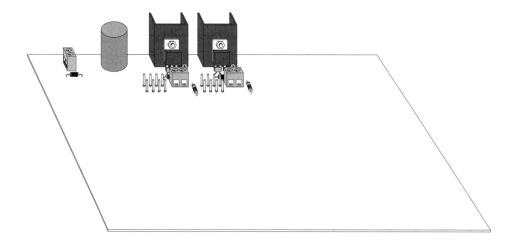

Figure A-9 Power Supply Circuit – components fitted.

Parallel Port Interface

Assembly

Fit and solder all the components listed in Table A-3 into their position as marked on the pcb overlay and as shown in Figure A-11 and Figure A-12.

Table A-3 Parallel Port Interface - Bill of Materials.

Quantity	Component Description	Lead Spacing or Footprint	Designator
2	0.1 µF ceramic monolithic capacitor	0.2 inch	C20, 22
4	470 Ω resistor	¼ W	R29, 30, 43, 44
16	10K resistor	¼ W	R78 - 93
2	74HC245 CMOS IC	DIL20	U6, 13
2	IC socket	20 pin	
1	D25 female right angle connector		J1
23	Pcb pin, 0.9 - 1.0 mm diameter		

Testing

Figure A-10 shows the schematic diagram for the interface circuitry. The two 'Buffer' ICs are used to protect the parallel port of the PC from damage should faults occur on the interface board.

Note: Ensure that both Buffers are fully disconnected from other devices (including the parallel port) prior to testing.

The voltage at the power supply pin of each logic 'Buffer' (74HC245, pin 20) should be approximately +5V (the same as that output by the +5V voltage regulator). If this is not the case, check:

- Incorrect IC orientation, faulty IC socket connections, short-circuits, open-circuits, a faulty IC, and the +5V internal power supply.

The input pins that control the direction of data flow and enable the output signals must be connected as follows:

- DIR input (pin 1) must be at +5V, and EN (pin 19) must be at 0V.

The input data lines to the Buffers have *pull-up* resistors fitted. This ensures correct interfacing to any TTL logic in the parallel port. The resistors also connect any unused input pins to a high logic state. Inputs not connected to a logic level can cause unpredictable circuit behaviour. These resistors will produce +5V input voltages at each pin when the interface cable to the PC is disconnected. The output data pins from the Buffers should have corresponding high logic levels. The four resistors connected to the D25 connector help protect the BASE+2 interface at the parallel port from damage.

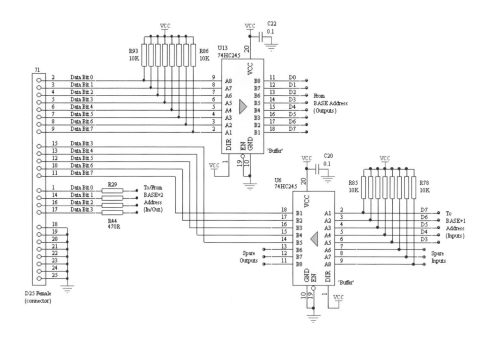

Figure A-10 Parallel Port Interface Circuit - schematic diagram.

Disconnect the PC interface cable when testing the following circuitry. Test one data input of the Buffer (74HC245, U13) at a time as follows:

The input data pin (shown as A1, A2, ..., A8) should be at +5V, and therefore, at a high logic state. The corresponding output pin (shown as B1, B2, ..., B8) should be at the same high logic state. Repeat this test for all other input pins.

Connect the input data pin (shown as A1, A2, ..., A8) to GND by first grounding one end of an interconnecting lead. Touch the other end of the lead to that pin's pull-up resistor lead that is in-line with the pin on the board. The corresponding output pin (B1, B2, ..., B8) should be at the same low logic state. Repeat for all other input data pins.

Test one data input of the Buffer (74HC245, U6) at a time as follows:

Connect one input data pin to GND by first grounding one end of an interconnecting lead and connecting the other end to an input pcb pin. The corresponding output pin should be at the same low logic state. Repeat for all other input data pins.

Swap the lead connection from GND to +5V and connect the other end to apply +5V to an input pcb pin. The corresponding output pin should be at the same high logic state. Repeat for all other input data pins.

If any of the above tests fail, check for:

  –   Incorrect IC orientation, short-circuits, open-circuits and a faulty IC.

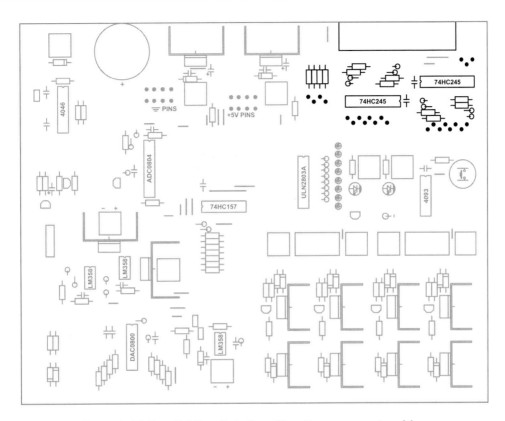

Figure A-11 Parallel Port Interface Circuit - component positions.

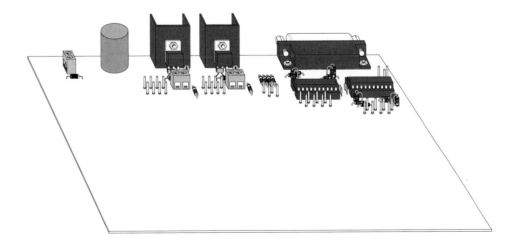

Figure A-12 Parallel Port Interface Circuit - components fitted.

LED Driver Circuitry

Assembly

Fit and solder all the components listed in Table A-4 into their position as marked on the pcb overlay and as shown in Figure A-14 and Figure A-15.

Table A-4 LED Driver Circuitry - Bill of Materials.

Quantity	Component Description	Lead Spacing or Footprint	Designator
8	330Ω resistor	¼ W	R1, ..., R10 (not inclusive)
8	Red LED 3mm diameter body		LED1-8
1	ULN2803A transistor array	DIL18	U3
1	IC socket	18 pin	
8	Pcb pin, 0.9 - 1.0 mm diameter		

Testing

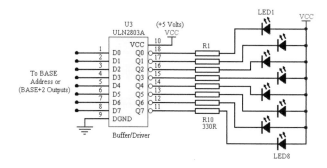

Figure A-13 LED Driver Circuit - schematic diagram.

Figure A-13 shows the schematic diagram of the LED Driver circuitry. This circuit block is ideal for testing logic levels of particular signals read or controlled by software – especially when writing and debugging a program. The circuit comprises one ULN2803A Driver IC along with eight associated resistors and LEDs. The IC contains eight separate darlington transistors, each one used to switch current through an output pin. Connecting logic level signals to the Driver will turn on and off the respective LEDs to indicate their logic state.

Note: assemble and test one series connected LED and resistor first – to check the correct polarity of the LED. Establishing LED polarity is discussed in the earlier section of this appendix titled "The Assembly Process".

The voltage level at the ULN2803A power pin (10) should be +5V. The LED terminals furthest from the resistors should also be at +5V. If this is not so, check:

- Incorrect IC orientation, faulty IC socket connections, short-circuits, open-circuits, a faulty IC or LED, and faulty +5V internal power supply.

The ULN2803A Driver operates as follows:

3. When a Driver input pin (D0, D1, ..., D7) is taken to a *high* logic level (use +5V), the corresponding output pin (Q0, Q1, ..., Q7) will be switched internally to ground voltage (0V). This will light the corresponding LED as current flows from VCC (+5V), through the LED, resistor and the Driver output pin to its internal ground.

4. When a Driver input is driven to a *low* logic level (use GND), the corresponding output pin connection to GND will be broken, interrupting current flow through the LED and resistor, extinguishing the LED.

Should any LED fail to light, check:

- Incorrect LED polarity, short-circuits, open-circuits and faulty LEDs or resistors.

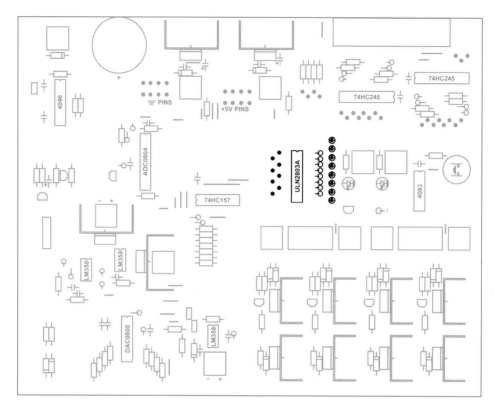

Figure A-14 LED Driver Circuit – component positions.

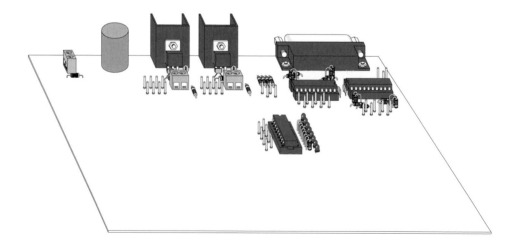

Figure A-15 LED Driver Circuit – components fitted.

Digital to Analog Converter Circuitry

The DAC circuitry is shown in Figure A-16 and comprises the DAC itself and a DAC Buffer Circuit.

Assembly

Fit and solder all the components listed in Table A-5 into their position as marked on the pcb overlay and as shown in Figure A-17 and Figure A-18.

Table A-5 Digital to Analog Converter - Bill of Materials.

Quantity	Component Description	Lead Spacing or Footprint	PCB Designator
-8V Supply:			
1	1N4004 diode		D3
1	2 way terminal block	5 mm pitch	J14
1	9V battery clip		
DAC cct:			
1	10 nF ceramic monolithic capacitor	0.2 inch	C16
2	0.1 µF ceramic monolithic capacitor	0.2 inch	C7, 8
11	10K resistor	¼ W	R39-41, 70-77
1	20K resistor	¼ W	R22
1	DAC0800 CMOS IC	DIL16	U8
1	IC socket	16 pin	
8	Pcb pin, 0.9 – 1.0 mm diameter		
2	2 pin header	0.1 inch	LINK1, LINK2
1	Jumper (fit to header LINK1 or LINK2)	0.1 inch	
DAC Buffer cct:			
2	0.1 µF ceramic monolithic capacitor	0.2 inch	C5, 6
2	10K resistor	¼ W	R26, 27
1	LM358 IC	DIL8	U10
1	IC socket	8 pin	
1	Pcb pin, 0.9 - 1.0 mm diameter		

Testing the DAC Circuit

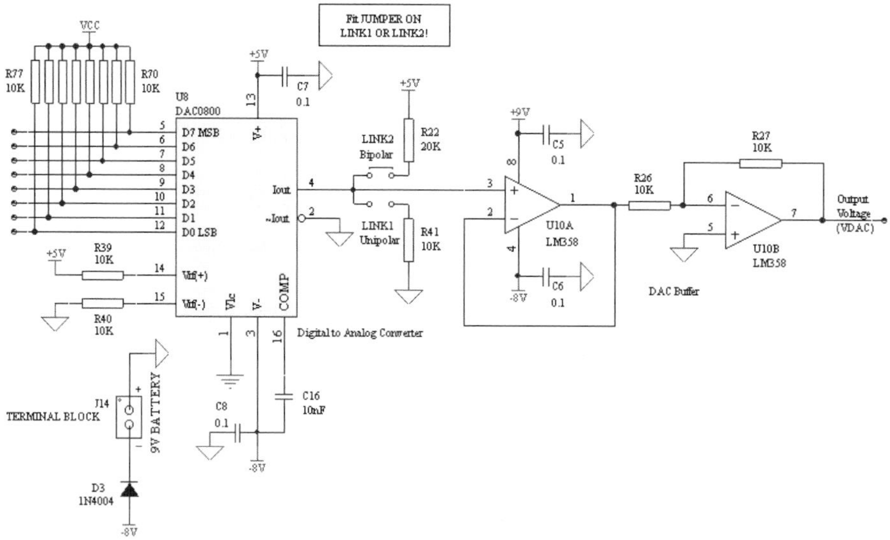

Figure A-16 DAC & DAC Buffer Circuit - schematic diagram.

Note: Ensure that a usable 9V battery is connected to the terminal block J14.

The voltage level at the DAC (U8) positive power supply pin V+ (13) should be at +5V, the DAC negative power supply pin V- (3) should be at approximately –8V. If not, check:

- Incorrect IC orientation, faulty IC socket connections, flat 9V battery, short-circuits, open-circuits, faulty DAC IC, and the +5V internal supply.

When all DAC logic input pins (D0, D1, ..., D7) are unconnected, all logic inputs should be pulled up to +5V. If this is not so, check for:

- Short-circuits, open-circuits, faulty resistors, and poor solder joints.

Fit the single Jumper to the Unipolar position, marked as LINK1. With all DAC logic inputs pulled to +5V (input pins unconnected), the output of the DAC (pin 4) should be at –5V. Conversely, when all DAC logic inputs are connected to GND (0V) the DAC should produce 0V. If this is not so, check for:

- Poor lead connections, short-circuits, open-circuits, faulty soldering of components, components having incorrect value, and faulty components.

Fit the jumper to the bipolar position (LINK2). With all DAC logic inputs pulled to +5V (input pins unconnected), the DAC output should be at –5V. With all DAC logic inputs connected to GND (0V), the DAC output should produce +5V.

Testing the Buffer Circuit

This circuitry buffers the output voltage generated by the DAC circuitry and inverts this voltage about zero volts to bring the DAC output voltage, VDAC, to a positive convention (increasing value of the DAC input byte produces an increasing voltage, VDAC).

The voltage level at the op-amp (U10) power supply pins should be +9V (pin 8) and approximately –8V (pin 4). If not, check:

 – Incorrect IC orientation, faulty IC socket connections, +9V internal power supply, flat 9V battery, short-circuits, open-circuits, and a faulty LM358 IC.

With the DAC output at –5V (all DAC logic inputs pulled to +5V), the non-inverting input of the 'Buffer' op-amp (U10, pin 3) should also be –5V. Likewise the inverting input and the output of the 'Buffer' op-amp (U10, pin 2 and pin 1 respectively) should be –5V. If not, check:

 – Short-circuits, open-circuits, faulty IC socket connections, and a faulty LM358 IC.

The 'inverter' op-amp's non-inverting input pin (U10, pin 5) and the inverting input pin (U10, pin 6) should both be at 0V. The 'inverter' op-amp output pin (U10, pin 7) should be at +5V. If not, check:

 – Short-circuits, open-circuits, faulty IC socket connections, incorrect value or faulty resistors R26, R27, and a faulty LM358 IC.

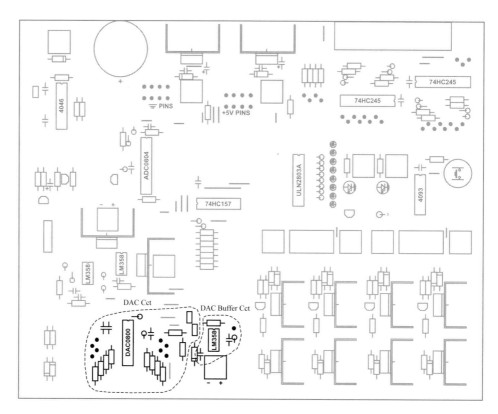

Figure A-17 DAC & DAC Buffer Circuit – component positions.

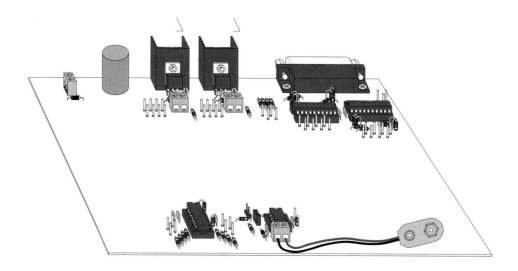

Figure A-18 DAC & DAC Buffer Circuit – components fitted.

Motor Control Circuitry

Assembly

Fit and solder all the components listed in Table A-6 into their position as marked on the pcb overlay and as shown in Figure A-20 and Figure A-21.

Table A-6 Motor Control Circuitry - Bill of Materials.

Quantity	Component Description	Lead Spacing or Footprint	Designator
4	4K7 resistor	¼ W	R34, 35, 51, 53
12	10K resistor	¼ W	R18-21, 24, 25, 45-50
8	1N4004 diode		D5-8, 11-14
4	24V zener diode	1W	ZD1, 2, 4, 5
4	BC547 npn transistor	TO-92	Q1, 2, 13, 14
4	BD649 npn darlington transistor (or equivalent)	TO-220	Q5, 6, 15, 16
4	BD650 pnp darlington transistor (or equivalent)	TO-220	Q8, 10, 17, 18
4	2 way terminal block	5mm pitch	J7, 8, 12, 13
2	4 way terminal block	5 mm pitch	J6, 11
8	Heatsink 20°C/W		HS5-12
8	Pcb pin, 0.9 – 1.0 mm diameter		
8	M3 screw 6-10 mm long (or equivalent)		
8	M3 nut		
8	M3 locking washer		

Testing

Figure A-19 shows the schematic diagram for the motor control circuitry, comprising two H-bridge circuits used for driving DC and stepper motors. Ensure wire links are fitted across each two-way terminal block before testing the circuitry. Connect an external DC power supply (less than 30V) capable of powering the motor(s) (or the +12V DC 1A power pack if suitable) to the four-way terminal block connections: Power Supply +ve connects to Vm1 and Vm2, Power Supply –ve connects to Vm1 GND and Vm2 GND.

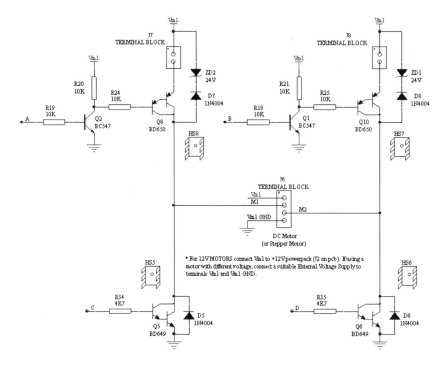

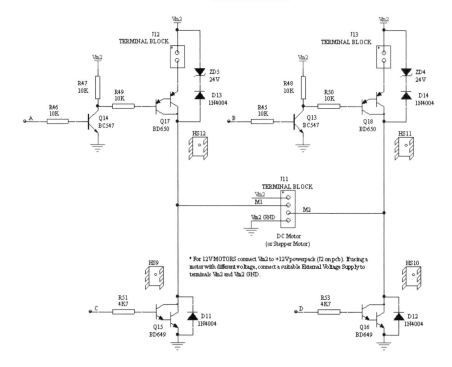

Figure A-19 Motor Control Circuit - schematic diagram.

The appropriate voltage from the motor power supply should be present at each of the respective two-way terminal blocks. If not, check:

–   Powerpack or external power supply wiring/operation, and wire links are fitted across two-way terminal blocks.

The four 'transistor switches' used in each H-bridge operate independently - each 'closes' when its logic input is taken to a high level. Each H-bridge circuit uses two types of transistor:

1. Two identical *lower* transistor switch circuits, each using a npn darlington transistor (e.g. Q5).

2. Two identical *upper* transistor switch circuits, each using a npn transistor (e.g. Q2) driving a pnp darlington transistor (e.g. Q8). Without the npn transistor, the logic input signal could not drive the pnp darlington transistor, since this darlington configuration needs to be controlled by an input voltage capable of rising close to the motor supply voltage (Vm1) to turn off the transistor. The npn transistor 'inverts' the logic input signal and can switch off at voltages up to Vm1. The 'inverted' signal then drives the pnp darlington transistor.

Test each of the transistor switches in the two H-bridge circuits in-turn. Fit a resistor, say 1KΩ (1000 Ω) as a light load in place of an actual motor across the four way terminal block contacts marked M1 and M2.

One lower npn darlington transistor and its diagonally opposing upper pnp darlington transistor circuitry will be tested at a time, all other logic inputs **must not** be connected. For example, logic input C and logic input B will be switched as follows:

When logic input C is wired to +5V (Q5 ON) and logic input B is wired to +5V (Q10 ON), the voltage at the four way terminal block contact, M1 should be equal to approximately 1V and M2 should be equal to approximately Vm1 minus 1V.

Repeat this test for the opposite transistor circuits in the H-bridge controlled by logic inputs A and D. This circuit test will give opposite voltages at terminal block contacts M1 (Vm1 minus approximately 1V) and M2 (approximately 1V). Should any of these tests fail, check:

–   Power supply operation, poor lead connections, missing wire links, short-circuits, open-circuits, faulty soldering of components, components having incorrect value or incorrect orientation, and faulty components.

Note that the BD649 and BD650 darlington transistors used in the H-bridge (shown on the schematic with two arrows) are capable of switching currents up to 8A. Other darlington transistors with similar current capacity and matching pin configuration can be used. A 24V zener diode in series with a regular diode is connected across the BD650 darlington transistors to absorb spikes of back-emf generated during switching. The diodes across the BD649 darlington transistors limit negative voltages across the transistor to one forward-biased diode drop approximately (0.7V).

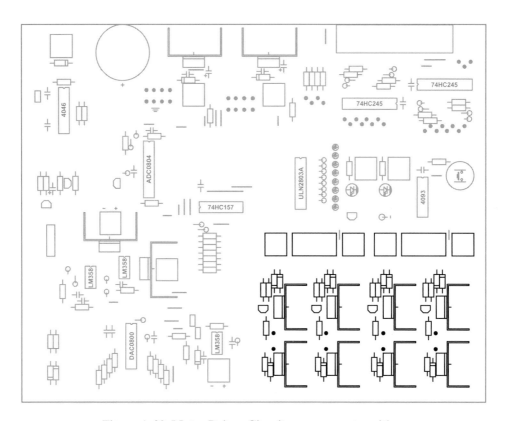

Figure A-20 Motor Driver Circuit – component positions.

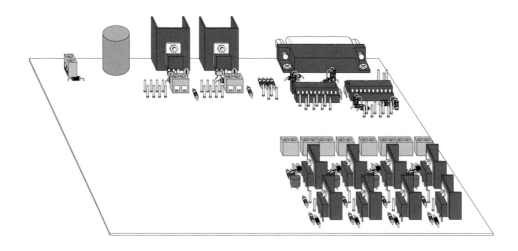

Figure A-21 Motor Driver Circuit – components fitted.

Voltage-Controlled Oscillator Circuitry

The VCO circuit uses the Thermistor circuit for several projects within the main text. Therefore, assembling and testing the Thermistor circuit has been included with the VCO material.

Assembly

Fit and solder all the components listed in Table A-7 into their position as marked on the pcb overlay and as shown in Figure A-23 and Figure A-24.

Table A-7 Voltage-Controlled Oscillator - Bill of Materials.

Quantity	Component Description	Lead Spacing or Footprint	Designator
VCO cct:			
1	1µF ceramic monolithic capacitor	0.2 inch	C18
1	100K resistor	¼ W	R11
1	1M resistor	¼ W	R13
1	4046 CMOS IC	DIL16	U4
1	IC socket	16 pin	U4
2	Pcb pin, 0.9 - 1.0 mm diameter		U4
Thermistor cct:			
1	0.1µF ceramic monolithic capacitor	0.2 inch	C9
1	100K - 470K resistor (suit thermistor - see main text Chapter 10)		R31
1	Thermistor		RT1

Testing the VCO Circuit

Figure A-22 shows the circuit diagram for the voltage-controlled oscillator (part of the phase lock-loop IC) and the thermistor circuitry. The VCO circuit outputs a digital signal having a square waveform at a frequency proportional to the voltage applied to its input pin.

The voltage at the power supply pin of the Phase-Lock Loop IC (U4, pin 16) should be equal to approximately +5V, the same as that output by the +5V voltage regulator. If not, check:

– Incorrect IC orientation, short-circuits, open-circuits, faulty IC socket connections, faulty 4046 IC, and the +5V power supply.

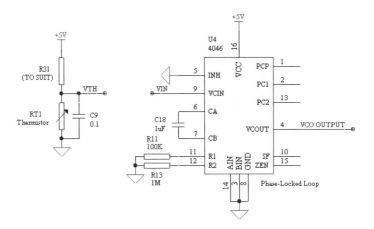

Figure A-22 VCO & Thermistor Circuit - schematic diagram.

Connect the VCO's voltage input pin VIN (9) to GND and connect an interconnect lead from the VCO output (pin 4) to one of the 8 LED Driver input pins. This should produce a slowly changing signal, evident by a LED flashing on and off. Connecting the VCO voltage input pin to +5V should produce an output pulse-train having a much higher frequency. If this is not so, check:

– Short-circuits, open-circuits, incorrect value of associated resistors and capacitors, faulty components, incorrect IC orientation, and faulty soldering.

Testing the Thermistor Circuit

This is a simple voltage divider circuit having a capacitor connected across its output to GND. The capacitor has been added to reduce the possible effects of any high-frequency noise coupled through from the +5V power supply.

Test for +5V at the end of resistor R31 (furthest from capacitor C9). If not present, check:

– +5V power supply, short-circuits, and open-circuits.

The thermistor will have a particular resistance at room temperature, producing a corresponding output voltage (VTH) from the voltage divider circuit. Hold the body of the thermistor between two fingers to warm it (change its resistance) and the output voltage should change. If not so, check:

– Inappropriate value of resistance for R31, faulty soldering of components, short-circuits, open-circuits, faulty resistor R31, faulty thermistor, or faulty capacitor.

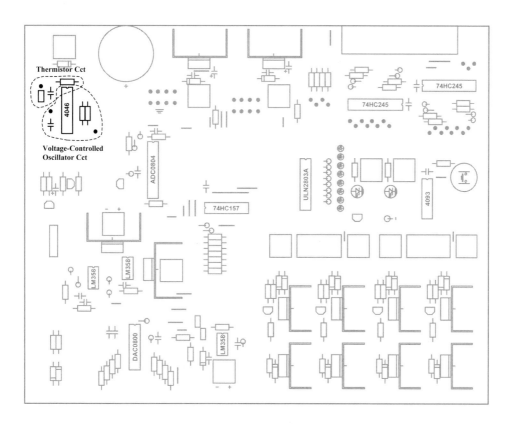

Figure A-23 VCO & Thermistor Circuit – component positions.

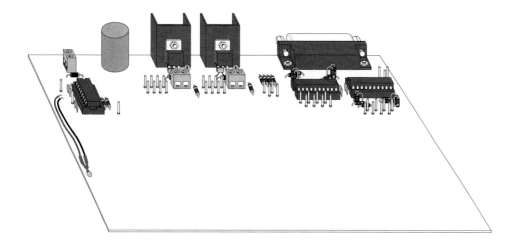

Figure A-24 VCO & Thermistor Circuit – components fitted.

Analog-to-Digital Converter Circuitry

Assembly

Fit and solder all components listed in Table A-8 into their position as marked on the pcb overlay and as shown in Figure A-26 and Figure A-27.

Table A-8 Analog-to-Digital Converter - Bill of Materials.

Quantity	Component Description	Lead Spacing or Footprint	Designator
1	150 pF ceramic capacitor	0.2 inch	C17
1	0.1 µF ceramic monolithic capacitor	0.2 inch	C11
1	100 Ω resistor	¼ W	R69
4	10K resistor	¼ W	R9, 15-17
1	ADC0804 CMOS IC	DIL20	U7
1	IC socket	20 pin	
13	Pcb pin, 0.9 - 1.0 mm diameter		

Testing

Figure A-25 shows the schematic for the Analog-to-Digital Converter (ADC) circuitry.

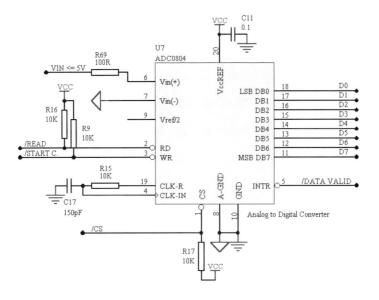

Figure A-25 ADC Circuit - schematic diagram.

This circuit uses an ADC0804 IC to sample an analog voltage (0 to +5V) at its input pin (VIN) and produce a digital output byte (0 to 255) that represents this analog voltage.

The voltage at the ADC0804 power pin (20) should be +5V. If not, check:

- +5V power supply, incorrect IC orientation, short-circuits, open-circuits, faulty IC socket connections, and a faulty ADC0804 IC.

If the IC has +5V present at its power pin as intended, but the IC is HOT, the device may be suffering from what is known as *CMOS SCR Latch-up*. This phenomena can occur with some CMOS devices, and has the effect of internally short-circuiting the power pin to GND. To overcome this problem, turn the power off to the interface board and allow the IC to cool, then re-apply power and check for normal cool temperature.

The three logic inputs to the ADC0804 (/READ, /START C., and /CS) should be pulled to +5V by the pull-up resistors R16, R9, and R17. If not, check:

- Faulty soldering of components, short-circuits, open-circuits, faulty resistors, incorrect IC orientation, and a faulty ADC0804 IC.

To perform the final test for the ADC, connect the /CS and /READ input pins to GND. Connect an interconnecting lead to the /START C. input, at this stage leaving the other end of the lead free.

Connect the ADC digital output pins (D0-D7) to the LED Driver Circuit to visually display their logic state.

Connect the analog input voltage (VIN) to GND. Briefly connect the /START C. input to GND to initiate a conversion. The resulting ADC output byte should be close to decimal 0, or binary 0000 0000.

Repeat this test with VIN connected to +5V. The resulting ADC output byte should be close to decimal 255, or binary 1111 1111.

Note that the most significant bit of the ADC0804 is DB7 (pin 11). This bit should remain high until the input voltage drops below half the input voltage range (2.5V).

If any of the above tests fail, check:

- Incorrect IC orientation, faulty soldering of components, short-circuits, open-circuits, faulty IC socket connections, and faulty components.

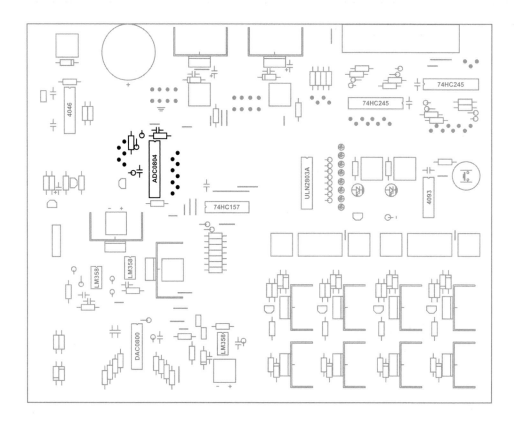

Figure A-26 ADC Circuit – component positions.

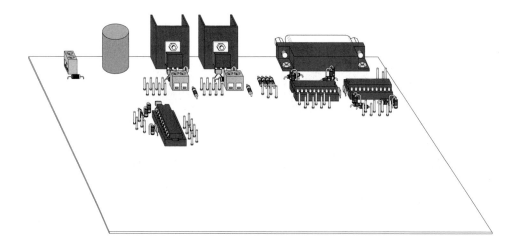

Figure A-27 ADC Circuit – components fitted.

Miscellaneous Circuitry

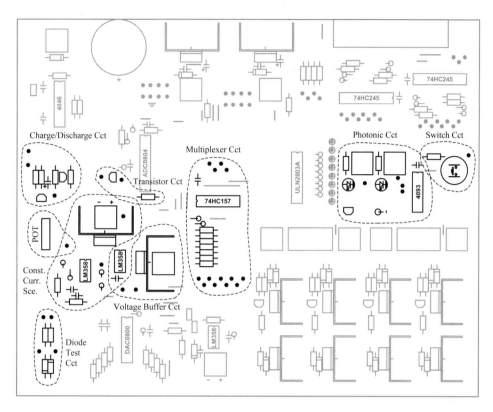

Figure A-28 Miscellaneous Circuits – component positions.

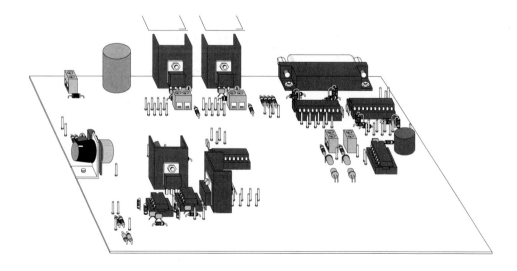

Figure A-29 Miscellaneous Circuits – components fitted.

Multiplexer Circuit

Assembly

Fit and solder all the components listed in Table A-9 into their position as marked on the pcb overlay and as shown in Figure A-28 and Figure A-29.

Table A-9 Multiplexer - Bill of Materials.

Quantity	Component Description	Lead Spacing or Footprint	Designator
1	0.1 µF ceramic monolithic capacitor	0.2 inch	C19
9	10K resistor	¼ W	R58-66
1	74HC157 CMOS IC	DIL16	U12
1	IC socket	16 pin	
13	Pcb pin, 0.9 - 1.0 mm diameter		

Testing

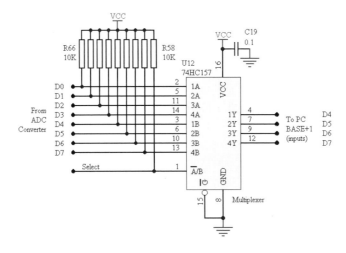

Figure A-30 Multiplexer Circuit - schematic diagram.

Figure A-30 shows the schematic for the multiplexer circuitry. This circuit splits eight bits (or digital signals) into two groups of four bits and allows one group at a time to be switched through to its four output pins. In this manner, a stream of eight-bit data can be transmitted as a series of two four-bit groups to another device, using only four signal transmission signal lines. The disadvantage is the doubling of time required to transmit when compared to the use of eight direct connections and no multiplexer.

The *Select* input (pin 1) to the multiplexer controls the connection of inputs 1A-4A or inputs 1B-4B to the outputs 1Y-4Y respectively. When *Select* is at a low logic level, A's inputs are switched to the output pins. Conversely when *Select* is at a high logic level, B's inputs are switched to the output pins.

The voltage at the power supply pin of the Multiplexer (74HC157, pin 16) should be equal to approximately +5V, the same as that output by the +5V voltage regulator. If not, check:

- Incorrect IC orientation, short-circuits, open-circuits, faulty IC socket connections, and a faulty 74HC157 IC.

Switch Test I

- Ensure all input pins (and output pins) are disconnected. This places all inputs (including Select) in a high state due to their pull-up resistors. This will switch the second set of four high-level inputs 1B-4B (inputs D4-D7) to their corresponding output pins 1Y-4Y (outputs D4-D7) that should all be in a high state.

- Connect the Select input to GND using an interconnect lead (input data pins disconnected) as before. This will switch the first set of four high-level inputs 1A-4A (inputs D0-D3) to their corresponding output pins 1Y-4Y (outputs D4-D7) that should all be in a high state.

If either of these tests show bits that are not as stated, then measure the voltages at each input data pin (disconnected as before). If voltages are not at +5V, check:

- Continuity (turn off power) across each resistor to its respective data bit (pcb pin). If this value is greater than the 10K resistance specified, then check the resistor(s) for open-circuits and faulty solder joints.

- Incorrect IC orientation, short-circuits, open-circuits, faulty IC socket connections, and a faulty 74HC157 IC.

Switch Test II

- Connect all four inputs 1A-4A (inputs D0-D3) to GND using interconnect leads. Ensure remaining four inputs 1B-4B (inputs D4-D7) are disconnected (pulled to a high state). Connect the Select input to GND using an interconnect lead. This will switch the four input bits 1A-4A to their corresponding output pins 1Y-4Y (outputs D4-D7). The output bits should all be in a low state. If not, check:

  - Incorrect IC orientation, poor connections, short-circuits, open-circuits, faulty IC socket connections, and a faulty 74HC157 IC.

- Connect the Select input to a high state (disconnect the interconnect lead). This will switch the other four high-level input bits 1B-4B (inputs D4-D7) to their corresponding output pins 1Y-4Y (outputs D4-D7). At this stage the output bits should be in a high state. Connect all four inputs 1B-4B (inputs D4-D7) to GND using interconnect leads. The output bits should now all be in a low state. If either of these two tests do not comply as stated, check:

  - Incorrect IC orientation, poor connections, short-circuits, open-circuits, faulty IC socket connections, and a faulty 74HC157 IC.

Adjustable Current Source Circuit

Assembly

Fit and solder all the components listed in Table A-10 into their position as marked on the pcb overlay and as shown in Figure A-28 and Figure A-29.

Table A-10 Adjustable Current Source - Bill of Materials.

Quantity	Component Description	Lead Spacing or Footprint	Designator
3	0.1 µF ceramic monolithic capacitor	0.2 inch	C4, 10, 23
1	3K resistor	¼ W	R67
4	10K resistor	¼ W	R28, 55-57
1	12K resistor	¼ W	R68
1	BD650 pnp darlington transistor (or equivalent)	TO-220	Q9
2	LM358 IC	DIL8	U9, 11
2	IC socket	8 pin	
2	2 way terminal block	5 mm pitch	J4, 16
1	Heatsink 12°C/W		HS4
2	Pcb pin, 0.9 - 1.0 mm diameter		
1	M3 screw 6-10 mm long (or equiv.)		
1	M3 nut		
1	M3 locking washer		

Testing

Figure A-31 shows the schematic for the adjustable current source circuitry. Adjustable current sources have many uses such as charging NiCad batteries and for electronic test and measurement purposes. This circuitry generates a current that is proportional to an input voltage (0 to +5V) such as that produced by VDAC. The resistor *Rcurr* (of appropriate value) is fitted across the terminal block J16 to set the range of current level that can be supplied.

The adjustable current source is implemented using three op-amp stages, each stage performing a different operation. The first op-amp circuit (U11B) scales the input voltage VDAC by 4/5ths, using a voltage divider circuit formed by resistors R67 and R68. This means that for a maximum input voltage of +5V, the resistive divider will give +4V at its output. Op-amp U11B buffers this voltage for use by the second stage of the circuitry (U11A).

The second stage uses op-amp U11A configured as a *non-inverting* amplifier, amplifying the voltage signal at its +ve (non-inverting) input terminal using a gain of two. Note: for a non-inverting amplifier configuration, the gain is equal to $1 + R_f$ /R, where R_f is the feedback resistor and R is the grounded resistor. A second voltage divider circuit (that uses resistors R28 and R57) generates the input voltage at the non-inverting terminal. When the output from the first stage (U11B, pin 7) is +4V, the input voltage at the non-inverting terminal of U11A will be +4.5V. This voltage is amplified by two to produce +9V at the output of U11A (pin 1).

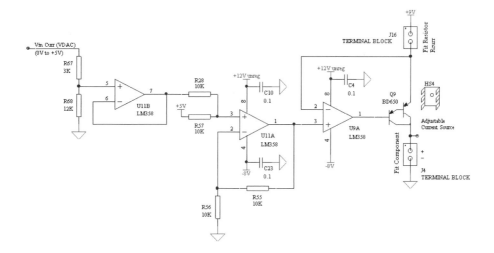

Figure A-31 Adjustable Current Source Circuit - schematic diagram.

The final stage in the current source circuitry uses op-amp U9A operating as an adjustable current source. It takes the output voltage from U11A and drives a pnp darlington transistor (Q9) such that the voltage at the emitter of this transistor (marked by an arrow and connected to the –ve input terminal) is equal to the voltage at the +ve input terminal. Therefore, when the previous stage outputs +9V, the voltage at the emitter side of the terminal block J16 will also be at +9V. This produces a net voltage of zero across the resistor Rcurr, meaning that no current flows through Rcurr, the transistor, and the component fitted across the terminal block J4.

When the input voltage (VDAC) to the first stage is 0V, the output of U11B will be 0V, the output of the second stage will be +5V and the voltage at the emitter side of terminal block J16 will also be +5V, producing a maximum voltage across Rcurr equal to 4V. The level of current generated by the four volts across Rcurr will depend on its resistance value. Select the value of Rcurr to set the maximum current range needed.

This adjustable current source can be used for many purposes, including the three following examples:

    i)   Virtual Ohmmeter - fit a resistor of correct value (Rcurr) across terminal block J16 to generate a corresponding range of current. The resistor to be measured is fitted across terminal block J4. A controlled voltage VDAC applied to input pin 'Vin Curr' will generate a current through the resistor being evaluated, to produce a voltage across it. This voltage can be measured using the ADC (or the VCO if its transfer function has been measured).

ii) NiCad Battery Charger - fit the battery across terminal block J4 with correct polarity and charge with an appropriate level of current for a suitable period of time (or until the battery has been charged to the required voltage).

iii) Transistor or Diode tester (plot characteristic curve) - connect the component being tested to terminal block J4 with correct polarity. Then control the current through the device while measuring the voltage generated across it.

To test this circuit, fit the resistors across each terminal block – try using a 390Ω resistor for Rcurr and a 470 Ω resistor across terminal block J4.

First test the voltage at both op-amp power supply pins (U11 and U9, pin 8). The voltage should be approximately +12V when using a +12V powerpack or +12V power supply. The negative voltage supply pins (U11 and U9, pin 4) should be approximately –8V.

Connect either 0V or +5V as the input voltage Vin Curr (described above using VDAC) and move through each stage, testing the input voltage and then the corresponding output voltage of that stage. Note that the op-amp +ve and –ve input terminals will be at the same voltage (within a few microvolts) if the op-amp is functioning properly.

Remembering that effectively zero current passes through the op-amp –ve input (and +ve input), the current through Rcurr will flow down and branch into the emitter of darlington transistor Q9. Zero volts connected to Vin Curr will generate 4V across Rcurr (390 Ω) and produce a current through Rcurr of approximately 10mA. Nearly all of this current will flow down through terminal block J4 (470 Ω) to generate slightly less than 4.7V across it. The very small amount of current that does not pass through the terminal block (470 Ω) passes through the emitter of the darlington transistor Q9 to the output of op-amp U9A. This current is used to drive the emitter voltage to match that voltage present on the op-amp +ve input pin.

If any faults are detected, check:
- +12V power supply, -8V power supply (9V battery), incorrect IC and transistor orientation, incorrect components, faulty soldering, short-circuits, open-circuits, faulty IC socket connections or faulty components.

Voltage Buffer Circuit (High Current)

Assembly

Fit and solder all the components listed in Table A-11 into their position as marked on the pcb overlay and as shown in Figure A-28 and Figure A-29.

Table A-11 Voltage Buffer (High Current) - Bill of Materials.

Quantity	Component Description	Lead Spacing or Footprint	Designator
1	10K resistor	¼ W	R23
1	BD649 npn darlington transistor (or equivalent)	TO-220	Q7
1	LM358 IC	DIL8	U9
1	IC socket	8 pin	
1	Heatsink 12°C/W		HS3
2	Pcb pin, 0.9 - 1.0 mm diameter		
1	M3 screw 6-10 mm long (or equivalent)		
1	M3 nut		
1	M3 locking washer		

Testing

Figure A-32 shows the schematic for the voltage buffer circuitry.

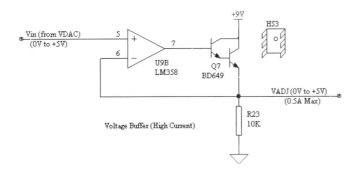

Figure A-32 Voltage Buffer Circuit (High Current) - schematic diagram.

This circuit takes an input voltage signal having low current drive capacity (say a few milliamps) and produces a matching output voltage capable of providing up to 0.5A of current.

The circuit uses op-amp U9B to drive an npn darlington transistor (Q7) such that the emitter (shown with an arrow) voltage matches the input voltage at the +ve input terminal. Resistor R23 is used to generate the feedback voltage and allow proper circuit operation. The op-amp automatically adjusts the drive current to the transistor to maintain a constant output voltage as changes in current draw take place at the output.

First test the voltage at both op-amp power supply pins. These voltages should be approximately +12V (pin 8) and approximately –8V (pin 4). Connect a resistive load (say a 100Ω resistor) across the output pin (VADJ) to GND. With the input voltage Vin at 0V, the output voltage VADJ should also be at 0V. Apply fixed DC voltages (up to a maximum of +5V) to the input, Vin. The output voltage (VADJ) should match the input voltage being applied. The Potentiometer is ideal for generating various voltages up to +5V.

Note: op-amp U9 uses the –8V power supply. Therefore, ensure the components listed in Table A-5 that make up the –8V supply are fitted, and a usable 9V battery is connected.

If any faults are detected, check:
- – +12V power supply and –8V power supply (9V battery), incorrect IC and transistor orientation, faulty soldering, short-circuits, open-circuits, faulty IC socket connections, and faulty components.

Charge/Discharge RC Circuit

Assembly

Fit and solder all the components listed in Table A-12 into their position as marked on the pcb overlay and as shown in Figure A-28 and Figure A-29.

Table A-12 Charge/Discharge RC Circuit - Bill of Materials.

Quantity	Component Description	Lead Spacing or Footprint	Designator
1	1μF, ≥ 16V tantalum electrolitic capacitor	0.2 inch	C15
2	4K7 resistor	¼ W	R36, 37
1	100K resistor	¼ W	R12
1	470K resistor	¼ W	R38
1	BC547 npn transistor	TO-92	Q3
1	BC557 pnp transistor	TO-92	Q12
3	Pcb pin, 0.9 - 1.0 mm diameter		

This circuitry is shown in Figure A-33 and is used to demonstrate the charging/discharging characteristics of a resistor/capacitor (RC) circuit. The ADC circuit can digitise the output signal produced by this circuit and then a waveform can be plotted.

Charging takes place when the /Charge input signal is activated by driving it from a high logic level to a low logic level while the Discharge input signal is kept inactive at low logic level. The voltage across capacitor C15 will increase during the charging process. The pnp transistor (Q12) controlled by /Charge, conducts when its base terminal (connected to R36) voltage drops at least approximately 0.7V lower than the emitter terminal (shown with an arrow). This allows current to flow through charging resistor R12 and charge up capacitor C15 to +5V.

Testing

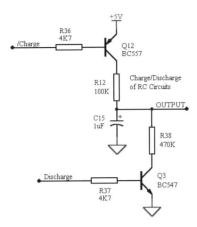

Figure A-33 Charge/Discharge RC Circuit - schematic diagram.

The npn transistor (Q3) is controlled by Discharge and will only conduct when its base terminal is at a voltage at least approximately 0.7V above its emitter terminal (connected to GND).

Discharging occurs when the /Charge input is driven to its inactive state (> 4.3V ensuring transistor Q12 will not conduct) and the Discharge input signal is activated using a logic-HIGH level. This causes transistor Q3 to conduct and allow charge stored on the capacitor to flow as a current through the discharge resistor (R38) and transistor (Q3) to GND. The voltage across the capacitor (C15) will drop to zero volts by the end of this process. Note that the discharge resistor (R38) is approximately five times the resistance value of the charge resistor (R12), producing different charge and discharge time constants (see Figure 13-6).

First test the voltage at the charge transistor (Q12) emitter pin (connected to the thick track on the pcb) - it should be equal to +5V. If not so, check:

- +5V power supply, incorrect transistor or capacitor orientation, faulty soldering, short-circuits, open-circuits, and faulty components.

Connect the /Charge and Discharge inputs as described above and observe circuit function. If the output voltages are not as described, check:

- Incorrect transistor or capacitor orientation, faulty soldering, short-circuits, open-circuits, and faulty components and/or drive signals used for /Charge and Discharge inputs that cannot supply the required voltages.

LED and Photodiode/Phototransistor Pair Circuit

Assembly

Fit and solder all the components listed in Table A-13 into their position as marked on the pcb overlay and as shown in Figure A-28 and Figure A-29.

Table A-13 LED and Photodiode/Phototransistor Pair - Bill of Materials.

Quantity	Component Description	Lead Spacing or Footprint	Designator
1	0.1μF ceramic monolithic capacitor	0.2 inch	C21
2	150Ω resistor	¼ W	R14, 54
2	LED (red or infrared)		LED9, 10
1	Photodiode (suit type of LED)		D9
1	Phototransistor (suit type of LED)		Q11
1	4093 CMOS IC	DIL14	U5
1	IC socket	14 pin	U5
2	2 way terminal block	5 mm pitch	J3, 15
4	Pcb pin, 0.9 - 1.0 mm diameter		

Testing

Figure A-34 shows the schematic for the circuirtry associated with the LED and photodiode/phototransistor pair. These circuits can be used to measure light level, as proximity sensors, for optical communication and for detecting rotational and linear position/speed.

Fit appropriate value resistors across terminal blocks J3 and J15 to bias (allow desired operation) the photodetectors, being either photodiodes or phototransistors. The value of these resistors will need to be determined through repeated trials. Note that the phototransistor can be replaced by a second photodiode. Try increasing resistor values by powers of ten, starting from 100R. Then fine tune for the most suitable value.

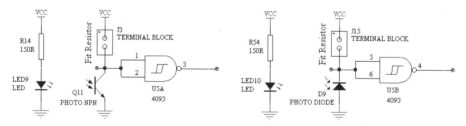

Photonic Emitter/Detectors - Motor Speed Feedback and Quadrature Detector.

Figure A-34 LED and Photodiode/Phototransistor Pair - schematic diagram.

First, test the supply voltage VCC (+5V) at all four resistors. If the voltage is not correct, check:

- +5V power supply, faulty soldering, short-circuits, open-circuits and faulty components.

Test the photo-response of the two photo-transceiver pairs by directing the LED light sources at the photo-detectors and observing the voltage increase at their pcb pin terminals as the light is blocked. Note that infra-red light is *not* blocked by paper. If these tests fail, check:

– incorrect LED or photodetector orientation, faulty soldering, short-circuits, open-circuits and faulty components.

The NAND gates (NOT AND) are used to produce digital signals from the analog voltages generated by the photodetectors. They have in-built hysteresis indicated by the symbol inside their outline. This hysteresis prevents the NAND gate output fluctuating with small changes of light level.

Switch Interface, Potentiometer, Diode, Zener Diode and Transistor Circuits

Assembly

Fit and solder all the components listed in Table A-14 into their position as marked on the pcb overlay and as shown in Figure A-28 and Figure A-29.

Table A-14 Switch Interface, Potentiometer, Diode, Zener Diode and Transistor Circuits - Bill of Materials.

Quantity	Component Description	Lead Spacing or Footprint	Designator
Switch Interface:			
1	2K7 resistor	¼ W	R8
1	SPST normally open keyboard pushbutton switch (or equivalent)	5 mm pitch	SW1
1	Pcb pin, 0.9 - 1.0 mm diameter		
Potentiometer:			
1	Knob for potentiometer		
1	1K 16 mm potentiometer	0.2W	POT1
1	Pcb pin, 0.9 - 1.0 mm diameter		
1	Fabricate POT right angle support bracket		
Semiconductor circuits:			
2	1K resistor	¼ W	R32, 33
1	Resistor (test with a range of values, from 100Ω to 10K)	¼ W	R42
1	1N4148 diode		D4
1	3V3 zener diode	1W	ZD3
1	BC547 npn transistor	TO-92	Q4
6	Pcb pin, 0.9 - 1.0 mm diameter		

Testing

Figure A-35 shows the schematic for the switch interface, potentiometer, diode, zener diode, and transistor circuitry.

Switch Interface:

When the switch is open the output will be low. Conversely, when the switch is closed the output will be high.

If the switch interface circuit does not function correctly, check:

- Faulty soldering, short-circuits, open-circuits, incorrect switch orientation, and faulty components.

Note: This switch circuit will need to be altered if it is to interface with TTL logic devices. TTL logic circuits require much greater current flow in and out of their input pins than do CMOS circuits. Should you use this circuit with TTL devices, lower the value of resistor R8 to approximately 330Ω.

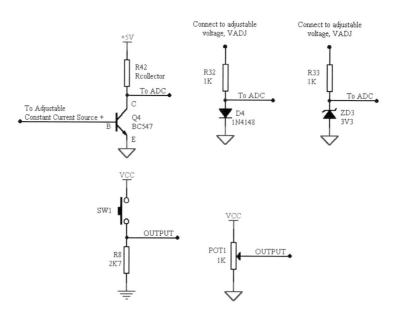

Figure A-35 Switch Interface, Potentiometer, Diode, Zener Diode, and Transistor Circuits - schematic diagram.

Potentiometer:

Ensure that the wiper terminal of the potentiometer is not connected to another circuit. As the knob of the potentiometer is rotated through its range, the output of its wiper terminal should produce voltages ranging from 0V to +5V.

Semiconductor circuits:

These circuits are used to observe the electrical characteristics of diodes, zener diodes, and bipolar transistors.

Diode D4 is driven by either an adjustable voltage source or current source while the voltage at its anode is measured to observe its basic electrical characteristics.

Zener diode ZD3 is also driven by an adjustable voltage source or current source while the voltage at its anode is measured to observe its basic electrical characteristics.

Bipolar npn transistor Q4 is driven by the adjustable current source (0 to 1mA using Rcurr at 3K9) while the voltage at its base (B) and collector (C) are measured to observe basic electrical characteristics. It can also be driven by a voltage source using a series connected resistor placed between the voltage source and the base pin.

Two sockets from an IC socket strip can be used to provide a socket at each end of resistor R42. This will ease the process of trialling different values of this resistor.

Note: the adjustable current source can be controlled using the output of the potentiometer POT1 connected to the input of the adjustable current source (Vin Curr).

Interface Board Bill of Materials

Quantity	Component Description	Lead Spacing or Footprint	PCB Designator
1	150 pF ceramic capacitor	0.2 inch	C17
1	10 nF ceramic monolithic capacitor	0.2 inch	C16
14	0.1μF ceramic monolithic capacitor	0.2 inch	C2-11, 19-23
1	1μF ceramic monolithic capacitor	0.2 inch	C18
3	1μF, ≥ 16V tantalum electrolitic capacitor	0.2 inch	C13-15
1	4700 μF, ≥16V electrolytic capacitor	RB 0.5 or 0.35 inch	C1
1	100 Ω resistor	¼ W	R69
2	150Ω resistor	¼ W	R14, 54
8	330Ω resistor	¼ W	R1,...,R10 not incl.
4	470 Ω resistor	¼ W	R29, 30, 43, 44
3	1K resistor	¼ W	R32, 33, 94
1	1K8 resistor	¼ W	R52
1	2K7 resistor	¼ W	R8
1	3K resistor	¼ W	R67
6	4K7 resistor	¼ W	R34-37, 51, 53
59	10K resistor	¼ W	R9, 15-17, 18-21, 23-28, 39-41, 45-50, 55-66, 70-93
1	12K resistor	¼ W	R68
1	20K resistor	¼ W	R22
2	100K resistor	¼ W	R11, 12
1	470K resistor	¼ W	R38
1	1M resistor	¼ W	R13
1	100K to 470K resistor (suit thermistor)		R31
1	resistor (range of values, say 100Ω to 10K)	¼ W	R42
1	1K 16 mm potentiometer	0.2W	POT1
1	Knob for potentiometer		
1	Thermistor		RT1
1	1N4148 diode		D4
12	1N4004 diode		D1-3, 5-8, 10-14
1	3V3 zener diode	1W	ZD3
4	24V zener diode	1W	ZD1, 2, 4, 5
8	Red LED 3mm		LED1-8
2	LED (red or infrared)		LED9, 10
1	Photodiode (suit type of LED)		D9
1	Phototransistor (suit type of LED)		Q11
6	BC547 npn transistor	TO-92	Q1-4, 13, 14

1	BC557 pnp transistor	TO-92	Q12
5	BD649 npn darlington transistor (or equivalent)	TO-220	Q5-7, 15, 16
5	BD650 pnp darlington transistor (or equivalent)	TO-220	Q8-10, 17, 18
1	4046 CMOS IC	DIL16	U4
1	4093 CMOS IC	DIL14	U5
1	74HC157 CMOS IC	DIL16	U12
2	74HC245 CMOS IC	DIL20	U6, 13
1	DAC0800 CMOS IC	DIL16	U8
3	LM358 IC	DIL8	U9, 10, 11
1	ADC0804 CMOS IC	DIL20	U7
1	LM7805CT	TO-220	U1
1	LM7809CT	TO-220	U2
1	ULN2803A transistor array	DIL18	U3
3	IC socket	20 pin	
1	IC socket	18 pin	
3	IC socket	16 pin	
2	IC socket	14 pin	
3	IC socket	8 pin	
2	2 pin header	0.1 inch	LINK1, LINK2
1	Jumper (across header)	0.1 inch	LINK1 or LINK2
12	2 way terminal block	5mm pitch	J2-4, 7-10, 12-16
2	4 way terminal block	5 mm pitch	J6, 11
1	9V battery clip		
1	SPST normally open keyboard pushbutton switch (or similar)	5 mm pitch	SW1
1	D25 female right angle connector		J1
1	D25 Male to D25 Male, one to one cable		
50	Pcb pin socket, suit pin 0.9 – 1.0 mm		
111	Pcb pin, 0.9 - 1.0 mm diameter		
7 m	Hookup wire (for interconnect cables)		
1m	Heatshrink tubing, 2.5 - 3 mm diameter		
4	Heatsink 12°C/W		HS1-4
8	Heatsink 20°C/W		HS5-12
1	Heatsink paste		
12	M3 screw 6-10 mm long (or equivalent), nut, locking washer		
4	Pcb rubber stick-on feet		
1	Power pack; +12V DC, 1A		
1	Fabricated right angle support bracket (POT)		

Appendix B - Software

C++ Keywords

Operator Precedence

ASCII Character Set

C++ Keywords

asm	private
auto	protected
break	public
case	register
catch	return
char	short
class	signed
const	sizeof
continue	static
default	struct
delete	switch
do	template
double	this
else	throw
enum	try
far	typedef
float	union
for	unsigned
friend	virtual
goto	void
if	volatile
inline	while
int	
interrupt	
long	
near	
new	
operator	

Operator Precedence

The table below shows the precedence of C++ operators. The highest priority is given to the operators on the first row and the lowest priority is given to the operator in the last row. Operators placed on the same row have the same priority. The operators are used from left to right, except for the rows marked with a †. These rows are used from right to left. For example, `a=b;` is used from right to left. That is, b's value is assigned to a.

Table B-1 Operator precedence.

Operators	
() [] -> :: .	
! ~ + - ++ -- & *	†
sizeof new delete	†
.* ->*	
* / %	
+ -	
<< >>	
< <= > >=	
== !=	
&	
^	
|	
&&	
||	
?:	†
= *= /= %= += -= &= ^= |= <<= >>=	†
,	

ASCII Character Set

NUL 0x00	DLE 0x10	SP 0x20	0 0x30	@ 0x40	P 0x50	` 0x60	p 0x70
SOH 0x01	CD1 0x11	! 0x21	1 0x31	A 0x41	Q 0x51	a 0x61	q 0X71
STX 0x02	DC2 0x12	" 0x22	2 0x32	B 0x42	R 0x52	b 0x62	r 0x72
ETX 0x03	DC3 0x13	# 0x23	3 0x33	C 0x43	S 0x53	c 0x63	s 0x73
EOT 0x04	DC4 0x14	$ 0x24	4 0x34	D 0x44	T 0x54	d 0x64	t 0x74
ENQ 0x05	NAK 0x15	% 0x25	5 0x35	E 0x45	U 0x55	e 0x65	u 0x75
ACK 0x06	SYN 0x16	& 0x26	6 0x36	F 0x46	V 0x56	f 0x66	v 0x76
BEL 0x07	ETB 0x17	' 0x27	7 0x37	G 0x47	W 0x57	g 0x67	w 0x77
BS 0x08	CAN 0x18	(0x28	8 0x38	H 0x48	X 0x58	h 0x68	x 0x78
HT 0x09	EM 0x19	) 0x29	9 0x39	I 0x49	Y 0x59	i 0x69	y 0x79
LF 0x0A	SUB 0x1A	* 0x2A	: 0x3A	J 0x4A	Z 0x5A	j 0x6A	z 0x7A
VT 0x0B	ESC 0x1B	+ 0x2B	; 0x3B	K 0x4B	[0x5B	k 0x6B	{ 0x7B
FF 0x0C	FS 0x1C	, 0x2C	< 0x3C	L 0x4C	/ 0x5C	l 0x6C	| 0x7C
CR 0x0D	GS 0x1D	- 0x2D	= 0x3D	M 0x4D	] 0x5D	m 0x6D	} 0x7D
SO 0x0E	RS 0x1E	. 0x2E	> 0x3E	N 0x4E	^ 0x5E	n 0x6E	~ 0x7E
SI 0x0F	US 0x1F	/ 0x2F	? 0x3F	O 0x4F	_ 0x5F	o 0x6F	DEL 0x7F

Index